AF614759

METHODS IN MOLECULAR BIOLOGY™

Series Editor
John M. Walker
School of Life Sciences
University of Hertfordshire
Hatfield, Hertfordshire, AL10 9AB, UK

For other titles published in this series, go to
www.springer.com/series/7651

Plant Transcription Factors

Methods and Protocols

Edited by

Ling Yuan

University of Kentucky, Lexington, KY, USA

Sharyn E. Perry

University of Kentucky, Lexington, KY, USA

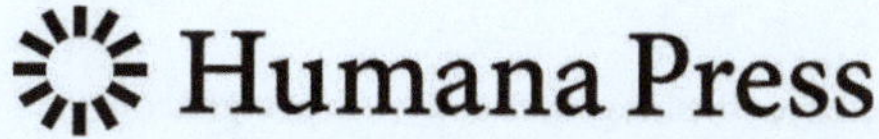

Editors
Ling Yuan
Department of Plant & Soil Sciences
College of Agriculture
University of Kentucky
Lexington, KY 40546-0236, USA
lyuan3@uky.edu

Sharyn E. Perry
Department of Plant & Soil Sciences
University of Kentucky
Lexington, KY 40546-0312, USA
sperr2@uky.edu

ISSN 1064-3745 e-ISSN 1940-6029
ISBN 978-1-61779-153-6 e-ISBN 978-1-61779-154-3
DOI 10.1007/978-1-61779-154-3
Springer New York Dordrecht Heidelberg London

Library of Congress Control Number: 2011931514

Printed on acid-free paper

Humana Press is part of Springer Science+Business Media (www.springer.com)

Preface

Transcription factors (TFs) are regulatory proteins that generally function by modulating local protein concentrations of limiting factors at the target promoters. TFs operate alone or in complexes with other molecules to activate or repress the recruitment of the basal transcriptional machinery to specific genes, thereby controlling transcription of DNA to mRNA. Close to 10% of higher plant genes encode TFs. The large numbers of TF genes in plant genomes are the result of higher rate of expansion in specific TF superfamilies compared to those in other biological kingdoms. TFs are pivotal in the regulation of plant development, reproduction, intercellular signaling, response to environment, cell cycle, and metabolism. Despite the importance of TFs and the rapid expansion of TF genes discovered through genomics and bioinformatics, only a few TFs have been subjected to extensive functional characterization. A monumental challenge facing plant molecular biologists is the mechanistic understanding of TF functions. This task is particularly challenging considering that TFs may bind to different *cis* elements throughout the genome with varying affinities and often function in a combinatorial fashion with other regulatory factors. Nevertheless, in recent years we have seen a significant advancement in the development of enabling technologies that facilitate the study of TFs.

The purpose of *Plant Transcription Factors* is to introduce the basic concepts and the detailed protocols of a series of commonly used tools for investigating plant TFs. In editing this volume, we envision that the readership will include seasoned plant molecular biologists as well as scientists new to the field of TFs. The chapters are contributed by many leading scientists in the respective areas. Sufficient experimental details are provided to minimize the need to consult additional references.

The organization of this volume first provides an initial discussion of select TF families in plants, with a focus on plant development. The chapters in Section II focus on approaches to identify a TF, either based on homology to TFs/TF families of interest, in association with a particular *cis* element, or via a phenotype (developmental, response to environment, metabolic consequence) of interest. Once identified, it is important to verify the function of the TF in control of a particular process or the ability to impact on gene expression (topics in Section III). Because TFs are most often members of large families, loss-of-function analysis can be hindered by redundancy. Chapter 5 introduces one approach that can address this redundancy.

TFs act in combinatorial fashion and particular complexes can have different consequences for gene expression. Therefore, it is important to identify protein interactions in which TFs are involved and how the interaction is mediated (Section IV, Chapters 8, 9, 10, and 11). Other post-transcriptional or post-translational control mechanisms of TFs occur and an example of this is covered in Chapter 12. Increasing examples of TFs that function non-cell-autonomously are being discovered and methods to assess intercellular trafficking are addressed in the chapter by Ahmad et al. (Chapter 13).

Section V examines TF interaction with DNA. Many TFs associate with DNA in a sequence-specific manner and Chapters 14 and 15 present approaches to characterize these DNA *cis* motifs. Finally a key to understanding how TFs actually impact on a process, be it developmental or response to a stimulus, involves determining what genes are directly controlled by the TF or TF complex of interest. Evaluation of the transcriptome in response to TF accumulation is one piece of this puzzle (Chapter 7), but assessment of direct interaction with regulatory regions of genes is also important and approaches to ascertain this are presented in Chapters 16, 17, and 18.

Finally, how can we use or even improve TFs to meet the current challenges we face in agriculture? Using TFs to modify specific metabolic pathways in order to improve crop quality is increasingly viewed as a powerful approach in plant biotechnology. However, because TFs commonly regulate more than one gene, to avoid undesirable (unintended) consequences it is critical that a TF is functionally characterized prior to use for the manipulation of plant pathways. Once understood, it is even possible to improve TFs and the final section of this volume discusses directed evolution to generate TFs that can more efficiently control desired processes.

These are exciting times in which to be investigating functions and roles of TFs. With the ever-increasing numbers of organisms with available genome sequence information and the technologies that come with this information, it is becoming possible to obtain a global or near-global view of TF function at a specific developmental stage or in response to a particular stimulus. A challenge in the future will be integrating all of this information to identify regulatory networks and mechanisms underpinning particular processes. We hope *Plant Transcription Factors* provides access of many necessary methods to all scientists who are interested in exploring the functions of TFs. We would like to thank Kathy Shen for her expert assistance in editing this volume. We also would like to thank all the authors who so enthusiastically contributed to this volume and acknowledge their patience and responsiveness to our requests for clarification.

Ling Yuan
Sharyn E. Perry

Contents

Contributors

MUNAWAR AHMAD • *Division of Applied Life Science (BK21 program), Environmental Biotechnology National Core Research Center, Plant Molecular Biology and Biotechnology Research Center (PMBBRC), Gyeongsang National University, Jinju, Korea*

GERCO C. ANGENENT • *Business Unit Bioscience, Plant Research International, Wageningen, The Netherlands; Laboratory of Molecular Biology, Wageningen University, Wageningen, The Netherlands*

KIM BOUTILIER • *Plant Research International, Wageningen, The Netherlands*

CHENGLIN CHAI • *Department of Plant Cellular and Molecular Biology, Plant Biotechnology Center, The Ohio State University, Columbus, OH, USA*

WON KYONG CHO • *Division of Applied Life Science (BK21 program), Environmental Biotechnology National Core Research Center, Plant Molecular Biology and Biotechnology Research Center (PMBBRC), Gyeongsang National University, Jinju, Korea*

STEFAN DE FOLTER • *Laboratorio Nacional de Genómica para la Biodiversidad (LANGEBIO), Centro de Investigación y de Estudios Avanzados del IPN (CINVESTAV-IPN), Irapuato, Guanajuato, México*

GEERT DE JAEGER • *Department of Plant Systems Biology, Flanders Interuniversity Institute for Biotechnology (VIB), Ghent University, Gent, Belgium*

DOMINIQUE EECKHOUT • *Department of Plant Systems Biology, Flanders Interuniversity Institute for Biotechnology (VIB), Ghent University, Gent, Belgium; Department of Plant Biotechnology and Genetics, Ghent University, Gent, Belgium*

VALÉRIE GAUDIN • *Institut Jean-Pierre Bourgin, UMR1318 INRA-AgroParisTech, INRA Centre de Versailles-Grignon, Versailles Cedex, France*

JAN GEERINCK • *Department of Plant Systems Biology, Flanders Interuniversity Institute for Biotechnology (VIB); Department of Plant Biotechnology and Genetics, Ghent University, Gent, Belgium*

SOPHIE GERMANN • *Institut National de la Santé et de la Recherche Médicale (Inserm), Centre Léon Bérard, Lyon Cedex, France*

DANIEL H. GONZALEZ • *Cátedra de Biología Celular y Molecular, Facultad de Bioquímica y Ciencias Biológicas, Instituto de Agrobiotecnología del Litoral (CONICET-UNL), Universidad Nacional del Litoral, Santa Fe, Argentina*

MAARTJE GORTE • *Molecular Genetics Group, Department of Biology, Faculty of Science, Utrecht University, Utrecht, The Netherlands*

ERICH GROTEWOLD • *Department of Plant Cellular and Molecular Biology, Plant Biotechnology Center, The Ohio State University, Columbus, OH, USA*

RENZE HEIDSTRA • *Molecular Genetics Group, Department of Biology, Faculty of Science, Utrecht University, Utrecht, The Netherlands*

ANNEKE HORSTMAN • *Plant Research International, Wageningen, The Netherlands*

LIJUN HUANG • *Division of Applied Life Science (BK21 program), Environmental Biotechnology National Core Research Center, Plant Molecular Biology and Biotechnology Research Center (PMBBRC), Gyeongsang National University, Jinju, Korea*

ENAMUL HUQ • *Section of Molecular Cell and Developmental Biology, Institute for Cellular and Molecular Biology, The University of Texas, Austin, TX, USA*

MIHO IKEDA • *Bioproduction Research Institute, National Institute of Advanced Industrial Science and Technology (AIST), Tsukuba, Ibaraki, Japan*
RICHARD G.H. IMMINK • *Plant Research International, Centre for BioSystems Genomics (CBSG), Wageningen, The Netherlands*
YUJI IWATA • *Huck Institutes of the Life Sciences, Pennsylvania State University, University Park, PA, USA*
KERSTIN KAUFMANN • *Laboratory of Molecular Biology, Wageningen University, Wageningen, The Netherlands; Business Unit Bioscience, Plant Research International, Wageningen, The Netherlands*
JAE-YEAN KIM • *Division of Applied Life Science (BK21 program), Environmental Biotechnology National Core Research Center, Plant Molecular Biology and Biotechnology Research Center (PMBBRC), Gyeongsang National University, Jinju, Korea*
NOZOMU KOIZUMI • *Graduate School of Life and Environmental Sciences, Osaka Prefecture University, Sakai, Osaka, Japan*
JASMINA KUREPA • *Department of Plant and Soil Sciences, University of Kentucky, Lexington, KY, USA*
MI-HYUN LEE • *Graduate School of Life and Environmental Sciences, Osaka Prefecture University, Sakai, Osaka, Japan*
JUN-JUN LIU • *Natural Resources Canada, Canadian Forest Services, Pacific Forestry Centre, Victoria, BC, Canada; Agriculture and Agri-Food Canada, Victoria, BC, Canada*
SERGIY LOPATO • *Australian Centre for Plant Functional Genomics, The University of Adelaide, Glen Osmond, SA, Australia*
NAYELLI MARSCH-MARTÍNEZ • *Laboratorio Nacional de Genómica para la Biodiversidad (LANGEBIO), Centro de Investigación y de Estudios Avanzados del IPN (CINVESTAV-IPN), Irapuato, Guanajuato, México*
KYOKO MATSUI • *Bioproduction Research Institute, National Institute of Advanced Industrial Science and Technology (AIST), Tsukuba, Ibaraki, Japan*
RAINER MELZER • *Department of Genetics, Friedrich Schiller University Jena, Jena, Germany*
NOBUTAKA MITSUDA • *Bioproduction Research Institute, National Institute of Advanced Industrial Science and Technology (AIST), Tsukuba, Ibaraki, Japan*
JOSE M. MUIÑO • *Business Unit Bioscience, Plant Research International, Wageningen, The Netherlands*
YUKARI NAGATOSHI • *Bioproduction Research Institute, National Institute of Advanced Industrial Science and Technology (AIST), Tsukuba, Ibaraki, Japan*
MASARU NAKATA • *Bioproduction Research Institute, National Institute of Advanced Industrial Science and Technology (AIST), Tsukuba, Ibaraki, Japan*
MASARU OHME-TAKAGI • *Bioproduction Research Institute, National Institute of Advanced Industrial Science and Technology (AIST), Tsukuba, Ibaraki, Japan*
YOSHIMI OSHIMA • *Bioproduction Research Institute, National Institute of Advanced Industrial Science and Technology (AIST), Tsukuba, Ibaraki, Japan*
ROBERT B. PAGE • *Department of Biology, Auburn University, Montgomery, AL, USA*
SITAKANTA PATTANAIK • *Department of Plant and Soil Sciences, Kentucky Tobacco Research & Development Center, University of Kentucky, Lexington, KY, USA*
SHARYN E. PERRY • *Department of Plant and Soil Sciences, University of Kentucky, Lexington, KY, USA*

GEERT PERSIAU • *Department of Plant Biotechnology and Genetics, Ghent University, Gent, Belgium; Department of Plant Systems Biology, Flanders Interuniversity Institute for Biotechnology (VIB), Ghent University, Gent, Belgium*

TATIANA PYVOVARENKO • *Australian Centre for Plant Functional Genomics, The University of Adelaide, Glen Osmond, SA, Australia*

YEONGGIL RIM • *Division of Applied Life Science (BK21 program), Environmental Biotechnology National Core Research Center, Plant Molecular Biology and Biotechnology Research Center (PMBBRC), Gyeongsang National University, Jinju, Korea*

JAN A. SMALLE • *Department of Plant and Soil Sciences, University of Kentucky, Lexington, KY, USA*

ARNOLD STROMBERG • *Department of Statistics, University of Kentucky, Lexington, KY, USA*

GÜNTER THEIßEN • *Department of Genetics, Friedrich Schiller University Jena, Jena, Germany*

EVELINE VAN DE SLIJKE • *Department of Plant Systems Biology, Flanders Interuniversity Institute for Biotechnology (VIB), Ghent University, Gent, Belgium; Department of Plant Biotechnology and Genetics, Ghent University, Gent, Belgium*

GERT VAN ISTERDAEL • *Department of Plant Systems Biology, Flanders Interuniversity Institute for Biotechnology (VIB), Ghent University, Gent, Belgium; Department of Plant Biotechnology and Genetics, Ghent University, Gent, Belgium*

JELLE VAN LEENE • *Department of Plant Systems Biology, Flanders Interuniversity Institute for Biotechnology (VIB), Ghent University, Gent, Belgium; Department of Plant Biotechnology and Genetics, Ghent University, Gent, Belgium*

IVANA L. VIOLA • *Cátedra de Biología Celular y Molecular, Facultad de Bioquímica y Ciencias Biológicas, Instituto de Agrobiotecnología del Litoral (CONICET-UNL), Universidad Nacional del Litoral, Santa Fe, Argentina*

JOSHUA R. WERKMAN • *Department of Plant and Soil Sciences, Kentucky Tobacco Research & Development Center, University of Kentucky, Lexington, KY, USA*

ERWIN WITTERS • *Department of Biology, Center for Proteome Analysis and Mass Spectrometry (CEPROMA), University of Antwerp, Antwerp, Belgium; Department of Ecophysiology, Biochemistry and Toxicology (EBT), University of Antwerp, Antwerp, Belgium; Flemish institute for Technological Research (VITO), Mol, Belgium*

YU XIANG • *Pacific Agri-Food Research Center, Agriculture and Agri-Food Canada, Summerland, BC, Canada*

ZIDIAN XIE • *Department of Plant Cellular and Molecular Biology, Plant Biotechnology Center, The Ohio State University, Columbus, OH, USA*

LING YUAN • *Department of Plant and Soil Sciences, Kentucky Tobacco Research & Development Center, University of Kentucky, Lexington, KY, USA*

YUMEI ZHENG • *Department of Plant and Soil Sciences, University of Kentucky, Lexington, KY, USA*

LING ZHU • *Section of Molecular Cell and Developmental Biology, Institute for Cellular and Molecular Biology, The University of Texas, Austin, TX, USA*

Section I

Introduction

Chapter 1

MADS and More: Transcription Factors That Shape the Plant

Rainer Melzer and Günter Theißen

Abstract

All major processes of life depend on differential gene expression, which is largely controlled by the activity of transcription factors (TFs). In plants many TFs are encoded by members of multigene families that expanded much more dramatically during land plant evolution than during the evolution of animals and fungi. Here we review typical features such as domain structure, DNA binding, and protein interactions of TFs from some families that have contributed to the development and evolution of plant-specific structures in especially important ways. Our survey includes the MADS-domain protein family involved in specifying meristem and organ identity; YABBY proteins controlling lamina outgrowth; TCP proteins controlling floral zygomorphy and apical dominance; and finally homeodomain proteins involved in stem-cell maintenance and many other processes. Common themes as well as interesting differences between these "molecular architects of plant body plans" will become apparent.

Key words: Transcription factor, MADS, YABBY, TCP, homeodomain, plant, body plan.

1. Introduction: Transcription Factors in Plants – It's a Family Business

All major processes of life, such as development and reproduction, depend on differential gene expression. To a large extent, the temporal and spatial expression of genes is regulated at the transcriptional level. Therefore, proteins that activate or repress transcription, termed transcription factors (TFs) if they bind directly to DNA, are of prime biological interest. TFs exert their function by binding to specific DNA sequences, termed *cis*-regulatory elements, in the promoters or enhancers of their target genes. They do so by a DNA-binding domain, which is almost always the most highly conserved part of the TFs. Therefore, it is often used to classify transcription factors and to assign them to different families or classes. In line with the unusual high conservation

L. Yuan, S.E. Perry (eds.), *Plant Transcription Factors*, Methods in Molecular Biology 754,
DOI 10.1007/978-1-61779-154-3_1, © Springer Science+Business Media, LLC 2011

of DNA-binding domains, those plant TFs that belong to families that are present in most eukaryotes usually do not share significant similarity with protein family members from the other kingdoms except in the respective DNA-binding domain.

Many TFs have a modular structure, comprising not only the DNA-binding domain but often also other distinct sections. Typical are domains for transcriptional activation or repression and for protein–protein interactions with TFs of the same or other families (1). Often TFs bind to DNA only as dimers, either homo- or heterodimers, in a sequence-specific way. The specific part of the *cis*-regulatory DNA sequence they bind to is usually a quite short stretch of typically only about 6–12 nucleotides.

Comparative bioinformatic analyses based on the availability of genome sequences of diverse land plants (embryophytes), ranging from the moss *Physcomitrella patens* (*P. patens*) to flowering plants (angiosperms) as distantly related as *Arabidopsis thaliana* (*A. thaliana*; thale cress) and *Oryza sativa* (*O. sativa*; rice), enabled a comprehensive identification of genes encoding for TFs (some little conserved putative TFs that have not been identified notwithstanding) (1, 2). It turned out that there are about 1600 genes encoding for TFs in the genome of *A. thaliana*, accounting for about 6% of the estimated 26,000 protein-coding genes (1). A similar figure has been determined for the number of genes encoding TFs in the genome of the distantly related flowering plant rice (1800 genes, 4.6%).

The TFs encoded by the *A. thaliana* genome can be classified into more than 40 major families (1). The three largest are the MYB superfamily, the basic helix-loop-helix (bHLH) family, and the APETALA2/ETHYLENE RESPONSIVE ELEMENT BINDING PROTEIN (AP2/EREBP) family, all having more than 120 members (1, 2). Single-copy genes encoding TFs appear to be rare, but include genes with important developmental functions such as the floral meristem identity gene *LEAFY* (*LFY*) and *SPOROCYTELESS/NOZZLE* (*SPL/NZL*), a "floral organ-building gene" with a central role in regulating anther cell differentiation. Similar numbers have been determined for plant genomes other than *A. thaliana* (2).

The high number of genes in many gene families in plants is in quite some contrast to the situation in other eukaryotes such as animals and fungi. For example, there are only 2–8 MADS-box genes in diverse animals that have been studied, ranging from nematode worms such as *Caenorhabditis elegans* and flies such as *Drosophila melanogaster* to several vertebrates, including humans (*Homo sapiens*), 1–4 genes in fungi including yeasts, but 23 genes in *P. patens*, 107 genes in *A. thaliana*, and 74 genes in *O. sativa* (2, 3). Obviously, many TF families shared between plants, animals, and fungi have undergone much more dramatic expansion in plants than in other eukaryotes (2). The reason for this is not simply a higher duplication rate of plant genomes, but because there is also a higher degree of expansion for TF genes compared to other plant genes (2). One reason is that plant TF genes have been preferentially retained after whole-genome duplications.

Several functional features of plant TFs could account for their higher expansion rate (2): (i) *Neofunctionalization*: Duplication of TF genes may have contributed to the evolution of regulatory novelties in an especially pronounced way in plants. Good cases in point may be developmental and morphological novelties of seed plants such as ovules, seeds, flowers, and fruits whose development evolved under the control of novel MADS-box gene subfamilies not present in non-seed plants (see below; (4)). (ii) *Subfunctionalization*: Gene expansion may be due to the ease of subfunctionalization among TF gene duplicates; the high number of partially redundant MADS-box genes may provide suitable examples (see below; (3)). (iii) *Dosage balance*: Due to requirements for dosage balance, duplications of all genes encoding proteins involved in protein complexes, as realized by whole-genome duplications, might be more tolerable than single gene duplications. Again, MADS-box genes may provide good cases in point, since at least some plant-specific subtypes (MIKC type) constitute DNA-bound multimeric complexes (floral quartets), a feature not known from animal or plant MADS-domain proteins (see below). Of course, these scenarios are not mutually exclusive and looking into the diversity of TFs in plants may provide examples for all of them.

With the several dozens of families comprising thousands of TF genes in any typical plant genome, a comprehensive review about all TFs in plants is by far beyond the scope of this chapter. Rather we try to focus on some basic features and biological functions of those TFs that are of special importance for the development of prominent traits of plant body plans, such as plant habit, or the symmetry or identity of plant organs. Therefore, we do not consider in more detail the "big three," i.e., MYB, bHLH, and AP2/EREBP, and some others such as WRKY proteins, because they are mainly involved in other biological processes such as stress response and control of secondary metabolism. We rather refer to excellent reviews that have recently been published, e.g., about AP2/EREBP proteins, bHLH proteins, and WRKY proteins (5–7).

Since our treatment is exemplary rather than comprehensive we apologize to all authors whose favorite gene family we ignored or whose work we could not cite due to space constraints. For the same reason, we mostly cite reviews rather than publications of original work. For anyone interested in a more global overview about the TFs of plants and other transcriptional regulators (TRs), we refer to several databases that have recently been established (8, 9).

2. MADS-Domain Proteins: Specifying Identity

In contrast to several other plant transcription factors, MADS-domain proteins are not exclusively found in plants, but rather in almost all eukaryotes (**Fig. 1.1**). They can be divided into type I and type II MADS-domain proteins, representatives of which were very likely present already in the most recent common ancestor of extant eukaryotes (10).

Even though animal and fungal MADS-domain proteins often fulfill essential functions, the number of MADS-box genes in these taxa is rather small. In contrast, the family of MADS-domain proteins experienced a tremendous expansion in plants, especially in seed plants (11). Though there is a growing body

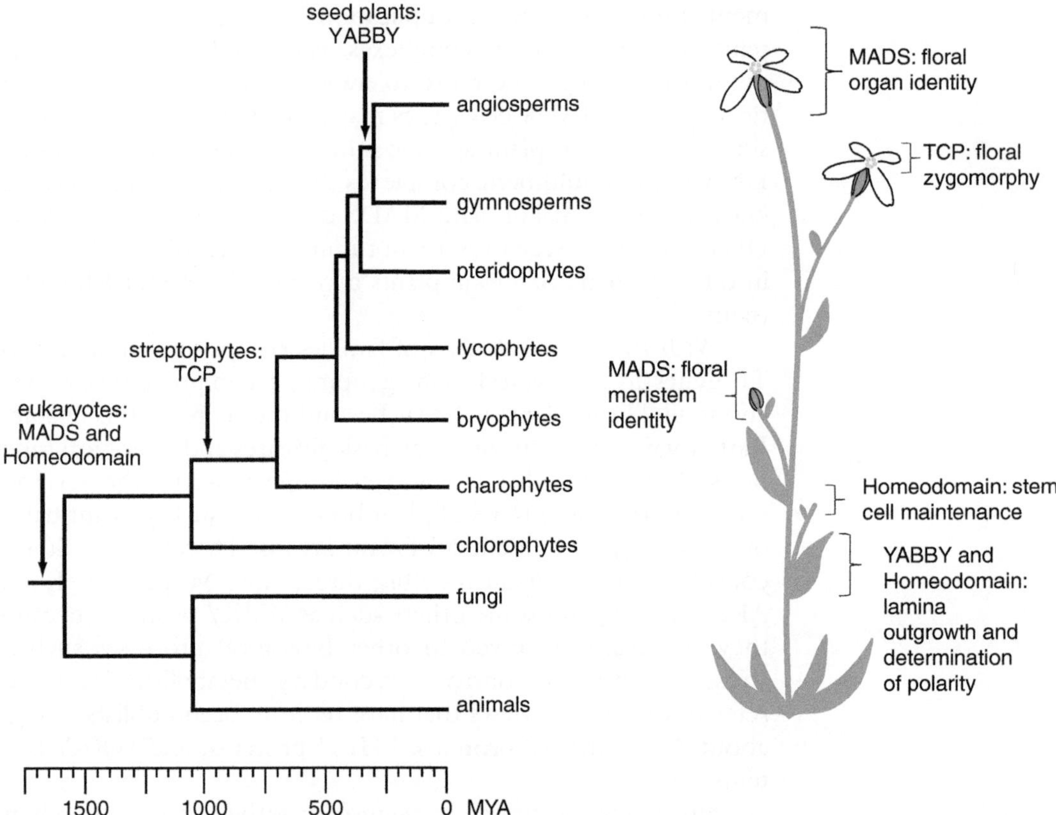

Fig. 1.1. Ancestry and functions of transcription factors found in plants. On the *left*, a very simplified phylogeny of animals, fungi, and plants is shown. The names of families of transcription factors at some internal nodes denote protein families that originated at the time interval represented by the respective branches of the phylogenetic tree, at the latest, as revealed by the presence of respective family members in extant taxa. The approximate age of the different taxa can be inferred from the timescale below the phylogeny (MYA, million years ago). On the *right*, typical and prominent functions of TF family members are illustrated on the picture of a generic plant. Note that the key words given are by no means meant to cover the whole spectrum of the functions of the respective family of transcription factors.

of evidence that some plant type I MADS-domain proteins are involved in female gametophyte development or early seed development (3, 12 and citations therein), the function of most plant type I proteins remains elusive. This is in contrast to type II MADS-domain proteins from plants, many of which fall into subfamilies that are functionally well characterized and highly conserved across all seed plants (13). Plant type II MADS-domain proteins possess a characteristic domain structure including the DNA-binding *M*ADS-domain, an *I*ntervening region, a *K*eratin-like domain, and a *C*-terminal domain (14). They are therefore often referred to as MIKC-type MADS-domain proteins. Many MIKC-type MADS-domain proteins are key regulators of plant development; we will focus on this subgroup of MADS-domain proteins here.

Many MIKC-type MADS-box genes are involved in reproductive development. The genetic interactions among these genes have been the subject of extensive research for the past 20 years (15). Most famous are the floral homeotic class A, B, C, D, and E MADS-box genes. Combinatorial interactions among these selector genes determine floral organ identities (for a review *see* (15, 16)). However, it is intriguing that MIKC-type genes are also involved in processes "upstream" and "downstream" of the developmental events of floral organ determination. For example, the flowering time genes *AGAMOUS-LIKE24* (*AGL24*) and *SUPRESSOR OF OVEREXPRESSION OF CONSTANS1* (*SOC1*) are involved in the switch from vegetative to reproductive development (17). They activate meristem identity genes like *APETALA1* (*AP1*) and *CAULIFLOWER* (*CAL*) that in turn activate the floral homeotic genes. Targets of floral homeotic genes again include MIKC-type genes like *SEEDSTICK* (*STK*), *SHATTERPROOF1* (*SHP1*), and *SHP2* that are involved in ovule and fruit development. In fact, every stage of reproductive development is at least partially controlled by an intricate network of MADS-box gene and MADS-domain protein interactions. Very often, the molecular basis of the genetic interactions is protein–protein interactions between the different MADS-domain proteins (14). Prominent examples are the floral homeotic class B proteins APETALA3 (AP3) and PISTILLATA (PI) that need to heterodimerize with each other to specify petal and stamen identity.

Depending on different interaction partners in space and time, many MADS-domain proteins can fulfill different functions. For example, the same proteins participate in different complexes to specify different floral organs or to regulate the transition from one stage of reproductive development to the other (18). AP1 is a good case in point here. By forming dimers with AGL24, AP1 first represses the class E floral homeotic gene *SEPALLATA3* (*SEP3*). However, as AP1 (possibly again in an AP1/AGL24 dimer) also represses *AGL24*, repression of *SEP3* is finally released (19, 20). Afterward, AP1 interacts with SEP3 and activates the floral homeotic genes *AGAMOUS*, *AP3*, and *SEP3* itself, thus ensuring that floral organ development proceeds.

Another important aspect of MADS-domain protein function that is nicely illustrated by this example is the importance of positive and probably also negative autoregulatory feedback loops. AP1/SEP3 dimers foster *SEP3* expression, whereas AP1/AGL24 dimers may lead to an inhibition of *AGL24* expression. Positive autoregulatory feedback loops have been demonstrated for a number of MADS-domain proteins. For example, they ensure continuous expression of organ identity genes even if the factors that initially activated expression of the respective genes are no longer present. Many of these feedback loops are assumed to be direct, i.e., the respective proteins bind to the promoters of the genes that encode them to sustain expression (21).

Negative autoregulatory feedback loops may help to facilitate the switch from one stage of reproductive development to another (21). The proposed inhibition of *AGL24* expression by an AP1/AGL24 heterodimer ensures that, once AP1 is activated and floral organ formation is initiated, flowering time genes are repressed. Positive as well as negative autoregulatory feedback loops are likely involved in sharpening the expression domains of MADS-box genes. Positive autoregulation enables the amplification of initially small differences in expression whereas negative autoregulation is a "built-in" control system that disables gene expression at the wrong place or time.

Yet another aspect of MADS-domain protein function is the ability to form tetrameric, DNA-bound complexes. In case of floral homeotic proteins, these DNA-bound tetramers are called "floral quartets" (22). Combinatorial quartet formation is assumed to determine floral organ identity. For example, AP3/PI and AP1/SEP3 heterodimers likely interact to form a DNA-bound tetramer that is involved in petal specification, whereas a tetramer composed of an AP3/PI and an AG/SEP3 heterodimer probably determines stamen identity. SEP3 and AP1, but neither AP3 nor PI nor AG possess transcription activation domains. Therefore, one function of tetramerization is probably to supply transcription activation potential to proteins that lack this capacity (23). Importantly, the tetramer is a dimer of two dimers that are bound to neighboring DNA sites and contact each other by looping the intervening DNA. This mode of DNA binding is cooperative, as one dimer helps the other dimer to bind to DNA. One important aspect of cooperative DNA binding is that small changes in protein concentration suffice to increase occupancy of target DNA sites substantially. Thus, cooperativity in DNA binding might enable MADS-domain proteins to function as "molecular switches" that bind to their target genes in an almost all-or-none fashion (24).

Experimental evidence for many of the properties described here is available for a few MIKC-type MADS-domain proteins only. However, combinatorial interactions, autoregulation, multimerization, and cooperative DNA binding may be features common to many MADS-domain proteins. These attributes might have been of critical importance for MADS-domain proteins to become key regulators in the molecular network controlling reproductive development. Strikingly, MIKC-type MADS-domain proteins fulfill their function as regulators of reproductive development predominantly by interacting with each other. Though accumulating experimental evidence suggests that they regulate transcription by interacting with chromatin-modifying complexes (surprisingly, so far especially complexes with repressive functions have been identified (25)), interactions between MIKC-type MADS-domain proteins and genuine transcription factors from other families have barely been reported. However, all MADS-domain proteins bind as dimers to DNA elements termed CArG boxes, short stretches of DNA with the consensus sequence 5′-CC(A/T)$_6$GG-3′ or very similar sequences. So far, there is little evidence that different complexes bind with different affinities to certain CArG boxes, raising the question of how MADS-domain proteins achieve target gene specificity (21, 26). Whether additional yet unidentified cofactors are required to confer target gene specificity or whether the specificity lies in so far unexplored properties of the MIKC-type MADS-domain protein complexes will be one major challenge of future research.

3. YABBY Proteins: Controlling Lamina Outgrowth

YABBY genes encode a small family of transcription factors that are found in gymnosperms and angiosperms, but not outside the seed plants (**Fig. 1.1**) (27). YABBY transcription factors possess a characteristic zinc finger domain close to the N-terminal end and a helix-loop-helix "YABBY" domain close to the C-terminal end (28). The YABBY domain is partially similar to the HMG domain of high-mobility group non-histone proteins (28). Knowledge on DNA binding of YABBY proteins is scarce. Studies on the YABBY protein FILAMENTOUS FLOWER (FIL) from *A. thaliana* suggest that it binds as a dimer to DNA in a non-sequence-specific manner and that the YABBY domain is involved in DNA binding (29). In contrast, the rice YABBY1 protein binds to a gibberellin responsive element in a DNA sequence-specific manner (30). Protein–protein interaction studies indicate that many

YABBY proteins interact with each other, but interactions with other transcription factors and co-repressor complexes have also been reported (31, 32).

The *A. thaliana* genome contains six YABBY genes. Of these, *FIL*, *YABBY2*, and *YABBY3* are specifically expressed in the abaxial domain of all above-ground lateral organs, whereas *CRABS CLAW* (*CRC*) expression is restricted to the abaxial domain of carpels and *INNER NO OUTER* (*INO*) is expressed in the outer integuments of the ovules (the expression pattern of *YABBY5* remains to be determined) (28, 32). As indicated by these expression patterns, the YABBY genes are involved to different extents in the development of certain organs. For example, *CRC* is involved in nectary specification and carpel polarity whereas *INO* is involved in ovule development. *FIL* is involved in floral organ development and also functions in leaf development. This was only uncovered by double mutant analysis, however, as *FIL* acts redundantly with *YAB3* in this pathway (28, 32). The distinct developmental functions of *A. thaliana* YABBY genes notwithstanding, one important commonality is that they all appear to be involved in the determination of abaxial cell identity of lateral organs (28, 32). In *A. thaliana*, loss-of-function mutants of YABBY genes show adaxialized phenotypes whereas ectopic expression of YABBY genes leads to abaxialization of lateral organs (28, 33). Based on expression patterns and mutant analyses of YABBY genes from other eudicots, it was thus proposed that a function in determination of abaxial cell identity is conserved in eudicots (28, 33).

The molecular evolution of YABBY genes is beginning to be unraveled. Five distinct clades of YABBY genes can be assigned in phylogenetic reconstructions, covering *CRC*-like, *INO*-like, *YAB5*-like, *YAB2*-like, and *FIL/YAB3*-like genes, respectively (34). Within these clades, orthologous genes can be found between eudicots, monocots, and basal angiosperms, suggesting that the most recent common ancestor of all angiosperms already possessed a minimum of five YABBY genes (27, 34).

Given the high degree of conservation of YABBY genes along with their polar expression in *A. thaliana* it came as a surprise that in monocots YABBY gene expression is not confined to abaxial domains (35). For example, YABBY genes from maize are expressed in the adaxial domain of developing leaf primordia (35). It was concluded that determining polarity of lateral organs is very likely not the ancestral function of YABBY genes in angiosperms (33). As loss of YABBY function in leaves also coincides with the loss of lamina outgrowth, one current view is that the ancestral YABBY function was not in specifying abaxial cell fate but rather in promoting lamina outgrowth along an adaxial–abaxial boundary (33). Genetic analyses meanwhile revealed that, with respect to polar development, YABBY genes are downstream of conserved polarity determinants like class III homeodomain-leucine zipper (HD-ZIP III) proteins. Interestingly, whereas *A. thaliana* YABBY proteins are repressed by HD-ZIP III proteins, HD-ZIP III proteins positively regulate maize YABBY genes (35). Thus, the regulation of YABBY genes by upstream factors may have diverged during evolution, leading to the divergent expression patterns of YABBY genes observed in different species (33).

Thus far, only a limited number of YABBY genes have been studied. It is therefore not clear to which extent the present observations can be generalized. However, the divergent expression patterns of YABBY genes provide a fascinating starting point for further studies on the diversification of gene functions.

4. TCP Proteins: Conferring Floral Zygomorphy and Apical Dominance

TCP proteins represent plant-specific TFs with a non-canonical bHLH motif involved in DNA binding and protein–protein interactions. Some family members have important functions in the control of cell proliferation in developing tissues and thus contribute in a significant way to plant shape. In one of the most spectacular and widely known TCP gene mutants the *CYCLOIDEA* (*CYC*) and *DICHOTOMA* (*DICH*) genes of snapdragon (*Antirrhinum majus*) have lost their function, leading to peloric, radially symmetrical (actinomorphic) rather than bilaterally symmetrical (zygomorphic) flowers. In another famous mutant the *TEOSINTE BRANCHED1* (*TB1*) gene of maize (*Zea mays* ssp. *mays*) is affected. Maize *tb1* loss-of-function mutants show excessive shoot branching. Thus *TB1* is a major determinant of apical dominance which distinguishes maize from its heavily branched wild ancestor, teosinte (*Zea mays* ssp. *parviglumis*). *TB1*, *CYC*, and the *PROLIFERATING CELL FACTORS1* and *2* (*PCF1* and *2*) from rice were the founding members that were used to define the TCP family (reviewed in (36)).

TCP proteins appear to be absent from the chlorophyte *Chlamydomonas reinhardtii*, but have been identified in some charophyte green algae as well as in mosses, lycophytes, and ferns, suggesting that they originated near the base of streptophytes about 700 mya (**Fig. 1.1**) (36). While in non-seed plants there appear to be about half a dozen TCP genes, more than 20 members have been identified in whole-genome analyses on diverse flowering plants (36).

Based on differences in their TCP domains, two types of TCP proteins have been distinguished, class I (also known as PCF and TCP-P class) and class II (TCP-C class), with class I including the PCF1 and 2 of rice and class II TB1 and CYC (36). Members of both classes have been identified in some green algae belonging to the streptophytes, and it is thus unclear whether class I or class II proteins resemble the ancestral TCPs more closely. While the class I proteins of angiosperms are relatively closely related, class II proteins can be further subdivided into the CINCINNATA (CIN) clade and the CYC/TB1 (or ECE) clade (36). Since CIN, but not CYC/TB1 clade members are present in mosses and lycophytes, the CIN clade appears to be more ancient than the CYC/TB1 clade. Further duplications led to a total of 24 TCP genes in *A. thaliana* including 13 class I, 8 *CIN*, and 3 *CYC/TB1* class genes and almost the same gene numbers (26 genes, thereof 13 class I, 10 CIN and 3 CYC/TB1) in *O. sativa* (36). Two gene duplications near the base of core eudicots gave rise to three types of genes, *CYC1*-like, *CYC2*-like, and *CYC3*-like genes, with *CYC1*-like genes having retained *TB1* roles in the control of shoot branching, and *CYC2*-like genes (including *CYCLOIDEA*) having key roles in the evolution of floral dorsoventral asymmetry (zygomorphy). The *CYC3* subclade comprises genes such as *BRANCHED2* (*BRC2*) from *A. thaliana*. These genes are expressed in branch and flower primordia and appear to play minor roles in the control of shoot branching; their function during flower development is still unclear (36).

Most class I single mutants that have been analyzed so far show only mild, if any, deviations from wild-type phenotype, probably due to functional redundancy between closely related genes. Concerning class II genes, *CIN* clade members appear to be involved in lateral organ development, as exemplified by the name giving *cin* mutant of snapdragon, which shows larger leaves with altered shape and crinkled surface. The reason is that in *cin* loss-of-function mutants, leaf cells keep on dividing for a longer period of time compared with wild-type plants. This is in line with the notion that class II TCP genes have generally a role in preventing growth and proliferation (36).

In contrast, CYC/TB1 clade members are involved in the development of axillary meristems giving rise to lateral shoots or flowers (36). As revealed by mutant analysis, CYC-type TFs determine dorsal identity within flowers; by interaction with MYB TFs and cell cycle genes, their activity may lead to alterations in size, shape, or number of organs and hence to the development of zygomorphic (dorsoventral asymmetric, bilaterally symmetric) flowers (37). There is evidence that the multiple independent origins of zygomorphic flowers in different lineages of core eudicots are based on the parallel recruitment of CYC homologues (37). Whether this parallel recruitment applies even to more distantly related angiosperms (such as the spectacularly zygomorphic flowers of orchids, which are monocotyledonous plants) remains to be seen. Recent data suggest that CYC-like genes do not only control the development of dorsal and lateral floral organs, but that in some species they are also involved in the abortion of ventral stamens (38). This reveals that CYC-like genes played a broader role in the evolution of flower structure than previously thought.

While CYC TFs may shape the symmetry of flowers, TB1 activity affects whole plant architecture as revealed by the unrestrained outgrowth of axillary meristems in *tb1* loss-of-function mutants (see above). One may hypothesize that TB1 acts as a negative regulator of meristem proliferation and organ growth in maize.

The bHLH motif of TCP proteins is 59 amino acids long. The basic part is highly conserved in all TCP proteins and is necessary but not sufficient for DNA binding. As revealed by random binding site selection experiments and electrophoretic mobility shift assays (EMSA), TCP proteins bind similar yet different consensus binding sites for class I and class II factors, i.e., sequences resembling 5′-GGNCCCAC-3′ and 5′-G(T/C)GGNCCC-3′, respectively (36). DNA binding appears to require dimer formation. TCP proteins preferentially form heterodimers, but as far as known only among members of the same class. For many TCP proteins, mono- or bipartite nuclear localization signals (NLS) have been identified and nuclear localization has been demonstrated experimentally. Some TCP proteins, however, might be targeted to the chloroplast and control the transcription of chloroplast genes.

Specific TCP proteins can act as transcriptional activators, repressors, and some maybe even as both. The molecular mechanisms by which TCP proteins control transcription are still poorly understood. Their trans-activation potential (if any) appears quite limited, and it appears quite likely that at least some TCP proteins require interaction with other proteins to control the transcription of target genes (36). Potential TCP binding sites in the DNA of target genes are often associated with other *cis*-acting elements, suggesting that TCP proteins function as part of multimeric complexes of TFs.

5. Homeodomain Proteins: The Check of All Traits

Homeodomain proteins are characterized by the homeodomain (HD), a typical DNA-binding domain of about 60 amino acids that is encoded by the homeobox (HB). The homeodomain folds into a characteristic three-dimensional structure containing three alpha-helices, of which the second and third form a helix-turn-helix motif. TFs that contain a homeodomain have been found in all eukaryotes that have been investigated so far, but not in prokaryotes (**Fig. 1.1**). Remarkably, most homeodomain proteins are able to bind to DNA as monomers with high affinity, by interactions made by the third helix (helix III, recognition helix) and a disordered N-terminal arm located beyond helix I (39).

Homeobox genes are widely known among biologists due to their essential contributions to the development of animal body plans, as first revealed by pioneering studies in the fruit fly (*D. melanogaster*). The first homeobox gene discovered in plants was *KNOTTED1* (*KN1*) from maize, named after a characteristic leaf phenotype generated by ectopic expression of the gene. Since then, numerous homeobox genes have been isolated from a wide variety of plants, and they revealed an involvement in almost all developmental processes. However, the mutants known in plant homeobox genes typically do not show homeotic transitions, i.e., the development of organs at wrong positions within a given body plan, implying that plant homeodomain TFs usually do not exert homeotic functions in specifying organ identity (39) (for an exciting exception see below).

Based on specific distinguishing features, such as sequence and location of the homeobox, association with other domains, size, and gene structure, the members of the superfamily have been classified into about half a dozen to 14 families, or classes, depending on the different authors. For example, Ariel et al. (39) define six different families, which are the HD-Zip proteins, in which the homeodomain is associated with a leucine zipper; the PHD finger proteins, which have a homeodomain associated with a finger domain; Bell proteins, which have a characteristic Bell domain; ZF-HD proteins, which have a zinc finger associated with a homeodomain; Wuschel-related (WOX) homeodomain proteins; and finally KNOTTED-related (KNOX) homeodomain proteins. There is evidence that at least some of these subfamilies have already been established before the separation of the lineages that led to animals/fungi and plants (39).

According to the most comprehensive analysis up to now, there are 110 homeobox genes in the flowering plants *A. thaliana* and rice, 66 homeobox genes in the moss *P. patens*, and 5 homeobox genes in the unicellular green alga *C. reinhardtii* (40), reflecting an expansion of the homeobox genes with land plant evolution as observed with other gene families encoding transcription factors. In the following, we have a closer look at two of the families of plant homeodomain TFs that contribute to plant shape in especially important ways.

HD-Zip TFs exhibit a combination of a homeodomain with a leucine zipper (Zip) immediately downstream of the HD which is unique to plants. HD-Zip proteins bind to DNA as dimers, with the Zip sequence acting as a dimerization motif. They function in a wide variety of processes, including stress response and development. Based on the sequence of the HD-Zip domain determining DNA-binding specificities, gene structures, additional conserved motifs, and function, HD-Zip proteins have been classified into four subfamilies, termed HD-Zip I–IV.

In *A. thaliana*, the HD-Zip I subfamily comprises 17 members. Most of them bind as dimers to the pseudopalindromic sequence 5′-CAAT(A/T)ATTG-3′. HD-Zip I proteins function as positive or negative regulators of development; they are involved in diverse processes such as response to abiotic stress or abscisic acid (ABA), de-etiolation, and blue-light perception signaling (39).

The HD-Zip II subfamily of *A. thaliana* comprises nine members. These proteins bind as dimers to the pseudopalindromic sequence 5′-CAAT(C/G)ATTG-3′. There is evidence that in an oxidative environment these proteins form high molecular weight multimers through intermolecular Cys–Cys bridges that cannot be transported into the nucleus. Thus, intriguingly, the redox state of the cell might regulate the activity of HD-Zip II proteins. The expression of HD-Zip II genes is generally affected by illumination conditions in photosynthetically active tissues, and they exert functions, e.g., in the shade avoidance response, which are well in line with this finding (39).

The HD-Zip III subfamily of *A. thaliana* is arguably the best characterized subfamily of HD-Zip TFs. It is composed of five members, termed PHABULOSA (PHB), PHAVOLUTA (PHV), REVOLUTA (REV), INCURVATA (ICU4)/CORONA (CNA), and ARABIDOPSIS THALIANA HOMEOBOX-8 (ATHB-8) (but most of them are also known under synonymous names). PHV binds in vitro to 5′-GTAAT(G/C)ATTAC-3′ with high affinity; the binding of the other proteins is not well studied. Members of the HD-Zip III subfamily are involved in different developmental processes, playing partially redundant, distinct, or even antagonistic roles. PHB, PHV, and REV, for example, have overlapping functions during embryogenesis and in the determination of leaf polarity (39). Recent investigations have shown that ectopic expression of HD-ZIP III TFs in the basal pole produces seedlings with a complete transformation of the root pole into a second shoot pole (41), implying a strong homeotic effect and thus identifying the first plant homeobox genes with a homeotic function. This way, HD-ZIP III proteins have been identified as master regulators of embryonic apical fate in *A. thaliana*.

The HD-Zip IV subfamily of *A. thaliana* comprises 16 members. They are also called the GLABRA2 (GL2) family after its founding member. HD-Zip IV proteins appear to bind to quite different DNA sequences that are characterized by the presence of a 5′-TAAA-3′ core. All HD-Zip IV proteins are expressed in the outer cell layer of plant organs. This is exactly where they exert their functions, e.g., by establishing cell fates in the epidermis, thus affecting traits such as anthocyanin accumulation and trichome formation (39).

WUSCHEL (WUS) from *A. thaliana* is the prototypic member of the WOX protein family. It is expressed in the cells of the organizing center of the shoot apical meristem, where it regulates the maintenance of stem cells. A closely related paralogue, WOX5, has a stem cell promoting function in roots. WUS has recently been shown to act as both a transcriptional repressor in stem cell regulation and an activator of floral patterning (42). WOX genes form a plant-specific clade and play important roles not only in stem cell maintenance but also in other key developmental processes such as embryonic patterning and organ formation. These functions are based on WOX protein activities in the promotion of cell division and/or the prevention of premature cell differentiation. WOX genes can be subdivided into three subclades, termed WUS (or modern), intermediate, and ancient (or WOX13-related) clades. The WUS clade is seed plant specific, and paralogous pairs of WUS and WOX5 even appear to be restricted to angiosperms, suggesting that discrete shoot and root stem cell promoting functions are an innovation that may have contributed to the rapid evolutionary "success" (e.g., in terms of species and ecological importance) of the flowering plants (43).

6. Outlook: Studying the TFs of Plants

As indicated above, TFs contribute in a way to plant development and evolution that can hardly be overestimated. Despite the fact that most of them come as members of a number of gene families, the diversity of plant TFs may appear overwhelming and puzzling at first glance. However, a closer look reveals typical themes, such as their sequence-specific binding to relatively short DNA motifs and the functional importance of dimerization or even multimerization. It is exactly these features that make TFs very well-suited for often quite simple yet powerful assays on protein–protein interactions (such as the Yeast Two-Hybrid (Y2H) assay), protein–DNA interactions (such as EMSA), and quite a number of more complex approaches, such as X-ray crystallography and NMR. But from the simplest to the most complex approach, the functional importance of plant TFs suggests that they well deserve all these efforts.

References

1. Riechmann, J. L. (2006) Transcription factors of *Arabidopsis* and rice: a genomic perspective. In: Grasser, K. D. (ed) Regulation of transcription in plants. Blackwell, Oxford. *Annu. Plant Rev.* **29**, 28–53.
2. Shiu, S.-H., Shih, M.-C., and Li, W.-H. (2005) Transcription factor families have much higher expansion rates in plants than in animals. *Plant Physiol.* **139**, 18–26.

3. Gramzow, L. and Theißen, G. (2010) A hitchhiker's guide to the MADS world of plants. *Genome Biol.* **11**, 214.
4. Theißen, G., Becker, A., Di Rosa, A., Kanno, A., Kim, J. T., Münster, T., Winter, K. U., and Saedler, H. (2000) A short history of MADS-box genes in plants. *Plant Mol. Biol.* **42**, 115–149.
5. Dietz, K.-J., Vogel, M. O., and Viehhauser, A. (2010) AP2/EREBP transcription factors are part of gene regulatory networks and integrate metabolic, hormonal and environmental signals in stress acclimation and retrograde signalling. *Protoplasma.* **245**, 3–14.
6. Pires, N. and Dolan, L. (2010) Origin and diversification of basic-helix-loop-helix proteins in plants. *Mol. Biol. Evol.* **27**, 862–874.
7. Rushton, P. J., Somsich, I. E., Ringler, P., and Shen, Q. J. (2010) WRKY transcription factors. *Trends Plant Sci.* **15**, 247–258.
8. Pérez-Rodriguez, P., Riano-Pachón, D. M., Guedes Correa, L. G., Rensing, S. A., Kersten, B., and Mueller-Roeber, B. (2010) PlnTFDB: updated content and new features of the plant transcription factor database. *Nucleic Acids Res.* **38**, D822–D827.
9. Richard, S., Lang, D., Reski, R., Frank, W., and Rensing, S. A. (2007) PlanTAPDB, a phylogeny-based resource of plant transcription-associated proteins. *Plant Physiol.* **143**, 1452–1466.
10. Gramzow, L., Ritz, M. S., and Theissen, G. (2010) On the origin of MADS-domain transcription factors. *Trends Genet.* **26**, 149–153.
11. Parenicova, L., de Folter, S., Kieffer, M., Horner, D. S., Favalli, C., Busscher, J., et al. (2003) Molecular and phylogenetic analyses of the complete MADS-box transcription factor family in *Arabidopsis*: new openings to the MADS world. *Plant Cell* **15**, 1538–1551.
12. Bemer, M., Gordon, J., Weterings, K., and Angenent, G. C. (2010) Divergence of recently duplicated M gamma-type MADS-box genes in *Petunia*. *Mol. Biol. Evol.* **27**, 481–495.
13. Becker, A. and Theißen, G. (2003) The major clades of MADS-box genes and their role in the development and evolution of flowering plants. *Mol. Phylogenet. Evol.* **29**, 464–489.
14. Kaufmann, K., Melzer, R., and Theißen, G. (2005) MIKC-type MADS-domain proteins: structural modularity, protein interactions and network evolution in land plants. *Gene* **347**, 183–198.
15. Causier, B., Schwarz-Sommer, Z., and Davies, B. (2010) Floral organ identity: 20 years of ABCs. *Semin. Cell Dev. Biol.* **21**, 73–79.
16. Krizek, B. A. and Fletcher, J. C. (2005) Molecular mechanisms of flower development: an armchair guide. *Nat. Rev. Genet.* **6**, 688–698.
17. Liu, C., Thong, Z. H., and Yu, H. (2009) Coming into bloom: the specification of floral meristems. *Development* **136**, 3379–3391.
18. Castillejo, C., Romera-Branchat, M., and Pelaz, S. (2005) A new role of the *Arabidopsis SEPALLATA3* gene revealed by its constitutive expression. *Plant J.* **43**, 586–596.
19. Gregis, V., Sessa, A., Dorca-Fornell, C., and Kater, M. M. (2009) The *Arabidopsis* floral meristem identity genes AP1, AGL24 and SVP directly repress class B and C floral homeotic genes. *Plant J.* **60**, 626–637.
20. Liu, C., Xi, W. Y., Shen, L. S., Tan, C. P., and Yu, H. (2009) Regulation of floral patterning by flowering time genes. *Dev. Cell* **16**, 711–722.
21. de Folter, S. and Angenent, G. C. (2006) Trans meets cis in MADS science. *Trends Plant Sci.* **11**, 224–231.
22. Theißen, G. (2001) Development of floral organ identity: stories from the MADS house. *Curr. Opin. Plant Biol.* **4**, 75–85.
23. Honma, T. and Goto, K. (2001) Complexes of MADS-box proteins are sufficient to convert leaves into floral organs. *Nature* **409**, 525–529.
24. Theißen, G. and Melzer, R. (2007) Molecular mechanisms underlying origin and diversification of the angiosperm flower. *Ann. Bot.* **100**, 603–619.
25. Liu, Z. C. and Mara, C. (2010) Regulatory mechanisms for floral homeotic gene expression. *Semin. Cell Dev. Biol.* **21**, 80–86.
26. Sablowski, R. (2010) Genes and functions controlled by floral organ identity genes. *Semin. Cell Dev. Biol.* **21**, 94–99.
27. Floyd, S. K. and Bowman, J. L. (2007) The ancestral developmental tool kit of land plants. *Int. J. Plant Sci.* **168**, 1–35.
28. Bowman, J. L. (2000) The YABBY gene family and abaxial cell fate. *Curr. Opin. Plant Biol.* **3**, 17–22.
29. Kanaya, E., Nakajima, N., and Okada, K. (2002) Non-sequence-specific DNA binding by the FILAMENTOUS FLOWER protein from *Arabidopsis thaliana* is reduced by EDTA. *J. Biol. Chem.* **277**, 11957–11964.
30. Dai, M. Q., Zhao, Y., Ma, Q. F., Hu, Y., Hedden, P. F., Zhang, Q., et al. (2007) The rice *YABBY1* gene is involved in the feedback regulation of gibberellin metabolism. *Plant Physiol.* **144**, 121–133.

31. Sieber, P., Petrascheck, M., Barberis, A., and Schneitz, K. (2004) Organ polarity in *Arabidopsis.* NOZZLE physically interacts with members of the YABBY family. *Plant Physiol.* **135**, 2172–2185.
32. Stahle, M. I., Kuehlich, J., Staron, L., von Arnim, A. G., and Golz, J. F. (2009) YABBYs and the transcriptional corepressors LEUNIG and LEUNIG_HOMOLOG maintain leaf polarity and meristem activity in *Arabidopsis. Plant Cell* **21**, 3105–3118.
33. Husbands, A. Y., Chitwood, D. H., Plavskin, Y., and Timmermans, M. C. P. (2009) Signals and prepatterns: new insights into organ polarity in plants. *Genes Dev.* **23**, 1986–1997.
34. Toriba, T., Harada, K., Takamura, A., Nakamura, H., Ichikawa, H., Suzaki, T., et al. (2007) Molecular characterization the YABBY gene family in *Oryza sativa* and expression analysis of *OsYABBY1. Mol. Genet. Genomics* **277**, 457–468.
35. Kidner, C. A. and Timmermans, M. C. P. (2007) Mixing and matching pathways in leaf polarity. *Curr. Opin. Plant Biol.* **10**, 13–20.
36. Martin-Trillo, M. and Cubas, P. (2009) TCP genes: a family snapshot ten years later. *Trends Plant Sci.* **15**, 31–39.
37. Preston, J. C. and Hileman, L. C. (2009) Developmental genetics of floral symmetry evolution. *Trends Plant Sci.* **14**, 147–154.
38. Hileman, L. C. and Cubas, P. (2009) An expanded evolutionary role for flower symmetry genes. *J. Biol.* **8**, 90.
39. Ariel, F. D., Manavella, P. A., Dezar, C. A., and Chan, R. L. (2007) The true story of the HD-Zip family. *Trends Plant Sci.* **12**, 419–426.
40. Mukherjee, K., Brocchieri, L., and Burglin, T. R. (2009) A comprehensive classification and evolutionary analysis of plant homeobox genes. *Mol. Biol. Evol.* **26**, 2775–2794.
41. Smith, Z. R. and Long, J. A. (2010) Control of Arabidopsis apical-basal embryo polarity by antagonistic transcription factors. *Nature* **464**, 423–426.
42. van der Graaff, E., Laux, T., and Rensing, S. A. (2009) The WUS homeobox-containing (WOX) protein family. *Genome Biol.* **10**, 248.
43. Nardmann, J., Reisewitz, P., and Werr, W. (2009) Discrete shoot and root stem cell-promoting WUS/WOX5 functions are an evolutionary innovation of angiosperms. *Mol. Biol. Evol.* **26**, 1745–1755.

Section II

Identification of Transcription Factors

Chapter 2

In Silico Mining and PCR-Based Approaches to Transcription Factor Discovery in Non-model Plants: Gene Discovery of the WRKY Transcription Factors in Conifers

Jun-Jun Liu and Yu Xiang

Abstract

WRKY transcription factors are key regulators of numerous biological processes in plant growth and development, as well as plant responses to abiotic and biotic stresses. Research on biological functions of plant WRKY genes has focused in the past on model plant species or species with largely characterized transcriptomes. However, a variety of non-model plants, such as forest conifers, are essential as feed, biofuel, and wood or for sustainable ecosystems. Identification of WRKY genes in these non-model plants is equally important for understanding the evolutionary and function-adaptive processes of this transcription factor family. Because of limited genomic information, the rarity of regulatory gene mRNAs in transcriptomes, and the sequence divergence to model organism genes, identification of transcription factors in non-model plants using methods similar to those generally used for model plants is difficult. This chapter describes a gene family discovery strategy for identification of WRKY transcription factors in conifers by a combination of in silico-based prediction and PCR-based experimental approaches. Compared to traditional cDNA library screening or EST sequencing at transcriptome scales, this integrated gene discovery strategy provides fast, simple, reliable, and specific methods to unveil the WRKY gene family at both genome and transcriptome levels in non-model plants.

Key words: Conifer, gene discovery, in silico mining, non-model plant, PCR-based gene cloning, WRKY transcription factor.

1. Introduction

WRKY proteins constitute a superfamily of plant transcription factors involved in plant growth, development, and plant interactions with environmental factors (1). These proteins contain one or two copies of a DNA-binding domain, designated as a WRKY domain, which is composed of about 50–60 amino acids with the N-terminal conserved motif WRKYGXK and a zinc

L. Yuan, S.E. Perry (eds.), *Plant Transcription Factors*, Methods in Molecular Biology 754,
DOI 10.1007/978-1-61779-154-3_2, © Springer Science+Business Media, LLC 2011

finger motif (C-X_{4-8}-C-X_{22-28}-H-X_{1-2}-H/C) at the C-terminus (1). WRKY proteins regulate transcript expression of downstream genes through their interaction with the *cis*-element W-box, (C/T)TGAC(T/C), localized in the promoter regions of the target genes (1–4). Large WRKY families have been identified in *Arabidopsis thaliana* and in rice (*Oryza sativa*) with 72 and 105 members, respectively (5, 6). Biological functions of a few WRKY genes have been characterized in angiosperms (1). Those co-expressed WRKY genes play their biological roles through co-regulatory networks in *Arabidopsis* and rice (7). All angiosperm plants analyzed to date have numerous WRKY genes, and angiosperm WRKY genes are classified into three distinct groups and five subgroups (I, IIa+b, IIc, IId+e, and III) based on the copy number of the WRKY domain, intron positions and phases, and the structural features of the zinc finger motif (1, 6). As large-scale EST data became available from a limited number of other plant species, several WRKY-homologous sequences have also been reported in lower plants (alga, moss, and ferns) and in conifers by searching DNA databases (1, 4–6, 8). This gives rise to questions about how evolutionary expansion has occurred for this gene superfamily in the plant kingdom and how WRKY proteins have adapted to various biological functions by structural differentiation.

In model plants, such as *Arabidopsis* and rice, identification of transcription factor families usually involves cDNA library construction, followed by DNA sequencing and gene annotation at the transcriptome scale. These methods are expensive and time-consuming and require a large amount of genomic information of the target species. It is difficult to apply a similar approach in non-model plants, such as conifers (which have the largest plant genomes), because of very limited genomic information in these species, scarcity of regulatory mRNAs in transcriptomes, and a specific spatiotemporal pattern of the expressed target genes. We recently demonstrated that a combination of an in silico-based bioinformatic prediction strategy with PCR-based experimental approaches could identify the WRKY family in western white pine (*Pinus monticola* Dougl. ex D. Don) (8). This simple, efficient, and inexpensive strategy provides a practical method for gene discovery of the WRKY superfamily in other non-model plants.

A BLAST search (9) of EST databases of the Gene Index Project, provided by the Computational Biology and Functional Genomics Laboratory at the Dana-Farber Cancer Institute and Harvard School of Public Health (http://compbio.dfci.harvard.edu/tgi/tgipage.html), identified 29 and 31 sequences in *Pinus taeda* and *Picea* species, respectively, with similarity to WRKY domains of the five subgroups from *Arabidopsis* and rice (**Table 2.1**). The amino acid motifs conserved among angiosperm and gymnosperm WRKY domains were determined by an alignment analysis and used to design a range of PCR primers

Table 2.1
Expressed members of the WRKY family in *Pinus* and *Picea* species

	Pinus taeda		*Picea*		
Group[a]	Gene No.	[EST sequence[b]]	Gene No.	[EST sequence[b]]	Species
I	1	[TC67599, TC10778, TC25556, TC31987, TC49174, TC5325]	1	[TC59872, TC2087, TC27577, EX331828]	*P. glauca*
I	2	[TC67819]	2	[TC53421, TC25771, DR588237]	*P. glauca*
I	3	[TC70314, TC11712, TC19803, TC37973, TC44417, TC6838]	3	[TC73002, ES665011]	*P. sitchensis*
I	4	[DR102534]		–	
IIa+b	5	[TC67611, TC29306, TC42772]	4	[TC72927, ES228321, ES228512, ES228759]	*P. abies*
IIa+b	6	[TC69912]	5	[EX371111]	*P. glauca*
IIa+b	7	[TC69221]	6	[TC59906, TC3742, TC20136, EX406580]	*P. glauca*
IIa+b	8	[CO198802]	7	[TC63325, TC24286, DR466155, CO207672]	*P. engelmannii* × *glauca*
IIa+b	9	[TC64319, TC45579]	8	[TC48570, ES261032, ES261221]	*P. sitchensis*
IIa+b	10	[TC75572, TC49622]	9	[TC49606, TC5670, TC37162, DR480521]	*P. sitchensis*
IIa+b	11	[TC62327]	10	[TC53881, TC17556, DR493091]	*P. sitchensis*
IIa+b	12	[TC68833, TC48794]		–	
IIa+b	13	[CO362998]		–	
IIc	14	[TC77638, TC44174]	11	[TC46708, EX396574, EX396867]	*P. glauca*

Table 2.1 (continued)

	Pinus taeda		*Picea*		
Group[a]	Gene No.	[EST sequence[b]]	Gene No.	[EST sequence[b]]	Species
IIc	15	[TC76195]	12	[TC57747, TC5412, TC28007, EX335523]	*P. glauca*
IIc	16	[TC69847]	13	[TC62191, TC23366, DV997756]	*P. glauca*
IIc	17	[TC77167]	14	[TC48310, TC30239, DR475708]	*P. engelmannii × glauca*
IIc	18	[TC71896]	15	[TC59853, TC2343, TC21165, DR475811]	*P. engelmannii × glauca*
IIc	19	[AW437880, ST73F02]	16	[ES260958]	*P. sitchensis*
IIc	20	[CF478959]	17	[ES871590]	*P. sitchensis*
IIc	21	[TC57307, TC20439, TC35508, TC47838][c]	18	[TC73769, TC6935, TC38386, FD731617]	*P. sitchensis*
IIc	22	[CF665435]		–	
IId+e	23	[DR694646, EST1084738]	19	[TC53444, ES875336]	*P. sitchensis*
IId+e	24	[TC59335, TC53663]	20	[TC52792, TC2588, TC20660, FD735167]	*P. sitchensis*
IId+e	25	[CF401103][d]	21	[DR503879]	*P. sitchensis*
IId+e	26	[TC66817, TC16416, TC26277, TC28002, TC41928, TC5870]	22	[DR495557]	*P. sitchensis*

Table 2.1 (continued)

	Pinus taeda		*Picea*		
Group[a]	Gene No.	[EST sequence[b]]	Gene No.	[EST sequence[b]]	Species
IId+e	27	[TC75206, TC36489, TC39130, TC48810, TC56850]	23	[DV982911]	*P. glauca*
IId+e	28	[TC75557]	24	[TC58586, EX386655, EX387020]	*P. glauca*
IId+e	29	[TC76519, TC55982]	25	[TC55698, EX328070, EX328428]	*P. glauca*
IId+e		–	26	[EX389580]	*P. glauca*
IId+e		–	27	[TC51682, EX307059]	*P. glauca*
IId+e		–	28	[TC80753, TC39381, DV986854]	*P. glauca*
IId+e		–	29	[TC66343, EX431002, EX402933]	*P. glauca*
IId+e		–	30	[TC83188, TC5604, DR581126]	*P. glauca*
IId+e		–	31	[DV979750]	*P. glauca*
III		–	32	[EX419977]	*P. glauca*

[a] Phylogenetic classification according to Zhang and Wang (6)
[b] GenBank accession numbers or the DFCI Gene Index (PGI) numbers (TCxxxxx)
[c] ORF interrupted by not processed intron sequence
[d] ORF interrupted by single nucleotide deletion

Table 2.2
PCR primer design for WRKY gene cloning

Primer name	Targeted a.a. motif	WRKY group	Nucleotide sequence (5′ to 3′)[a]	Reference/Accession No.[b]
WRKY-FP1	WRKYGQK	All groups	TGGMGIAARTAYGGNCARA	(11)
WRKY-FP2	RWRKYGQK	All groups	CGMTGGCGBAARTATGGACARAA	This study
WRKY-RP0	TEIVYKG	Group I	CCYTTGTAMACWAT TTCMGT	(8)/AK226301
WRKY-RP1	TTYEG(Q/V)H(N/T)H	Group I, IIa+b	TGRKTRTGYWSICCYTCRTAIGT	(11)
WRKY-RP2	SYLGRHNH	Group IIc	TGRTTRTGYCTNCCNAGRTANCT	(8)/DR024229
WRKY-RP3	TY(T/E)G(E/D)HNH	Group IId+e	TGRTTRTGYTCICCIKYRTAIGT	(8)/NM_122748, DAA05104
WRKY-RP4	TYXGEHTC	Group III	CAIGTRTGYTCICCIIIRTAIGT	(8)/DAA05110, ABF95809
WRKY-RP5	TYIGEHTC	Group III	CANGTRTGYTCNCCNATRTANGT	(8)/DAA05110
WRKY-RP6	TYYGHHTC	Group III	CANGTRTGRTGNCCRTARTANGT	(8)/CK109883
WRKY-RP7	TYSGVHSC	Group III	CARCTRTGNACNCCRCTRTANGT	This study, EX419977

[a]Degenerate IUB group codes: R=A+G, Y=C+T, M=A+C, K=G+T, W=A+T, S=G+C, D=G+A+T, B=T+C+G, N=A+T+C+G
[b]Primers were synthesized according to reference or designed in the present study based on amino acid sequences as indicated GenBank accession numbers

(**Table 2.2**). Using these WRKY domain primers, reverse transcription-polymerase chain reaction (RT-PCR) and rapid amplification of cDNA ends (RACE) were then performed to amplify expressed WRKY domain sequences and their respective full-length coding regions in various tissues of white pine. The primers were also used to amplify genomic DNA sequences of the WRKY domains and to detect DNA polymorphism for genetic mapping of the WRKY family in white pine populations. We identified 83 members of the *P. monticola* WRKY family (*PmWRKY*) (8). Using the neighbor-joining (NJ) method in the MEGA4 software package (10), the identified conifer WRKY genes were classified into groups I, IIa+b, IIc, IId+e, and III by a phylogenetic analysis at both nucleotide and amino acid sequence levels.

The WRKY gene superfamily is highly complex and divergent, suggesting its application for development of functional gene markers in genetic mapping (11, 12). A genetic mapping strategy modified from amplified fragment length polymorphism (AFLP) was used for genetic mapping of analogs of disease resistance genes (13–15) and transposable elements (16). We developed WRKY-AFLP markers by replacing a typical AFLP selective primer with one of the WRKY primers in the AFLP standard protocol. This WRKY-AFLP approach revealed 17–35% polymorphic bands, similar to the 26–40% found using a regular AFLP protocol in white pine populations (8). The identified WRKY-AFLP markers will advance our understanding of gene organization and evolution of the WRKY transcription factor family in a conifer species. Our research approach demonstrates a comprehensive and high-quality census of the WRKY transcription factors encoded within the white pine genome. The results provide a solid foundation for further systematic characterization of PmWRKY transcription factors at the level of either single genes or gene families.

2. Materials

2.1. Plant Tissue Samples

1. *Tissue for genomic DNA extraction.* In this study white pine seedlings were grown in the greenhouse and needle samples were collected and stored at –20°C.
2. *Tissue for RNA extraction.* In this study, samples of needles, stems, and roots were collected from pine seedlings, frozen in liquid nitrogen immediately, and stored at –80°C.

2.2. Genomic DNA Extraction

1. DNeasy Plant Mini Kit (Qiagen, Mississauga, ON, Canada).
2. NanoDrop 1000 UV-Vis Spectrophotometer (NanoDrop Technologies, Wilmington, DE, USA).

3. Agarose gel electrophoresis reagents and apparatus for DNA analysis.
4. Gel Doc 2000 image analysis system (Bio-Rad, Philadelphia, PA, USA).

2.3. Total RNA Extraction

1. RNA extraction buffer: 0.1 M Tris–HCl, pH 7.4, 0.5 M NaCl, 50 mM EDTA, 2% SDS, 2% PVP-40, 10 mM β-mercaptoethanol (added freshly before use). Autoclaving is not necessary.
2. Potassium acetate (KOAc) 3 M solution, pH 5.5, treated with diethylpyrocarbonate (DEPC), then autoclaved.
3. Phenol/chloroform/isoamyl alcohol (IAA) (25:24:1).
4. Chloroform/IAA (24:1).
5. Lithium chloride (LiCl) 8 M and 2 M solutions, treated with DEPC, then autoclaved.
6. Ethanol (70%), diluted from RNase-free ethanol with DEPC-treated ddH_2O.
7. TE buffer: 10 mM Tris–HCl, pH 7.5, 1 mM EDTA. Prepare from RNase-free chemicals and ddH_2O.
8. DEPC-treated ddH_2O.
9. Agarose gel electrophoresis reagents and apparatus for RNA analysis.

2.4. cDNA Synthesis

1. RQ1 RNase-free DNase (Promega, Madison, WI, USA).
2. Plant RNeasy extraction kit (Qiagen).
3. SMART cDNA library construction kit (Clontech, Palo Alto, CA, USA).
4. RNase-free ddH_2O.

2.5. WRKY Polymorphism by a Modified AFLP Method

1. AFLP analysis system I kit (Invitrogen, Carlsbad, CA, USA).
2. Polyacrylamide gel electrophoresis (PAGE) apparatus: adjustable nucleic acid sequencer (CBS Scientific Co., Del Mar, CA, USA) and PowerPac HV power supply (Bio-Rad, Hercules, CA, USA).
3. TBE buffer (5X): 0.445 M Tris, 0.445 M boric acid, 0.01 M EDTA, pH 8.0. Dissolve 54 g Tris and 27.5 g boric acid and add 20 mL of 0.5 M EDTA (pH 8.0) to a total volume of 1 L.
4. Formamide dye: 98% formamide, 10 mM EDTA, 0.025% bromophenol blue, and 0.025% xylene cyanol.
5. 40% acrylamide–bisacrylamide solution (20:1) (take appropriate safety precautions with these neurotoxic chemicals).

6. N,N,N',N'-tetramethylethylenediamine (TEMED) (Bio-Rad).
7. Ammonium persulfate (AP), 10% solution, freshly prepared for daily use although it may be stable for 1 week in darkness at 4°C.
8. Gel fixing solution: 10% (v/v) acetic acid.
9. Gel staining solution: 200 μL of 37% formaldehyde in 200 mL of 0.1% $AgNO_3$.
10. Gel developing solution: 200 μL of 37% formaldehyde and 20 μL of 2.0% (w/v) $Na_2S_2O_3 \cdot 5H_2O$ in 200 mL of 2.5% (w/v) Na_2CO_3.
11. Gel preserving solution: 20% ethanol, 20% isopropanol, and 10% glycerol.
12. Gel diffusion buffer: 0.5 M ammonium acetate, 1 mM EDTA, pH 8.0, 0.1% SDS.
13. QIAquick gel extraction kit (Qiagen).

2.6. PCR Cloning and Sequencing of Amplified DNA Fragments

1. Perkin-Elmer Thermocycler (Perkin-Elmer Applied Biosystems, Foster City, CA, USA).
2. Taq PCR Master Mix Kit (Qiagen).
3. Hot Start DNA Polymerase mix, dNTP solution (10 mM), Hot Start polymerase buffer.
4. PCR primers (10 μM): WRKY domain primers (**Table 2.2**), Clontech's CDS III/3′ PCR primer, Clontech's 5′ PCR primers, and Invitrogen's AFLP primers.
5. Reagents and apparatus for DNA electrophoresis.
6. MinElute PCR purification kit (Qiagen).
7. pGEM®-T Easy Vector System (Promega).
8. LB medium for bacterium culture.
9. *Escherichia coli* competent cells.
10. Isopropyl-beta-thiogalactopyranoside (IPTG) (Sigma), 0.1 M solution. Dissolve 0.238 g IPTG in 10 mL of water and sterilize by filtration. Store at –20°C.
11. 5-Bromo-4-chloro-indoly-β-D-galactoside (X-gal) (Sigma), 20 mg/mL solution in dimethylsulfoxide (DMSO) or dimethylformamide, but not in ddH_2O. Sterilize by filtration and wrap in foil for protection from light. Store at –20°C.
12. Plasmid mini kit (Qiagen).
13. Restriction enzymes and buffers (Invitrogen).

14. Reagents and apparatus for DNA sequencing.
15. Standard DNA ladders: 1 kb DNA ladder (Gibco BRL) and 25 bp DNA ladder (Invitrogen).

3. Methods

3.1. In Silico Data Mining and Design of PCR Primers

1. Select representatives of each WRKY subgroup (I, IIa+b, IIc, IId+e, and III) from genomes of *Arabidopsis* and rice for BLAST searches. Use a similar approach for other transcription factor families of interest.
2. Search EST databases of the Gene Index Project as in silico resources, provided by the Computational Biology and Functional Genomics Laboratory at the Dana-Farber Cancer Institute (DFCI) and Harvard School of Public Health (http://compbio.dfci.harvard.edu/tgi/tgipage.html).
3. Perform a tBlastn search through the database of the DFCI Gene Index Project online (http://compbio.dfci.harvard.edu/tgi/cgi-bin/tgi/Blast/index.cgi). For this study the DFCI Pine Gene Index (Release 6.0, July 19, 2005, total 45,557 output sequences) or the DFCI Spruce Gene Index (Release 3.0, July 11, 2008, total 80,494 output sequences) was searched with the representative WRKY domain sequences of the five subgroups from *Arabidopsis* or rice.
4. Retrieve all WRKY-homologous sequences from the Gene Index Project databases and store them in a local computer.
5. Perform alignment analysis of nucleotide or putative amino acid sequences online with the Clustal W network service at the European Bioinformatics Institute (EBI, Cambridge, UK; http://www.ebi.ac.uk/Tools/clustalw2/index.html) and identify those sequences with identities of 98% or more as the same gene. (The WRKY genes identified by this in silico data mining method are listed in **Table 2.1** with 29 WRKY family members in *P. taeda* and 31 members in *Picea* species.)
6. Construct phylogenetic trees using the neighbor-joining (NJ) method in the MEGA4 software package (10) based on sequence alignment analysis of the WRKY domains from a variety of plant taxa for classification of WRKY genes into distinct clusters; here the reliability of each tree is established by conducting 1,000 neighbor-joining bootstrap sampling steps. The gene clusters in the phylogenetic tree correspond to distinct WRKY gene groups (I, IIa+b, IIc, IId+e, and III) (*see* **Note 1**).

7. Search for conserved amino acid motifs in the WRKY domains based on alignment analysis of WRKY domain sequences from both angiosperms and gymnosperms; design PCR primers according to the identified amino acid motifs as listed in **Table 2.2** (*see* **Note 2**).

3.2. Genomic DNA PCR for WRKY Gene Discovery

1. Extract genomic DNA from plant tissues (white pine needles in this study) using a DNeasy plant mini kit following the manufacturer's instructions.
2. Determine the quality and concentration of extracted genomic DNA using a NanoDrop 1000 UV-Vis Spectrophotometer and verify the spectrophotometric results by 0.8% agarose gel electrophoresis.
3. Adjust genomic DNA concentration to 10 ng/μL with ddH_2O or TE buffer and store the genomic DNA samples at –20°C until further analysis.
4. Perform genomic DNA PCR using a Taq PCR Master Mix Kit (*see* **Note 3**) with one forward primer (FP1 or FP2) in combination with each of seven reverse primers (RP1–RP7) as listed in **Table 2.2**.
5. Prepare a PCR reaction in a total volume of 25 μL with 12.5 μL of Taq PCR Master Mix (2X), 1 μL of each primer (10 μM), 2 μL of genomic DNA (10 ng/μL), and 8.5 μL of ddH_2O (*see* **Note 4**).
6. Run the PCR on a Perkin-Elmer Thermocycler with an initial denaturation step at 94°C for 3 min, followed by 35 cycles of denaturation at 94°C for 40 s, primer annealing at 42–60°C for 1 min, and primer extension at 72°C for 1.5 min, and a final 10-min extension at 72°C (*see* **Note 5**).
7. Continue to **Section 3.5** for genomic PCR product separation and vector cloning. An example of genomic DNA amplification from WRKY domain primers is shown in **Fig. 2.1a**. Genomic PCR fragments (or smear regions) larger than 160 bp are subjected to further PCR fragment purification and vector cloning (*see* **Note 6**).

3.3. RT-PCR and RACE for Expression Profiling of WRKY Genes

3.3.1. RNA Isolation

1. Grind the tissue sample (~1.0 g of white pine tissue in this study) into a fine powder in liquid nitrogen and transfer the tissue powder into 10 mL of RNA extraction buffer (*see* **Note 7**).
2. Shake the extraction mixture thoroughly and incubate it at 65°C for 20–30 min.
3. Remove plant cellular debris by centrifugation at 12,000×*g* for 15 min at room temperature.

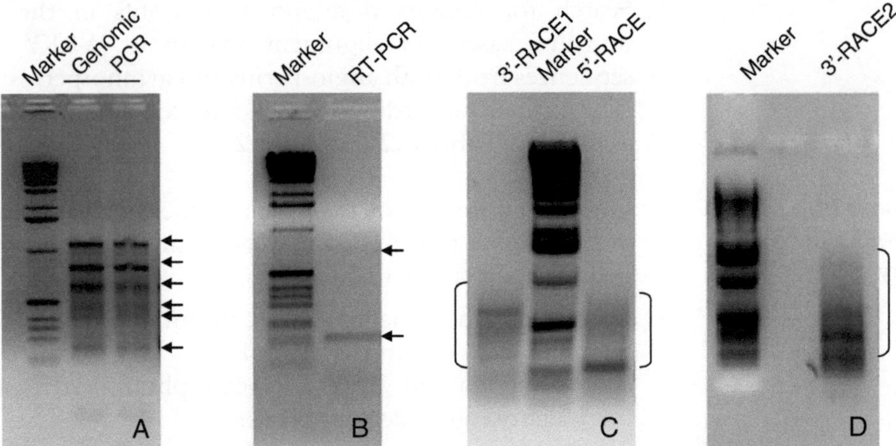

Fig. 2.1. Agarose gel electrophoresis of PCR products using WRKY domain primers. Marker standard 1 kb DNA ladder (Gibco BRL) is included in each gel. (**a**) Genomic DNA amplification using primers WRKY-FP1 and RP1. Six amplified fragments range from 0.25 to 1.15 kb and are shown by *arrows*. (**b**) cDNA fragments amplified by RT-PCR from total RNAs pooled from vegetative organs (shoots, stems, needles, and roots) using primers WRKY-FP1 and RP1. Two fragments are indicated by arrows with sizes about 0.16 and 0.70 kb. (**c**) PCR products from 5′RACE using WRKY-RP1 primer and Clontech's 5′ PCR primer and 3′RACE using WRKY-FP1 primer and Clontech's CDS III/3′ PCR primer with an annealing temperature of 45°C. (**d**) PCR products from 3′RACE using WRKY-FP2 primer and Clontech's CDS III/3′ PCR primer with an annealing temperature of 60°C.

4. Add 1/3 volume of 3 M KOAc, pH 5.5, to the supernatant and mix well; then incubate on ice for 30 min.
5. Remove the pellet with centrifugation at 12,000×*g* at 4°C for 10 min (*see* **Note 8**).
6. Extract the supernatant once with an equal volume of phenol/chloroform/IAA (25:24:1) and extract the supernatant once again with an equal volume of chloroform/IAA (24:1).
7. Precipitate total RNA from the supernatant by adding 1/3 volume of 8 M LiCl and incubate overnight at 4°C (*see* **Note 9**).
8. Recover total RNA by centrifugation at 12,000×*g* at 4°C for 30 min.
9. Wash total RNA pellet once with 2 M LiCl and then wash it once again with 70% ethanol.
10. Air-dry the RNA pellet and re-suspend it in 50 μL of TE buffer or DEPC-treated ddH_2O, then store RNA samples at –80°C until further analysis.
11. Measure RNA concentration by UV spectrophotometry for each individual tissue sample, verify RNA integrity by

agarose gel electrophoresis, and visualize the gel by ethidium bromide staining.

3.3.2. cDNA Synthesis

A common strategy for plant gene discovery is reverse transcription of plant messenger RNA into cDNA, followed by construction of a plant cDNA library or PCR cloning of targeted cDNA using gene-specific primers. A cap-switching approach is widely used for generating cDNA with high potential for full-length cDNA cloning from a small amount of mRNA or total RNA. The first strand (ss) cDNA is synthesized using a modified oligo d(T) primer. As the newly synthesized ss-cDNA extends to the 5′-cap structure at the end of the mRNA templates, a short non-template sequence, called a cap-switch oligonucleotide, is integrated to the cDNA 3′-end by the terminal transferase activity of the MMLV reverse transcriptase (17, 18). Therefore, additional sequences are attached at both 3′- and 5′-ends of the synthesized ss-cDNA that can be used for PCR primers and for the generation of double strand (ds) cDNA by PCR amplification. This cap-switching technology is very useful for cDNA library construction as well as for cDNA template production in targeted gene cloning by RT-PCR or RACE and is utilized in a variety of commercial kits. The SMART™ (*s*witching *m*echanism *a*t 5′ end of *R*NA *t*emplate) cDNA library construction kit (Clontech) is the one we use here.

1. Treat total RNA samples with RQ1 RNase-Free DNase to eliminate any contaminating genomic DNA, following the manufacturer's instructions.
2. Remove the enzyme from RNA samples by RNA re-purification using a Plant RNeasy extraction kit, following the manufacturer's instructions.
3. Verify total RNA integrity by agarose gel electrophoresis and measure RNA concentration using a NanoDrop 1000 UV-Vis Spectrophotometer.
4. Reverse-transcribe total RNA templates into ss-cDNA and synthesize ds-cDNA using the SMART cDNA library construction kit, following the manufacturer's instructions (*see* **Note 10**).
5. Dilute both ss-cDNA and ds-cDNA reactions appropriately and store them at –20°C for use as templates in RT-PCR and RACE.

3.3.3. RT-PCR for Amplification of WRKY Domain Sequences

The procedures are similar to genomic DNA PCR for *WRKY* discovery described in steps 4–6 of **Section 3.2**, but the PCR templates are replaced with diluted ss-cDNA or ds-cDNA. An example of RT-PCR amplification of two expressed WRKY domain fragments is shown in **Fig. 2.1b**. Usually one (~0.16 kb) or two cDNA fragments (~0.16 and ~0.70 kb) are amplified using each of the 14 combinations of the WRKY domain primers listed in **Table 2.2**. The two RT-PCR fragments, or smear regions with similar sizes, are subjected to further PCR fragment purification and vector cloning (*see* **Note 11**).

3.3.4. RACE for cDNA Cloning of Full-Length Coding Regions

1. Mix the following reagents in a sterile 0.2 mL tube: 5 μL of Hot Start polymerase buffer (10X), 1 μL of dNTP solution (each 10 mM), 2.5 μL of WRKY primer (FP1 or FP2 for 3′RACE or one of RP0–RP7 for 5′RACE), 1 μL of one of Clontech's primers (CDS III 3′ PCR primer for 3′RACE or 5′ PCR primer for 5′RACE), 2.5 units of Hot Start polymerase mix, 1 μL of diluted ss-cDNA or ds-cDNA (from step 5 of **Section 3.3.2**), and ddH_2O to a total volume of 50 μL (*see* **Note 12**).
2. Heat the mixture in a thermocycler at 95°C for 3 min, then run 30 cycles of RACE using a PCR program as follows: template denaturation at 95°C for 20 s, primer annealing at 45–68°C for 1 min, and primer extension at 72°C for 3 min (*see* **Note 13**).
3. Continue to **Section 3.5** for RACE product separation and vector cloning. Examples of RACE using WRKY domain primers are shown in **Fig. 2.1c, d**. The RACE products are usually in a continuous size range, so at least a few hundred recombinant plasmids are necessarily selected for DNA sequencing. A sampling of recombinant clones from one RACE experiment is shown in **Fig. 2.2.**

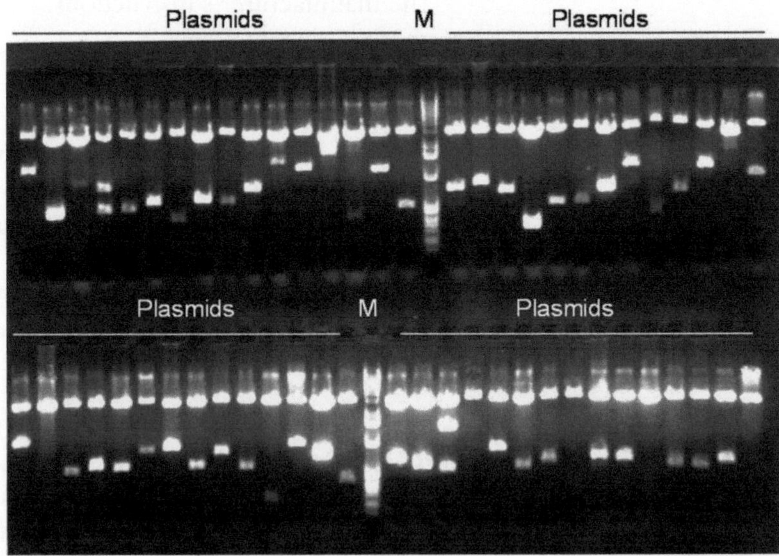

Fig. 2.2. Restriction enzyme analysis of the recombinant plasmids. 3′RACE products as shown in **Fig. 2.1d** were ligated into the pGEM-T Easy Vector (Promega). Plasmids were digested with *EcoR*I to show insert size variation.

3.4. A Modified AFLP for WRKY Polymorphism Detection

To detect DNA polymorphism and map WRKY genes, a modified AFLP method was developed based on a method described previously (8, 13). In this method, the WRKY-AFLP markers are first detected by PAGE followed by silver staining. The polymorphic DNA fragments of interest are then extracted, re-amplified, and sequenced to confirm their gene identities.

3.4.1. A Modified AFLP Protocol

Genomic DNA digestion and adapter ligation are performed using the AFLP analysis system I kit or another commercial AFLP kit following the manufacturer's instructions.

1. Add the genomic DNA (250 ng) with 5 μL of reaction buffer (5X), 2 μL of *Eco*RI/*Mse*I, and ddH$_2$O to a final volume of 25 μL.
2. Incubate at 37°C for 4 h, then at 70°C for 15 min to inactivate restriction enzymes.
3. Add 24 μL of adapter ligation solution and 1 μL of T4 DNA ligase to the 25 μL of genomic DNA digestion mixture for adapter ligation.
4. Incubate adapter ligation mixture at room temperature (20–22°C) for 2 h.
5. Dilute the ligation mix in a 1:10 ratio as follows: add 10 μL of ligation mix to 90 μL of TE buffer.
6. Store diluted and undiluted ligation mixes at –20°C.
7. Prepare AFLP pre-amplification reactions in 0.2 mL PCR tubes by adding the following: 5 μL of diluted template genomic DNA, 5 μL of PCR buffer plus Mg^{++} (10X), 2.5 μL of *Eco*RI-AC primer (10 μM), 2.5 μL of *Mse*I-CC primer (10 μM), 5 μL of dNTPs (each 2 mM), 1 μL of Taq DNA polymerase (5U/μL), and ddH$_2$O to a total volume of 50 μL (*see* **Note 14**).
8. Run AFLP pre-amplification with the following conditions: pre-denaturation at 94°C for 3 min, followed by 40 cycles of denaturation at 94°C for 30 s, annealing at 56°C for 60 s, and primer extension at 72°C for 60 s with a final extension at 72°C for 7 min.
9. Dilute pre-amplification PCR reactions to a ratio of 1:50 or 1:100 and store both diluted and undiluted reactions at –20°C until further analysis.
10. Perform WRKY-AFLP profiling by selective AFLP amplification reactions: prepare the reaction mixtures in 0.2 mL PCR tubes by adding 1 μL of the diluted pre-amplification reaction, 12.5 μL of 2X Taq Master Mix, 1 μL of *Eco*RI-ACNN or *Mse*I-CCNN primer (10 μM), 1 μL of WRKY primers (10 μM), and ddH$_2$O to a total volume of 25 μL (*see* **Notes 14** and **15**).

11. Run AFLP selective amplification under the following PCR conditions: pre-denaturation at 94°C for 3 min, followed by 30 cycles of denaturation at 94°C for 30 s, annealing at 56°C for 60 s, and primer extension at 72°C for 60 s with a final extension at 72°C for 7 min (*see* **Note 16**).

3.4.2. PAGE Separation of DNA Fragments

An adjustable nucleic acid sequencer (CBS Scientific) is used for PAGE analysis here. Advanced DNA sequencer systems can be used for high-throughput AFLP profiling.

1. Prepare a 6% polyacrylamide gel (0.75 mm thick, size 30 × 20 cm with 60 wells) with 60 mL of gel solution by adding the following: 27 g urea, 9 mL of 40% acrylamide–bisacrylamide stock solution, 6 mL of 5X TBE, 15 mL of ddH_2O, 0.5 mL of 10% APS, and 0.03 mL of TEMED.
2. Set up the adjustable nucleic acid sequencer apparatus and add 800 mL of 0.5X TBE buffer to the upper and lower chambers of the gel unit.
3. Run pre-electrophoresis for 20–30 min at 500–700 V for stabilization of the gel temperature at 50°C.
4. Prepare DNA samples while the gel is pre-running by adding equal volumes (25 μL) of formamide dye to each selective PCR reaction and mix well.
5. Denature DNA samples by incubation at 94°C for 3 min and keep the denatured DNA samples on ice until sample loading on the gel.
6. Load 10 μL of the denatured DNA samples onto the gel including one well at both sides with 50 ng of a 25 bp DNA standard for estimation of the amplified DNA fragment sizes.
7. Run AFLP electrophoresis with 0.5X TBE buffer until the xylene cyanol band is within 2–3 cm of the bottom edge of the gel.

3.4.3. PAGE Silver Staining

1. Fix the gel for 30 min by adding 200 mL of 10% acetic acid (v/v) (gel fixing solution) for each gel.
2. Rinse the gel in ddH_2O three times, each for 2 min; then stain the gel with 200 mL of gel staining solution for 20–30 min.
3. Rinse the gel on both surfaces with ddH_2O for 10–20 s.
4. Incubate the gel at room temperature in 200 mL of pre-cooled (4°C) gel developing solution until DNA bands are clearly visible (2–10 min).
5. Stop the developing reaction by incubating the gel in 10% acetic acid solution for 2–3 min, then rinse the gel with ddH_2O.

6. Scan the AFLP profile and save the image into a computer using the Quantity One software package (Bio-Rad). An example of a resulting WRKY-AFLP profile is shown in **Fig. 2.3**.
7. Incubate the gel in gel preserving solution overnight with slight shaking; then dry the gel for long-term storage if necessary.

3.4.4. Elution of DNA Fragments from Polyacrylamide Gel

Following treatment of a gel with gel diffusion buffer, a QIAquick gel extraction kit is used to elute DNA fragments from the polyacrylamide gel, following the manufacturer's instructions.

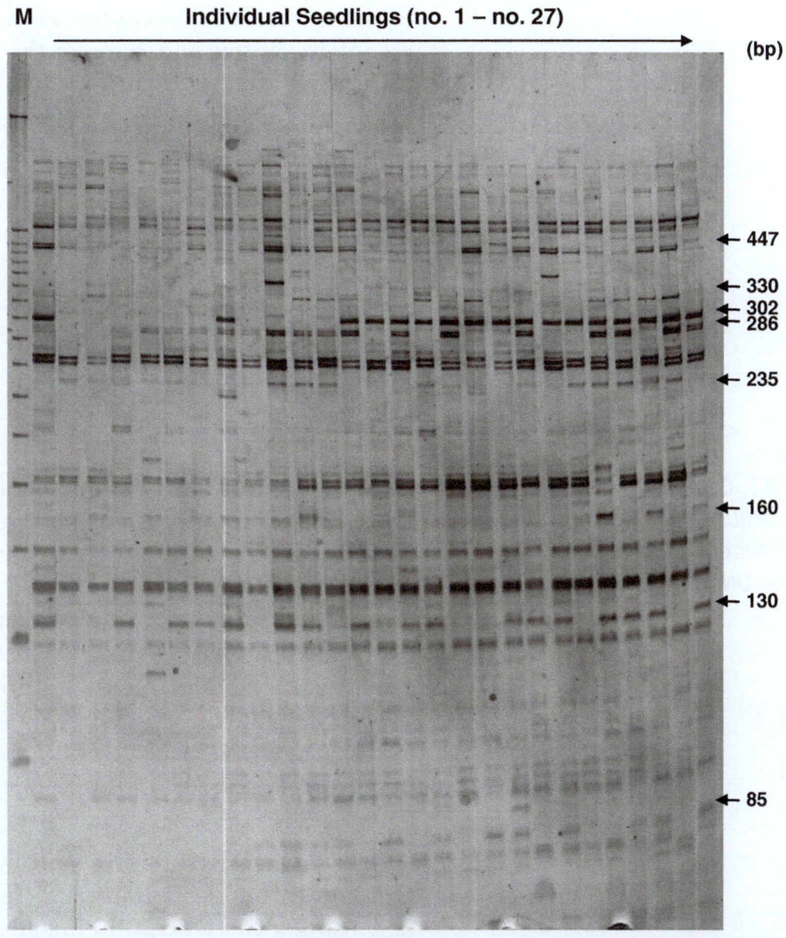

Fig. 2.3. An example of the white pine WRKY-AFLP polymorphic profiles. Marker standard 25 bp DNA ladder (M) is included in the left well. AFLP selective amplification was performed using one WRKY primer (RP0) coupled with one AFLP selective primer (*Eco*RI-ACTC). Eight WRKY-AFLP markers ranging from 85 to 447 bp are indicated by *arrows*. Amplicons of 27 individual seedlings (no. 1–no. 27) from a seed family are shown here.

1. Excise the gel slice containing the DNA band of interest from the polyacrylamide gel.
2. Weigh the gel slice and add 1–2 volumes (v/w) of gel diffusion buffer into the gel slice tube.
3. Incubate at 50°C for 30 min and centrifuge the sample tube at 12,000×*g* for 1 min.
4. Transfer the supernatant to a new tube and process it immediately with a QIAquick gel extraction kit.
5. Add three volumes of QG buffer to the supernatant and ensure the color of the mixture is yellow, which indicates the correct pH.
6. Apply the sample mixture to a QIAquick spin column for a quick spin (30–60 s). The DNA fragment is now bound to the QIAquick spin column. Wash the column with 0.75 mL of PE buffer and remove the PE buffer trace by centrifugation at 12,000×g for 2 min.
7. Add 50 μL of EB buffer to the column for a quick spin to elute DNA.
8. Re-amplify the eluted DNA fragment using the same pair of primers and the same PCR conditions used for WRKY-AFLP profiling.
9. Verify the size of the re-amplified DNA fragment with PAGE.
10. Continue to **Section 3.5** for DNA fragment separation and vector cloning.

3.5. Cloning, Sequencing, and Identity Confirmation of DNA Fragments

Cloning of DNA fragments amplified from genomic DNA or cDNA is necessary to confirm gene identity by DNA sequencing and sequence data analysis. There are a variety of commercial kits available for DNA fragment purification from agarose gel, PCR cloning, and subsequent DNA sequencing. Researchers can choose their own preferred kits and methods. We use Qiagen's gel extraction kit for DNA fragment purification and the pGEM-T Easy Vector for PCR cloning.

1. Perform agarose gel electrophoresis to separate PCR products from genomic DNA amplification, RT-PCR, RACE, or WRKY-AFLP.
2. Visualize the electrophoresis profiles of PCR fragments by ethidium bromide staining and record PCR profiles digitally using a Gel Doc 2000 image analysis system.
3. Excise agarose gel slices containing the DNA fragments of interest under UV light (*see* **Note 17**).
4. Extract each DNA fragment from the agarose gel slices using a gel extraction kit, such as Qiagen's QIAquick gel

extraction kit or the MinElute PCR purification kit (*see* **Note 18**), following the manufacturer's instructions.

5. Ligate each purified DNA fragment (or smear region) into Promega's pGEM-T Easy Vector at room temperature or 4°C (*see* **Note 19**), following the manufacturer's instructions.
6. Transform the ligation mixture into appropriate *E. coli* competent cells using white-blue selection with β-galactosidase under X-gal and IPTG in the LB-agar medium.
7. Select white *E. coli* colonies, grow overnight, and extract plasmids using a Plasmid mini kit.
8. Digest plasmids with *Eco*RI to verify insert sizes of recombinant clones using agarose gel electrophoresis.
9. Sequence inserts of recombinant plasmids with vector primers (T7 and SP6) and additional gene-specific primers if long inserts are present in the recombinant plasmids (*see* **Fig. 2.2**).
10. Analyze the DNA sequence data and annotate WRKY genes as described in **Section 3.1**.

4. Notes

1. This gene classification based on phylogenetic analysis is consistent with other group-specific features of the WRKY family, such as intron positions and phases when genomic sequences are available.
2. The PCR forward primers FP1 and FP2, targeted at the N-terminal region of the WRKY domain, are universal for both angiosperms and gymnosperms. In contrast, the C-terminal regions of the WRKY domains are less conserved, with group specificity or even gene specificity. Numerous reverse primers are required to cover a range of potential members of the WRKY family in a non-model plant. Reverse primers RP0–RP7 in **Table 2.2** are only a few oligonucleotides used in white pine. Other reverse primers with more diverse amino acid sequences may be needed to target the C-terminal regions of the WRKY domains in other non-model plants.
3. Researchers may wish to use another PCR kit or other thermostable DNA polymerases for genomic DNA amplification. For example, Hot Start DNA polymerase helps eliminate non-specific DNA amplification. *Pfu* DNA polymerase

is a better choice if nucleotide misincorporation is a concern in PCR cloning. *Tth* DNA polymerase may improve PCR results if there is a problem resulting from a high degree of secondary structure in the templates.

4. Because white pine has one of the largest genomes (~28.25 pg/1C), we used 20–50 ng of genomic DNA for WRKY gene amplification. The amount of genomic DNA template is adjustable, depending on the genome sizes of plant species, copy numbers of the targeted gene, and quality of the purified DNA samples.
5. Annealing temperature (Tm) could be adjusted from 42 to 60°C in the case of white pine with related primer combinations. A higher Tm will not only decrease non-specific amplification but also decrease the cloning coverage of the gene family. The primer extension time may be longer if a large intron (for example, 1 kb or longer) is predicted to be in the WRKY domain sequence.
6. An intron is predicted in the genomic DNA sequences of the WRKY domains. Because of the uncertainty regarding the size of this WRKY domain intron in uncharacterized plant genomes, we suggest collecting all amplified DNA fragments larger than 160 bp (a predicted size for WRKY domain mRNA) for genomic DNA cloning of the WRKY domains. For each genomic DNA fragment, multiple recombinant clones are needed for DNA sequencing to check gene diversity within a specific fragment because each DNA band probably contains PCR products amplified from multiple genes with very similar sizes.
7. No commercial RNA extraction kit is suitable for extraction of RNA from conifers. The method of conifer RNA extraction we describe here is adapted from a protocol reported previously (19, 20) and is effective for total RNA extraction from various conifer tissues. At this step, a high tissue-buffer ratio (w/v=1:10) helps eliminate polysaccharides in subsequent steps.
8. Many types of plant samples are rich in polysaccharides and polyphenols that are easily co-purified with RNA; this can seriously reduce RNA quality. Before the final RNA precipitation with 8 M LiCl or ethanol, this step of using 1/3 volume of 3 M KOAc, pH 5.5, is crucial to precipitate the sugars, although some RNA may be also lost. As an indicator of the increased purity, the RNA sample should have a higher ratio of A_{260}/A_{280}, which far outweighs the loss of RNA to the polysaccharide plugs.
9. If RNA extraction must be finished in 1 day, total RNA could be precipitated optionally by adding three volumes of

95% ethanol. Contaminating genomic DNA in total RNA usually is not a serious problem because each total RNA sample will be treated by RNase-free DNase before reverse transcription for cDNA synthesis. However, alcohol precipitation may result in co-purification of polysaccharides and polyphenolics along with RNA and genomic DNA, especially for plant tissues rich in secondary metabolites.

10. Because only a few cDNA syntheses are available from this kit, a mixture of total RNA pooled from multiple tissue samples is preferably used as the reverse-transcription template. In our study, total RNA was pooled equally from different tissues (roots, stems, needles, and immature male and female cones) or needle tissues with different treatments (wounding, disease infection, etc.) for optimal discovery of the expressed WRKY genes.

11. Even if an RT-PCR fragment is very sharp on an agarose gel, it is necessary to select at least 10 recombinant clones for DNA sequencing. If DNA sequencing reveals high gene diversity in a particular DNA fragment, a larger number of recombinant clones from the fragment should be sequenced in case it contains additional gene members (*see* **Note 6**).

12. When a ds-cDNA mixture is used as a RACE template, only 0.1–1.0 ng of ds-cDNA is added in an amplification reaction. In addition, Hot Start DNA polymerase is preferred for RACE because there is non-specific amplification in most cases (*see* **Note 3**). To increase the potential of cloning multiple genes simultaneously, we used degenerate WRKY domain primers (**Table 2.2**) to perform 3′RACE PCR. Once the sequence of the 3′-untranslated region is determined, a gene-specific primer could be designed for 5′RACE to clone a specific WRKY gene. A special commercial RACE kit may be used for this work.

13. A higher annealing temperature is recommended, especially for 5′RACE using WRKY reverse primers with high degeneracy, which easily cause non-specific amplification in RACE (*see* **Note 5**).

14. There are two options to modify the standard AFLP procedure for WRKY-AFLP profiling. One option is to modify AFLP at the step of AFLP selective amplification. Another option is to modify AFLP at the step of AFLP pre-amplification where one WRKY primer and one standard AFLP primer (*Eco*RI-AC or *Mse*I-CC) are used. Then, the same WRKY primer and one of 32 standard AFLP primers (*Eco*RI-ACNN or *Mse*I-CCNN) should be used at the next step for AFLP selective amplification.

15. When a WRKY primer is considered for WRKY-AFLP profiling, it will be designed to have an annealing temperature consistent with that of regular AFLP primers. In addition, for each of the 10 WRKY primers listed in **Table 2.2**, there are 32 primer combinations from standard AFLP selective primers (16 *Eco*RI-ACNN and 16 *Mse*I-CCNN). A selection of 320 primer combinations provides the WRKY-AFLP profiling with a high potential to discover WRKY polymorphism in a population.
16. The number of cycles in the AFLP selective amplification step may be adjusted according to the final dilution of pre-amplified products and at which step (pre-amplification or selective amplification) the WRKY primer is integrated into the AFLP protocol.
17. Ensure exposure to ultraviolet (UV) light is as brief as possible when DNA fragments are cut out from an agarose gel. The PCR product is preferably visualized only with a long-wave UV source. Otherwise, the efficiency of DNA cloning can decrease greatly because of UV-caused DNA damage.
18. There are various types of commercial kits for DNA fragment purification from agarose gel. Qiagen's MinElute PCR purification kit uses a small volume (10 μL) to elute a DNA fragment from a column at a relatively high concentration, which potentially increases efficiency of subsequent DNA ligation and cloning.
19. DNA ligation buffer (2X) contains ATP that might be degraded during temperature fluctuations. To avoid this, make single-use aliquots of the ligation buffer. Incubation of DNA ligation reactions overnight at 4°C often produces the maximum number of transformants.

Acknowledgments

The authors would like to thank Steve Glover for his critical review on the manuscript. This work was supported in part by Natural Resources Canada, the Canadian Forest Service, and Agriculture and Agri-Food Canada, the Pacific Agri-Food Research Centre.

References

1. Eulgem, T., Rushton, P. J., Robatzek, S., and Somssich, I. E. (2000) The WRKY superfamily of plant transcription factors. *Trends Plant Sci.* **5**, 199–206.
2. Rushton, P. J., Torres, J. T., Parniske, M., Wernert, P., Hahlbrock, K., and Somssich, I. E. (1996) Interaction of elicitor-induced DNA-binding proteins with elicitor response

elements in the promoters of parsley PR1 genes. *EMBO J.* **15(20)**, 5690–5700.

3. Du, L. and Chen, Z. (2000) Identification of genes encoding receptor-like protein kinases as possible targets of pathogen- and salicylic acid-induced WRKY DNA-binding proteins in *Arabidopsis*. *Plant J.* **24(6)**, 837–847.
4. Ülker, B. and Somssich, I. E. (2004) WRKY transcription factors: from DNA binding towards biological function. *Curr. Opin. Plant Biol.* 7, 491–498.
5. Wu, K. L., Guo, Z. J., Wang, H. H., and Li, J. (2005) The WRKY family of transcription factors in rice and *Arabidopsis* and their origins. *DNA Res.* **12**, 9–26.
6. Zhang, Y. and Wang, L. (2005) The WRKY transcription factor superfamily: its origin in eukaryotes and expansion in plants. *BMC Evol. Biol.* **5**, 1.
7. Berri, S., Abbruscato, P., Faivre-Rampant, O., Brasileiro, A. C. M., Fumasoni, I., Satoh, K., Kikuchi, S., Mizzi, L., Morandini, P., Pè, M. E., and Piffanelli, P. (2009) Characterization of WRKY co-regulatory networks in rice and Arabidopsis. *BMC Plant Biol.* **9**, 120.
8. Liu, J.-J. and Ekramoddoullah, A. K. M. (2009) Identification and characterization of the WRKY transcription factor family in *Pinus monticola*. *Genome* **52**, 77–88.
9. Altschul, S. F., Madden, T. L., Schaffer, A. A., Zhang, J., Zhang, Z., Miller, W., and Lipman, D. J. (1999) Gapped BLAST and PSI-BLAST: a new generation of protein database search programs. *Nucleic Acids Res.* **25**, 3389–3402.
10. Kumar, S., Tamura, K., and Nei, M. (2004) MEGA3: integrated software for molecular evolutionary genetics analysis and sequence alignment. *Brief. Bioinform.* **5**, 150–163.
11. Borrone, J. W., Kuhn, D. N., and Schnell, R. J. (2004) Isolation, characterization, and development of WRKY genes as useful genetic markers in *Theobroma cacao*. *Theor. Appl. Genet.* **109**, 495–507.
12. Trognitz, F., Manosalva, P., Gysin, R., Niño-Liu, D., Simon, R., del Herrera, M. R., Trognitz, B., Ghislain, M., and Nelson, R. (2002) Plant defense genes associated with quantitative resistance to potato late blight in *Solanum phureja* × Dihaploid *S. tuberosum* hybrids. *Mol. Plant Microbe Interact.* **15**, 587–597.
13. Hayes, A. J. and Maroof, M. A. (2000) Targeted resistance gene mapping in soybean using modified AFLPs. *Theor. Appl. Genet.* **100**, 1279–1283.
14. Liu, J.-J. and Ekramoddoullah, A. K. M. (2007) The CC-NBS-LRR subfamily in *Pinus monticola*: targeted identification, gene expression, and genetic linkage with resistance to *Cronartium ribicola*. *Phytopathology* **97(6)**, 728–763.
15. Liu, J.-J. and Ekramoddoullah, A. K. M. (2008) Development of LRR polymorphism, AFLP, and SCAR markers to the *Cronartium ribicola* resistance gene *Cr2* in western white pine (*Pinus monticola*). *Tree Genet. Genomes* **4**, 601–610.
16. Teunissen, H. A., Rep, M., Houterman, P. M., Cornelissen, B. J., and Haring, M. A. (2003) Construction of a mitotic linkage map of *Fusarium oxysporum* based on Foxy-AFLPs. *Mol. Genet. Genomes* **269**, 215–226.
17. Schmidt, W. M. and Mueller, M. W. (1999) CapSelect: a highly sensitive method for 5′cap-dependent enrichment of full-length cDNA in PCR-mediated analysis of mRNAs. *Nucleic Acids Res.* **27**, e31.
18. Schramm, G., Bruchhaus, I., and Roeder, T. (2000) A simple and reliable 5′-RACE approach. *Nucleic Acids Res.* **28**, e96.
19. Liu, J.-J., Goh, C.-J., Loh, C.-S., Liu, P., and Pua, E.-C. (1998) A method for isolation of total RNA from fruit tissues of banana. *Plant Mol. Biol. Rep.* **16**, 87a–f.
20. Liu, J.-J., Ekramoddoullah, A. K. M., and Yu, X. (2003) Differential expression of multiple PR10 proteins in western white pine following wounding, fungal infection and cold-hardening. *Physiol. Plant* **119**, 544–553.

Chapter 3

Isolation of Plant Transcription Factors Using a Yeast One-Hybrid System

Tatiana Pyvovarenko and Sergiy Lopato

Abstract

The yeast one-hybrid (Y1H) system is a powerful tool for the identification and isolation of cDNAs of transcription factors using promoter segments or regulatory elements as baits. Here we propose an adaptation of the Y1H system for identification and cloning of transcription factors using Matchmaker (Clontech) Y2H cDNA libraries. The method is a modification of the standard one-hybrid screening protocol, utilising a mating step to introduce the library and reporter constructs into the same cell. This extends the compatibility of Matchmaker cDNA libraries from yeast two-hybrid screens to one-hybrid screens. Libraries were successfully prepared from wheat, barley and maize grain, spike, leaf and root tissues from plants subjected to several environmental stresses. Using this method, we have isolated more than 50 cDNAs encoding transcriptional factors from several different families.

Key words: Yeast one-hybrid, transcription factors, *cis*-elements, cDNA libraries, wheat, barley, maize.

1. Introduction

Plant development and responses to different environmental stimuli and stresses are determined by differential gene expression and can be regulated at many different levels. A major mechanism of differential gene expression is transcriptional regulation (1) which involves the binding of transcription factors to DNA *cis*-elements located in gene promoters and/or introns. For example, more than 5% of the genes in *Arabidopsis* have been identified as transcription factors (2).

A powerful method for the identification and cloning of transcription factors using protein–DNA interactions is the yeast one-hybrid (Y1H) system. The Y1H system is a variant of the yeast

L. Yuan, S.E. Perry (eds.), *Plant Transcription Factors*, Methods in Molecular Biology 754,
DOI 10.1007/978-1-61779-154-3_3, © Springer Science+Business Media, LLC 2011

two-hybrid system (Y2H) (2–4). It is based on the domain structure of yeast transcription factors (e.g. GAL4), which consist of a transcription activation domain (AD) and a DNA-binding domain (BD). Exchange of the GAL4 BD for the full- or partial-length sequence of a plant transcription factor with its own, intact DNA-binding domain creates a new hybrid protein able to recognise specific plant *cis*-elements. A cDNA library which expresses a fusion protein of a constitutive (yeast) activation domain and a variable DNA-binding domain encoded by plant cDNAs can provide a pool of hybrid transcription factors for screening.

In the Y1H method, a segment of a plant promoter or tandem copies of putative transcription regulatory DNA elements are cloned in a reporter vector, upstream of a minimal yeast promoter and followed by the reporter/selection gene. In the modified version we describe here, part of the reporter vector is introduced into the genomic DNA of yeast via homologous recombination with a non-essential gene (5), thus generating a new reporter yeast strain containing the plant promoter/*cis*-element and reporter/selection gene. This stabilises the Y1H screen by bringing protein–DNA interactions into the nucleus and often ensuring correct protein folding and modifications (6). The new Y1H reporter strain is then transformed with a plant cDNA library vector, and interaction between target DNA and the hybrid protein is detected by reporter/selection gene expression (7).

The Y1H system enables known *cis*-elements as well as non-characterised fragments of promoters with predicted *cis*-elements to be used to search for specifically interacting DNA-binding proteins in expressional cDNA libraries (7, 8). It also enables the identification of families of factors with conserved DNA-binding domains which can interact with the same *cis*-activation DNA element (9–11). The potential to screen several millions of independent colonies simultaneously makes the Y1H system extremely sensitive, allowing the cloning of very low abundance transcription factors which may be absent from even very large EST databases (12). There is a high level of conservation of *cis*-elements across species, and *cis*-elements from well-characterised plants, like *Arabidopsis* and rice, can often be used successfully for the isolation of transcription factors from other species for which neither genomic nor EST data are yet available (9).

2. Materials

2.1. Generation of Yeast Reporter Strains

1. 10X annealing buffer: 100 mM Tris–HCl, pH 7.5, 1 M NaCl, 10 mM EDTA and partially complementary oligonucleotides representing the bait sequence (*see* **Table 3.1** for examples).

Table 3.1
Oligonucleotides used for the generation of bait sequences. These primer sets have been successfully used in Y1H screens of cDNA libraries prepared from wheat, barley and maize. Letters in *bold* represent several repeats of *cis*-elements and *underlined* sequences represent single *cis*-element

Cis-element repeat	Forward primer sequence	Reverse primer sequence	Ref.	Number of isolated TFs	Donor plant
HD-Zip class I and II	GGCCGCCAATAATTGCAATGAT-TGCAATTATTGCAATCATTGA	CTAGTCAATGATTGCAATAAT-TGCAATCATTGCAATTATTGGC	(16, 17)	6	Wheat and maize
HD-Zip class IV	GGCCGCCATTAAATGCATTAAA-TGCATTAAATGCATTAAATGA	CTAGTCATTTAATGCATTTAAT-GCATTTAATGCATTTAATGGC	(18)	2	Wheat
KN1-HD	GGCCGCTGACAGGTTGACAGCT-TGACAGGTTGACAGCTTGA CAG-GTTGACAGCTA	CTAGTAGCTGTCAACCTGTCAA-GCTGTCAACCTGTCAAGCTGTC-AACCTGTCAGC	(19)	2	Wheat
OsE2Fa	GGCCGCTTTTCCCGCCTTTTTT-CCCGCCTTTTTTCCCGCCTTA	CTAGTAAGGCGGGAAAAAAGG-CGGGAAAAAAGGCGGGAAAAGC	(20)	1	Wheat
DRE, *AtERD1*	GGCCGCTACCGACATTACCGAC-ATTACCGACATTACCGACATA	CTAGTATGTCGGTAATGTCGG-TAATGTCGGTAATGTCGGTAGC	(21)	6	Wheat, barley and maize
DRE, *ZmRab17*	GGCCGCGAACCGAGACGTGG-CGGGCCGGGCCACCGACGCG-AACCGAGACGTGGCGGGCCG-GGCCACCGACGCA	CTAGTGCGTCGGTGGCCCGGCC-CGCCACGTCTCGGTTCGCGTCG-GTGGCCCGGCCCGCCACGTCTC-GGTTCGC	(22)	3	Wheat and barley
CRT	GGCCGCGCCGACGCCGACGCC-GACGCCGACGCCGACA	CTAGTGTCGGCGTCGGCGTCGGC-GTCGGCGTCGGCGC	(23)	6	Wheat and barley
GCC-box, Nt	GGCCGCAGTGCCAAAAGCCGC-CACACCCCTAGTGCCAAAAGC-CGCCACACCCCTA	CTAGTAGGGGTGTGGCGGCTTT-TGGCACTAGGGGTGTGGCGGC-TTTTGGCACTGC	(24)	6	Wheat and maize
ABRE+ MYC (CACGTG)	GGCCGCGACACGTGGCGACAC-GTGGCGACACGTGGCA	CTAGTGCCACGTGTCGCCACGT-GTCGCCACGTGTCGC	(25)	12 bZIP and 3 MYC TFs	Wheat, barley and maize
MYB, *TaMYB80*	GGCCGCGGGAATATCCCGGG-AATATCCCGGGAATATCCCG-GGAATATCCCA	CTAGTGGGATATTCCCGGGATAT-TCCCGGGATATTCCCGGGATATT CCCGC	(26)	3	Wheat and maize

2.1.1. Vector Construction and Verification

2. The pINT1-HIS3NB (Acc. AY061966) vector: Available for non-commercial purposes on written request to Dr. PDF Ouwerkerk (e-mail: ouwerkerk@rulbim.leidenuniv.nl or p.b.f.ouwerkerk.2@umail.leidenuniv.nl).
3. DNA cleanup and gel extraction kits: NucleoSpin Extract II (Macherey-Nagel GmbH & Co, Duren, Germany) or QIAquick Gel Extraction Kit (Qiagen, Germany) or similar.
4. Plasmid DNA isolation kit: QiQprep Spin Miniprep Kit (Qiagen, Germany).
5. 10X ligation buffer and T4 DNA Ligase.
6. One Shot Competent *Escherichia coli* (Invitrogen, Carlsbad, CA, USA) or similar quality homemade competent cells.
7. LB medium: 1% Bacto Tryptone, 0.5% Bacto Yeast Extract, 1% NaCl, adjust pH to 7.0 and autoclave to sterilise.
8. 50 mg/mL kanamycin stock solution. Store aliquots at –20°C.
9. 10X TAE buffer: Dissolve 242 g of Tris base in 0.5 L of water, add 100 mL of 0.5 M EDTA, pH 8.0, adjust pH to 8.0 with glacial acetic acid (approx. 55 mL). Add water to 1 L.
10. 2% agarose gel in 1X TAE buffer.
11. *Not*I and *Spe*I restriction enzymes.
12. pINTseq and pINTseqr primers (**Table 3.2**) for PCR test for inserts.

Table 3.2
Primers used in protocols for yeast one-hybrid screening

Name of the primer	Primer sequence
pINTseq	TCACGCCGATAACGTAG
pINTseqr	CGTTTATCTTGCCTGCTC
5′ PCR primer	TTCCACCCAAGCAGTGGTATCAACGCAGAGTGG
3′ PCR primer	GTATCGATGCCCACCCTCTAGAGGCCGAGGCGGCCGACA
ADLD FOR	CTATTCGATGATGAAGATACCCCACCAAACCC
ADLD REV	AGTGAACTTGCGGGGTTTTTCAGTATCTACGAT
T7	TAATACGACTCACTATAGGC
2HRev	AGATGGTGCACGATGCACAG

2.1.2. Integration of the Construct into the PDC6 Locus of Yeast Genomic DNA via Double Crossover

1. Yeast strain Y187: *MATα ura3-52 his3-Δ200 ade2-101 trp1-901 leu2-3,112* met⁻ *gal4Δ gal80Δ URA3::GAL1*$_{UAS}$-*GAL1*$_{TATA}$-*lacZ MEL1*. This strain is provided with Clontech's Matchmaker Library Construction Kit 3.
2. YPDA liquid medium: 2% Difco peptone, 1% Yeast Extract and 15 mL of 0.2% adenine hemisulphate solution (filter sterilised) per 1 L of medium. Add water to 950 mL, autoclave, cool and add glucose to a final concentration of 2% (50 mL of a sterile 40% stock solution).
3. YPDA agar plates: YPDA containing 2% agar. Cool to about 55°C and add glucose to a final concentration of 2% and kanamycin, when required, to a final concentration of 25 mg/L. Mix and pour on plates. These can be stored at 4°C for up to 6 weeks.
4. YPDA-G418 plates: YPDA agar plates with 150 mg/L of G418 (Geneticin) (Sigma, St. Louis, MO, USA) added from a stock solution of 15% G418 (150 mg/mL) after cooling to 55°C. Plates can be stored at 4°C for up to 2 months. Avoid skin and eye contact with G418.
5. 50% (w/v) polyethylene glycol (PEG) stock solution: 50% PEG 3350 (Sigma, St. Louis, MO, USA) in water. Sterilise by autoclaving and store at room temperature.
6. 10X LiAc: 1 M lithium acetate. Adjust to pH 7.5 with acetic acid and autoclave. Store at room temperature. Avoid skin and eye contact with solutions containing LiAc.
7. 10X TE buffer: 0.1 M Tris–HCl, 10 mM EDTA, pH 7.5. Autoclave and store at room temperature.
8. TE/LiAc: 1 part of 10X TE buffer, 1 part of 10X LiAc solution and 8 parts of deionised water.
9. 40% (w/v) PEG solution (PEG/TE/LiAc): 8 parts of 50% (w/v) PEG, 1 part of 10X TE buffer and 1 part of 10X LiAc solution. Prepare fresh before use.
10. Herring Testes Carrier DNA, 10 mg/mL (Clontech, TAKARA BIO, Mountain View, CA, USA): Must be denatured by boiling for 10 min in a water bath and kept on ice until use.
11. Dimethylsulphoxide (DMSO) (Sigma, St. Louis, MO, USA).
12. 50% glycerol in water: Mix 100 mL of glycerol with 100 mL of deionised water and autoclave. Store at room temperature.

2.2. cDNA Library Construction

1. TRIzol or TRIzol-like reagent: Mix 380 mL of phenol equilibrated to pH 4.3, 118.16 g of guanidine thiocyanate,

76.12 g of ammonium thiocyanate, 33.4 mL of 3 M sodium acetate pH 4.5–5 and 50 mL of glycerol. Add water to 1 L. Store at 4°C.

2.2.1. Total RNA Isolation

2. Chloroform.
3. Isopropyl alcohol.
4. 75% ethanol in autoclaved deionised water.
5. RNase-free water.

2.2.2. RNA Quality Examination by Electrophoresis

1. Agarose.
2. 37% formaldehyde. Formaldehyde is toxic: All steps which involve the use of formaldehyde should be performed in a fume hood.
3. 100% deionised formamide.
4. 1% ethidium bromide (EthBr).
5. RNase-free or autoclaved deionised water.
6. Double-autoclaved plastic pipette tips and microfuge tubes.
7. 10X MOPS/EDTA buffer: Mix 42 g of MOPS, 10.88 g of NaAc·3H_2O, 20 mL of 0.5 M EDTA, pH 8.0 (or 3.72 g EDTA), and water to a final volume of 1 L, adjust pH to 7 with NaOH, autoclave and store at room temperature in darkness.
8. Master mix for gel electrophoresis of samples: 3 μL of 10X MOPS/EDTA, 5.25 μL of 37% formaldehyde, 15 μL of formamide and 0.2 μL of 1% EthBr for each sample.
9. Loading dye (6X Ficoll Dye): 15% Ficoll type 4000, 0.25% bromophenol blue and 0.25% xylene cyanol FF.

2.2.3. Poly A+ mRNA Isolation

1. Dynabeads mRNA Purification Kit (Invitrogen, Dynal, Oslo, Norway).
2. Autoclaved deionised water.
3. 10 mM Tris–HCl, pH 7.5.

2.2.4. Synthesis of the First-Strand cDNA Using Oligo(dT) Primer

1. BD Matchmaker Library Construction and Screening Kit (Clontech, TAKARA BIO, Mountain View, CA, USA).
2. dNTP mix: 10 mM of each dATP, dCTP, dGTP and dTTP. Store in aliquots at –20°C.
3. 20 mM DTT (dithiothreitol). Store in aliquots at –20°C.

2.2.5. Synthesis of the Second-Strand cDNA and Enrichment with Long cDNAs

1. Advantage 2 PCR Kit (Clontech, TAKARA BIO, Mountain View, CA, USA).
2. CHROMA SPIN TE-1000 columns (Clontech, TAKARA BIO, Mountain View, CA, USA).

3. 3 M NaAc, pH 4.8.
4. 95% ethanol cooled to –20 °C.

2.2.6. Preparation of Yeast (AH109) Competent Cells for Transformation

1. Yeast strain AH109: *MATa*, *trp1-901*, *leu2-3*, *112*, *ura3-52*, *his3-200*, *gal4D*, *gal80D*, *LYS::GAL1*$_{UAS}$-*GAL1*$_{TATA}$-*HIS3*, *GAL2*$_{UAS}$-*GAL2*$_{TATA}$-*ADE2*, *URA3::MEL1*$_{UAS}$-*MEL1*$_{TATA}$-*lacZ*, *MEL1*. This strain is provided with Clontech's Matchmaker Library Construction Kit.
2. YPDA liquid medium and agar plates, as described in steps 2–4 of **Section 2.1.2**.
3. TE/LiAc, as described in step 8 of **Section 2.1.2**.

2.2.7. Transformation of Yeast AH109 with ds cDNA and pGADT7-Rec

1. pGADT7-Rec vector: Provided as *Sma*I-digested linear DNA with BD Matchmaker Library Construction and Screening Kit (Clontech, TAKARA BIO, Mountain View, CA, USA).
2. Herring Testes Carrier DNA (Clontech, TAKARA BIO, Mountain View, CA, USA).
3. YPD Plus Liquid Medium (Clontech, TAKARA BIO, Mountain View, CA, USA).
4. YPDA agar plates (*see* steps 3 and 4 of **Section 2.1.2**).
5. –Leu DO supplement (Clontech, TAKARA BIO, Mountain View, CA, USA).
6. SD (–Leu) agar plates: 0.67% yeast nitrogen base without amino acids (BD Difco, Sparks, MD, USA), 2% agar, 850 mL of deionised water and 0.069% of –Leu DO supplement. Adjust pH to 5.8 if necessary and sterilise by autoclaving. Cool to about 55°C and add glucose to 2%, adjust the final volume to 1 L, mix and pour on plates. These can be stored at room temperature/4°C for up to several weeks.
7. YPDA Freezing medium: YPDA liquid medium (*see* step 2 of **Section 2.1.2**) containing 25% glycerol.
8. Steps 9 and 11 of **Section 2.1.2**.
9. 0.9% NaCl.

2.3. cDNA Library Screen

1. 2X YPDA liquid medium: Prepare as described for YPDA (*see* step 2 of **Section 2.1.2**), but double the concentration of all components.
2. 1 M 3-amino-1,2,4-triazole (3-AT). Do not autoclave. Filter-sterilise the solution and store at 4°C.
3. –Leu, –His DO supplement (Clontech, TAKARA BIO, Mountain View, CA, USA).
4. SD (–Leu, –His) containing 5–10 mM 3-AT: 0.67% yeast nitrogen base without amino acids, 2% agar, 850 mL of deionised water and 100 mL of 0.067% of a –Leu, –His

DO supplement. Check pH and adjust solution to pH 5.8 if necessary. Sterilise by autoclaving. Cool to about 55°C and add glucose to a final concentration of 2% and 3-AT to the required final concentration (usually 5 mM). Adjust the final volume to 1 L, mix and pour on large (15 cm) plates. Plates can be stored for up to 2 months at 4°C.

5. Kanamycin.

2.4. Further Characterisation of Positive Clones

2.4.1. Plasmid DNA Isolation from Positive Yeast Clones

1. Y-DER Yeast DNA Extraction Reagent Kit (Pierce Chemical, Rockford, IL, USA).
2. 70% ethanol.
3. Isopropyl alcohol.

2.4.2. Characterisation of cDNA Inserts by PCR and Restriction Digestion

1. *Hae*III or similar restriction enzyme.
2. 2% agarose gel in 1X TAE buffer (*see* steps 9 and 10 of **Section 2.1.1**).
3. ADLD FOR and ADLD REV primers (**Table 3.2**).

2.4.3. Transformation of Bacterial Cells and Isolation of Pure Plasmids

1. SOB medium: Mix 2% Tryptone/Peptone, 0.5% Yeast Extract, 10 mM NaCl and 2.5 mM KCl and autoclave. Add filter-sterilised $MgSO_4$ and $MgCl_2$ to a final concentration of 10 mM of each.
2. SOC medium: SOB medium supplemented with 20 mM glucose.
3. NEB10-beta electro-competent *E. coli* (New England BioLabs) or similar strain.
4. Gene Pulser Cuvettes, 1 mm between electrodes (Bio-Rad Laboratories, CA, USA).
5. Gene Pulser Xcell (Bio-Rad Laboratories, CA, USA).
6. Plasmid DNA Isolation Kit and QiQprep Spin Miniprep Kit (Qiagen, Germany).
7. LB-Ampicillin plates.
8. T7 and/or 2HRev primer(s) (**Table 3.2**).

3. Methods

The protocol involves cloning a plant-derived *cis*-element of interest into the vector construct, generating a reporter yeast strain by homologous recombination in yeast, preparation and screening of a plant cDNA library in the reporter yeast strain and recovery and analysis of cloned cDNA sequences.

It is possible to generate a number of reporter yeast strains simultaneously for later use. In our work, oligonucleotides with complementary sequences for each tandem of *cis*-elements (**Table 3.1**) were designed to create protruding ends for cloning into *Not*I–*Spe*I sites of the pINT1-HIS3NB binary vector. Linearised vector is integrated into yeast genomic DNA according to a previously described protocol (6). The resulting collection of reporter yeast strains can be stored at –80°C as glycerol cultures for later screening of cDNA libraries.

cDNA libraries are prepared using the BD Matchmaker Library Construction and Screening Kit (Clontech, TAKARA BIO, Mountain View, CA, USA), which provides a simple method for the construction of cDNA expression libraries for Y2H screening. The library construction protocol integrates the efficiency of SMART cDNA synthesis technology (4, 13, 14) with the simplicity of cloning using homologous recombination in yeast (15). Collections of cDNA libraries can be stored at –80°C in 1.5 mL aliquots (200–300 per library). No impact on library quality has been observed after 5–6 years of storage. Each aliquot can potentially be used for library amplification.

To perform the Y1H screen, each yeast reporter strain, based on Y187 (*MAT*α), is mated with the Yeast Matchmaker System cDNA library prepared in the yeast strain AH109 (*MAT*a). This approach permits the use in Y1H screens of cDNA libraries in yeast rather than having to revert to plasmid-based libraries and saves a great deal of time and labour.

An additional advantage of this method is that the same cDNA library can be subsequently used for further characterisation of any isolated transcription factor genes. After recloning of selected transcription factor cDNAs into the pGBKT7 vector (Clontech, TAKARA BIO, Mountain View, CA, USA), Y2H screening can then be used for identification of interacting partners.

3.1. Generation of Yeast Reporter Strains

3.1.1. Vector Construction and Verification

1. Prepare Y1H reporter constructs by cloning tandems containing 3–4 repeats of the *cis*-element of interest (*see* **Table 3.1** for DNA oligonucleotides used for the preparation of successful *cis*-elements in our work) (*see* **Note 1**). Mix together 45 μL of each of the pair of partially complementary oligonucleotides (at 100 μM concentration) and add 10 μL of 10X annealing buffer. Incubate the mix at 95°C for 10 min and slowly cool, first to room temperature and then transfer to ice. Use the resulting overhangs for cloning the *cis*-element repeats into the *Spe*I–*Not*I sites of the pINT1-HIS3NB vector.
2. Digest 2.5–5 μg of pINT1-HIS3NB vector with *Not*I and *Spe*I in a 20 μL reaction volume for 3 h at 37°C. Purify the

linearised vector using DNA electrophoresis in 2% agarose and a gel extraction kit (e.g. NucleoSpin Extract II or similar). Elute in a minimal amount (e.g. 20 μL) of water.

3. Mix together on ice 1 μL of linearised vector, 3 μL of *cis*-element repeat DNA with overhangs, 1 μL of 10X ligation buffer, 4 μL of water and 1 μL of T4 DNA Ligase and incubate the mix at 16°C overnight.
4. Gently mix the entire ligation reaction (10 μL) with 100 μL of chemically competent *E. coli* cells (e.g. One Shot Competent *E. coli*, Invitrogen, or similar quality 'home-made' competent cells). Incubate for 15 min on ice, 2 min in a 42°C water bath and again 2 min on ice. Add 1 mL of LB and incubate for 1 h at 37°C with shaking. Briefly spin down cells, resuspend them in 50 μL of LB medium and spread 5 and 45 μL of bacteria on two LB plates containing 50 μg/mL kanamycin. Incubate plates overnight at 37°C.
5. For each reporter construct, isolate plasmid DNA from three to five colonies using any suitable kit/method for plasmid isolation. Use 0.5 μL (50–250 ng) of each isolated plasmid and uncut pINT1-HIS3NB vector (negative control) as templates for PCR amplification with pINTseq and pINTseqr primers derived from the vector sequence (**Table 3.2**) (*see* **Note 2**).
6. Compare the lengths of obtained PCR products using DNA electrophoresis in 2% agarose. An example of this is shown in **Fig. 3.1a**. Select plasmids with PCR products which are

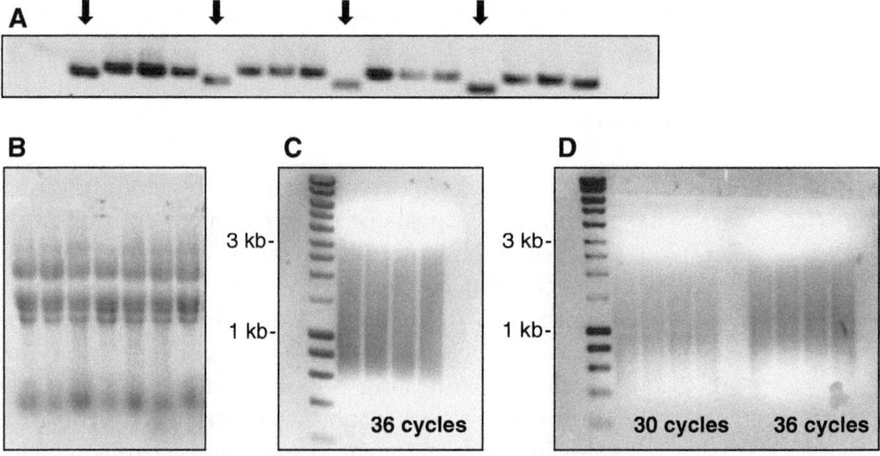

Fig. 3.1. Analyses of DNA and RNA quality in reporter vector and library construction. (**a**) PCR-based identification of pINT1-HIS3NB vector containing cloned inserts of *cis*-element repeats. PCR products from uncut control vector are shown with *arrows*. (**b**) An example of high-quality total RNA isolated from flag leaf of wheat, following RNA electrophoresis. (**c**, **d**) Size and amount of ds cDNA as assessed by DNA electrophoresis: (**c**) cDNA prepared from the wheat flag leaves and spikes of plants subjected to drought plus heat stress; (**d**) cDNA prepared from a 'difficult' material, roots of wheat plants grown in soil and subjected to drought. Number of PCR cycles is indicated under respective lanes.

longer than the product obtained from the uncut vector control and verify their sequences using either pINTseq or pINTseqr primers (**Table 3.2**).

7. Isolate 0.3–0.5 μg of confirmed plasmid fragment containing the *cis*-element, reporter gene and recombination sequences by restriction digestion with *Nco*I and *Sac*I. Check for completeness of the reaction by gel electrophoresis with 1 μL of the digested sample. If restriction digestion is complete, purify the resulting mix of DNA fragments using any suitable PCR/DNA purification kit (*see* **Note 3**).

3.1.2. Integration of the Construct into the PDC6 Locus of Yeast Genomic DNA via Double Crossover

1. Set up an overnight yeast culture (Y187 or similar strain) by transferring a few large colonies into 100 mL of YPDA in a 250 mL conical flask. Incubate at 30°C overnight with shaking. Pellet the cells by centrifugation in 50 mL plastic tubes at 700×*g*. Discard the supernatant.
2. Resuspend yeast pellets in 50 mL of fresh YPDA, return to the 250 mL flask and incubate at 30°C with shaking for an additional 2–3 h. Pellet cells by centrifugation (700×*g*, 5 min), resuspend in 50 mL of fresh deionised water and pellet cells again. Resuspend the final yeast pellet in 2 mL of TE/LiAc. Cells can be kept at room temperature for several hours without significant loss in competency.
3. Directly before yeast transformation, denature the carrier DNA by boiling for 10 min and then transfer to ice. Prepare PEG/TE/LiAc solution (step 9 in **Section 2.1.2**) in a quantity sufficient for use in the next step.
4. Mix 0.3–0.5 μg of linearised pINT1-HIS3NB construct with 25 μg of carrier DNA in a maximum total volume of 10 μL. Add 100 μL of the yeast suspension and 600 μL of freshly prepared 40% PEG/TE/LiAc solution. Add 70 μL of DMSO and vortex to mix.
5. Incubate the mix at 42°C for 15 min and then place on ice for 2 min. Harvest cells by centrifugation (2 min, 13,000×*g*), discard the supernatant and rinse tube with 400 μL of distilled water to remove traces of DMSO. Centrifuge again (2 min, 13,000×*g*), resuspend cells in 1 mL of YPDA medium and incubate for 3–6 h at 30°C with shaking.
6. Spin cells down in a microcentrifuge at 13,000×*g*, discard the supernatant and resuspend the yeast pellet in 100 μL of 1X TE, pH 7.5, by pipetting up and down. Plate an aliquot of 20 μL and the remainder of the cell suspension onto two YPDA-G418 plates, respectively. Incubate for 2–5 days at 30°C. Pick and re-streak large, white G418-resistant colonies (*see* **Note 4**) on new YPDA-G418 plates and incubate for a further 2–3 days at 30°C. These plates

may be stored at 4°C and streaked onto fresh plates every 1–2 months. For longer storage, prepare glycerol cultures.

7. For the preparation of glycerol cultures, inoculate a single colony in 4–5 mL of YPDA-G418 liquid medium and grow cells in a small flask overnight at 30°C with shaking. Add an equal volume of 50% glycerol and incubate for 30 min at 30°C with shaking. Store as 1 mL aliquots at –80°C (*see* **Note 5**).

3.2. cDNA Library Construction

*3.2.1. Total RNA Isolation (See **Note 6**)*

1. Transfer 50–100 mg of frozen powdered tissue to a plastic 1.5 mL microfuge tube on ice. Immediately add 1 mL of Trizol reagent and agitate on a daisy wheel rotor at room temperature for 5 min (*see* **Note 7**).
2. Centrifuge at 13,000×*g* for 10 min at 4°C. Transfer the supernatant to a clean microfuge tube and add 200 μL of chloroform. Shake tubes vigorously by hand for 15 s and incubate for 2–3 min at room temperature.
3. Centrifuge at 13,000×*g* for 15 min at 4°C (RNA remains in the colourless, upper aqueous phase). Transfer the aqueous phase to a fresh tube and add 500 μL of isopropyl alcohol. Incubate at room temperature for 10 min.
4. Centrifuge at 13,000×*g* for 10 min at 4°C (RNA forms a pellet on the side and bottom of the tube, white to brown in colour). Remove the supernatant and add 1 mL of 75% ethanol.
5. Mix the sample by vortexing and centrifuge at 7,400×*g* for 5 min at 4°C. Remove all traces of ethanol and leave tubes open in the fume hood for the RNA pellet to dry (the pellet will change from white to clear).
6. Resuspend RNA in 30–100 μL of RNase-free water and incubate for 10 min at 60°C. Store RNA at –80°C. Take an aliquot prior to freezing for determination of the quality and quantity of RNA obtained, using a spectrophotometer and by gel electrophoresis as described below.

3.2.2. RNA Quality Examination by Electrophoresis

1. For preparation of a 1% gel, weigh out 1.6 g of agarose and add 144 mL of autoclaved deionised water and 20 mL of 10X MOPS/EDTA buffer. Heat the mixture in a microwave oven to completely dissolve agarose. Cool to 55–60°C, add 6 mL of 37% formaldehyde (*see* **Note 8**) in the fume hood, mix and pour the gel. Leave the gel at room temperature to set and then pre-run in 1X MOPS/EDTA buffer at 60 V for 30 min.
2. Prepare a master mix for the required number of samples. For each sample you will need 3 μL of 10X MOPS/EDTA buffer, 5.25 μL of 37% formaldehyde, 15 μL of 100% for-

mamide and 0.2 μL of 1% EthBr. For each sample, mix together 8 μL of RNA (~5–8 μg) and 23.5 μL of master mix. Heat at 65–70°C for 10–15 min, cool on ice, add 1.5 μL of loading dye and load samples onto the gel. Run the gel at 40 V for 20 min and then at 60 V for 2.5 h.

3. Photograph the gel using a UV transilluminator (an example is shown in **Fig. 3.1b**). If the concentration and quality of the RNA are satisfactory, proceed with poly A+ mRNA isolation (*see* **Note 9**).

3.2.3. Poly A+ mRNA Isolation

To isolate poly A+ mRNA we used the Dynabeads mRNA Purification Kit (Invitrogen, Dynal, Oslo, Norway), essentially according to the supplied manual.

1. Adjust the volume of total RNA to 100 μL with distilled, RNase-free water (*see* **Note 10**). Heat sample to 65°C for 2 min, then place on ice.
2. Transfer 200 μL of well-suspended Dynabeads to a microcentrifuge tube and place the tube on a magnet for 30 s. Remove the supernatant, remove tube from the magnet and add 100 μL of Binding Buffer (supplied with the Dynabeads mRNA Purification Kit). Put the tube back onto the magnet and remove the supernatant. Remove tube from the magnet.
3. Add 100 μL of Binding Buffer and 100 μL of total RNA. Mix thoroughly and rotate on a roller for 3–5 min at room temperature. Place tube on the magnet until the solution is clear. Remove the supernatant.
4. Remove the tube from the magnet and wash the mRNA–bead complex twice with 200 μL of washing buffer B (supplied with the Dynabeads mRNA Purification Kit). With the help of the magnet remove all supernatant between each wash step.
5. Add 5–10 μL (*see* **Note 11**) of 10 mM Tris–HCl, pH 7.5, heat at 65–80°C for 2 min and place the tube immediately on the magnet. Transfer the eluted mRNA to a new tube and continue with the next step or store at –80°C.

3.2.4. Synthesis of the First-Strand cDNA Using Oligo(dT) Primer

1. Combine the following reagents in a PCR tube: 2 μL (100–600 ng) of poly A+ RNA sample, 1 μL of CDS III Primer (component of the Matchmaker Kit) and 1 μL of deionised water. Mix the contents and briefly spin. Incubate at 72°C for 2 min, cool on ice for 2 min and spin briefly.
2. Add the following to the reaction tube: 2 μL of 5X First-Strand Buffer, 1 μL of 20 mM DTT, 1 μL of 10 mM dNTP mix and 1 μL of MMLV Reverse Transcriptase (all are components of the Matchmaker Kit). Mix gently by tapping, spin briefly and incubate at 42°C for 10 min.

3. Add 1 μL of BD SMART III Oligonucleotide (component of the Matchmaker Kit) and incubate at 42°C for 1 h in a hot-lid thermocycler. Place the tube at 75°C for 10 min to terminate first-strand synthesis.
4. Cool the tube to room temperature and add 1 μL of RNase H (component of the Matchmaker Kit). Incubate at 37°C for 20 min. Store at –20°C or proceed with second-strand synthesis.

3.2.5. Synthesis of the Second-Strand cDNA and Enrichment with Long cDNAs

1. Mix the following components in a PCR tube: 2 μL of first-strand cDNA, 70 μL of deionised water, 10 μL of 10X Advantage 2 PCR buffer, 2 μL of 50X dNTP mix, 2 μL of 5′ PCR primer, 2 μL of 3′ PCR primer (*see* **Table 3.2**), 10 μL of 10X GC-Melt Solution and 2 μL of 50X Advantage 2 Polymerase Mix (except for the first-strand cDNA sample and primers, all are components of the Advantage 2 PCR Kit). Mix gently by flicking the tube. Centrifuge briefly. Cap the tube and place in a preheated (95°C) hot-lid thermal cycler.
2. Use the following conditions for double-stranded cDNA synthesis: 95°C for 30 s, 30–36 cycles (*see* **Note 12**) of 95°C for 10 s, 68°C for 6 min and termination at 68°C for 5 min. Remove a 5 μL aliquot of the reaction for analysis by gel electrophoresis. The sample can be stored at –20°C or purified immediately.
3. To enrich the pool of cDNAs for longer cDNAs, invert a CHROMA SPIN TE-1000 column (*see* **Note 13**) several times to resuspend the matrix. Snap off the breakaway end. Place the end of the spin column in a 2 mL microcentrifuge tube, remove the cap and centrifuge (700×*g*, 5 min).
4. Remove the spin column and collection tube from the centrifuge and discard the collection tube and the column equilibration buffer. Place the spin column in a second 2 mL tube. Carefully and slowly apply the ds cDNA sample (95 μL per column) to the centre of the matrix. Do not allow the sample to flow along the inner wall of the column.
5. Centrifuge at 700×*g* for 5 min. The purified sample is now at the bottom of the collection tube. Combine duplicate experimental samples in a single tube. Add 0.1 volume of 3 M NaAc, pH 4.8, and 2.5 volumes of 95% ethanol cooled to –20°C. Mix gently by rocking the tube back and forth. Incubate the sample overnight at –20°C to precipitate the cDNA.
6. Centrifuge (13,000×*g*, 20 min) the sample at room temperature and carefully remove the supernatant with a pipette. Spin briefly to collect any residual liquid and remove all

remaining liquid. Allow the tube to air-dry at room temperature for 10 min. Resuspend the pellet in 20 μL of deionised water. Store the sample at –20°C until the yeast transformation.

3.2.6. Preparation of Yeast (AH109) Competent Cells for Transformation

1. Streak a YPDA agar plate from a frozen stock of the AH109 yeast strain and incubate the plate upside down at 30°C until colonies appear (about 3 days). Inoculate a single colony (≤4 weeks old, 2–3 mm in diameter) into 3 mL of YPDA medium in a sterile 15 mL centrifuge tube and incubate at 30°C for 8 h with shaking. Transfer 5 μL of the culture to a 250 mL flask containing 50 mL YPDA medium and incubate at 30°C with shaking at 230–250 rpm for 16–20 h.
2. Centrifuge at 700×*g* for 5 min at room temperature. Discard the supernatant and resuspend the pellet in 100 mL of YPDA. Incubate at 30°C for 3–5 h (until OD_{600} = 0.4–0.5).
3. Centrifuge at 700×*g* for 5 min. Discard the supernatant and resuspend pellet in 60 mL of sterile, deionised water. Centrifuge at 700×*g* for 5 min, discard the supernatant and resuspend the pellet in 3 mL of TE/LiAc solution. Split the suspension between two 1.5 mL microcentrifuge tubes.
4. Centrifuge each tube at 13,000×*g* for 15 s. Discard the supernatant and resuspend each pellet in 600 μL of TE/LiAc solution. Cells may be left at room temperature for several hours without significant loss in competency.

3.2.7. Transformation of Yeast AH109 with ds cDNA and pGADT7-Rec

1. In a sterile 15 mL tube, combine the following: 20 μL of ds cDNA (after CHROMA SPIN purification), 6 μL of pGADT7-Rec (0.5 μg/μL, component of the Matchmaker Kit) and 20 μL of denatured Herring Testes Carrier DNA. Add 600 μL of competent cells to the DNA and mix gently by vortexing. Add 2.5 mL of PEG/TE/LiAc solution and mix gently by vortexing. Incubate at 30°C for 45 min, gently mixing the cells every 15 min.
2. Add 160 μL of DMSO, mix and then place the tube in a 42°C water bath for 20 min. Mix cells every 10 min. Centrifuge at 700×*g* for 5 min, discard the supernatant and resuspend in 3 mL of YPD Plus Liquid Medium. Incubate at 30°C without shaking for 90 min.
3. Centrifuge at 700×*g* for 5 min, discard the supernatant and resuspend in 30 mL of 0.9% NaCl solution. Spread 150 μL of 1:10, 1:100, 1:1,000 and 1:10,000 dilutions onto YPDA plates and SD (–Leu) plates to check for transformation efficiency. Spread 150 μL aliquots onto 200 large SD (–Leu) plates (*see* **Note 14**). Incubate the plates upside down

at 30°C without shaking for about 7 days. The expected number of independent colonies is 5,000–10,000 or more per plate.

4. When colonies have grown sufficiently (5–7 days), resuspend cells using a sterile spreader and 5 mL of YPDA Freezing medium for every four to five plates. Collect cell suspension into a sterile 2 L flask. Wash the same plates again using another 5 mL of YPDA Freezing medium and add to the flask. Incubate the cells for 2 h at 30°C with shaking. Aliquot 1.5 mL of the cell suspension in sterile 2 mL microcentrifuge tubes and store at –80°C.

3.3. cDNA Library Screen

1. Prepare 50 mL of overnight culture of each reporter strain in YPDA medium (alternatively, thaw 1 mL of glycerol culture in a water bath at room temperature and dilute to 50 mL with YPDA). Pellet cells by centrifuging for 5 min at 700×*g*. Resuspend the pellet in 50 mL of 2X YPDA medium containing 25 μg/mL kanamycin, mix with 1.5 mL of glycerol culture of the cDNA library thawed in a water bath at room temperature and incubate in a 2 L conical flask at 30°C overnight with slow (30–50 rpm) rotation (*see* **Note 15**).
2. Harvest yeast cells after 17–24 h of mating by centrifugation at 700×*g* and wash with 50 mL of 1X TE buffer, pH 8.0, containing 25 μg/mL kanamycin. Centrifuge again at 700×*g*, resuspend the pellet in 5 mL of the 1X TE buffer and spread onto plates with selective medium (–Leu, –His) containing 5–10 mM 3-amino-1,2,4-triazole (3-AT) to reduce possible leaky expression of the *HIS3* gene (*see* **Note 16**). Positive clones usually appear on plates after 2–5 days of incubation at 30°C. Plates can be stored for several weeks at 4°C.

3.4. Further Characterisation of Positive Clones

3.4.1. Plasmid DNA Isolation from Positive Yeast Clones (See Note 17)

1. Pick a single yeast colony with a toothpick and resuspend in 80 μL of Y-DER Reagent (Y-DER Yeast DNA Extraction Reagent Kit) in a 1.5 mL plastic tube. Incubate at 65°C for 10 min.
2. Centrifuge cells at 13,000×*g* for 5 min, carefully discard the supernatant using a pipette and resuspend the pellet in 64 μL of DNA Releasing Reagent A (Y-DER Kit). Add 64 μL of DNA Releasing Reagent B (Y-DER Kit), vortex briefly and incubate for 10 min at 65°C.
3. Add 32 μL of Protein Removal Reagent (Y-DER Kit) to the mixture and invert several times. Centrifuge for 5 min at 13,000×*g* and transfer the supernatant to a new 1.5 mL centrifuge tube.

4. Add 96 μL of isopropyl alcohol, mix and precipitate DNA by centrifugation (13,000×*g*, 10 min). Carefully remove the supernatant, then add 200 μL of 70% ethanol, invert the tube several times (do not vortex) and centrifuge again (13,000×*g*, 5 min). Carefully discard the supernatant with a pipette and dry the pellet at 37°C for 15 min.
5. Resuspend the pellet in 10 μL of deionised sterile water and store at −20°C. Use 1 μL of the yeast DNA solution for estimation of insert size by PCR (**Section 3.4.2**) and 0.5–1 μL for electroporation of bacterial cells (**Section 3.4.3**).

3.4.2. Characterisation of cDNA Inserts, by PCR and Restriction Digestion

1. Use ADLD FOR and ADLD REV primers derived from the pGADT7 vector (*see* **Table 3.2**), and 1 μL of each isolated yeast DNA as a template, in 20 μL reaction volumes for PCR amplification. Determine the length of the inserts by loading 1–5 μL of each PCR reaction on a 1.6% agarose gel for DNA electrophoresis. Group together inserts with the same length.
2. Use restriction analysis with *Hae*III or a similar enzyme for further analysis of inserts of the same size. Mix 5–7 μL of the PCR reaction with 2 μL of the appropriate restriction buffer and 1–2 μL of restriction enzyme in a 20 μL reaction volume and incubate at 37°C for 3 h. Load 15–20 μL of each reaction onto a 1.6–2% agarose gel and perform DNA electrophoresis. For bacterial transformation by electroporation, select one to two independent yeast DNA preparations as representatives of each group of inserts.

3.4.3. Transformation of Bacterial Cells and Isolation of Pure Plasmids

1. Add 0.5–1 μL of yeast DNA to 10 μL of electro-competent bacterial cells (*E. coli* strains NEB10-beta or DH5α). Mix gently with a pipette tip and keep on ice.
2. Prepare SOC medium by adding 0.7 mL of 50% glucose solution to 100 mL of SOB medium.
3. Transfer mix of DNA and electro-competent bacterial cells to a 1 mm cuvette and apply 1,800 V and 25 μF using a Gene Pulser Xcell. Immediately add 1 mL of SOB medium to the cuvette, resuspend the bacterial cells by pipetting and transfer the suspension to a 1.5 mL tube. Incubate tubes at 37°C for 1 h with shaking.
4. Pellet the bacterial cells for 5 min at 13,000×*g*, discard the supernatant, add 100 μL of SOB medium and plate onto LB plates containing 100 μg/mL ampicillin. Incubate plates at 37°C overnight.
5. Isolate plasmid DNA from two colonies (for each plate) using a plasmid isolation kit and sequence the cDNA

inserts using T7 and/or 2HRev primers (*see* **Table 3.2** and **Note 18**).

6. Use publically available DNA databases to identify isolated cDNAs.

4. Notes

1. The use of repeats of *cis*-elements helps to optimise the distance between the plant regulatory element and the minimal promoter of the yeast GAL4 gene.
2. In order to simplify electrophoretic analysis of the small-sized inserts, pINTsec and pINTsecr PCR primers were designed to anneal to approximately 50 bp on either side of the cloning sites, giving an amplicon from the empty vector of about 100 bp. Size differences between amplicons from empty vector and vector with the cloned insert can be detected using 2% agarose gels **(Fig. 3.1a)**. Plasmids with inserts can subsequently be analysed by sequencing using the same primers.
3. Linearisation and PCR purification of the vector before recombination improves recombination rate. *Bbe*I, *Ehe*I or *Nar*I restriction sites of the vector can be used in place of *Nco*I and *Asc*I, *Xcm*I or *Age*I in place of *Sac*I. Gel electrophoresis will indicate the presence of two fragments following digestion. However, it is not necessary to isolate the appropriate fragment from the gel. The second part of the plasmid is not able to recombine with yeast genomic DNA or to be amplified autonomously in yeast cells and consequently will be rapidly degraded.
4. Do not choose small, semi-transparent colonies as these are usually the result of spontaneous mitochondrial mutations. They cannot be used for the cDNA library screen.
5. If desired, yeast glycerol cultures can be quickly frozen in dry ice before being transferred to –80°C storage. However, we have observed no differences between the viability of yeast cells frozen on dry ice prior to or transferred directly to storage at –80°C. The use of liquid nitrogen instead of dry ice gives poor results and should be avoided.
6. This technique for RNA extraction is suitable for most plant tissues. We have successfully used it for wheat, barley, rice and *Arabidopsis* root and shoot material. Slight modifications are recommended when using polysaccharide-rich tissues (e.g. grain or anthers): after completing the

total RNA extraction protocol, heat the RNA to 65°C for 5 min. Cool tube to room temperature and briefly centrifuge (13,000×*g*, 15 s). Transfer the supernatant to a clean tube and incubate at –20°C for 5 min. Centrifuge briefly again (13,000×*g*, 15 s), then transfer the supernatant containing the RNA to a clean tube.

7. Avoid thawing of frozen material prior to the addition of Trizol. In the presence of Trizol reagent, RNA is protected from RNase degradation. To prevent RNA degradation, all non-disposable glass- and plasticware must be RNase-free. Glass items can be baked at 150°C for 4 h, and plastic items can be autoclaved or soaked for 10 min in 0.5 M NaOH, followed by thorough rinsing with sterile water. Always keep RNA samples on ice and return promptly to –80°C for storage.
8. Formaldehyde is toxic: gel casting and electrophoresis should be performed in a fume hood.
9. Total RNA can be used directly for the preparation of cDNA libraries. However, according to our experience, the quality of such libraries (based on complexity and average length of inserts) is much lower than the quality of libraries prepared from poly A+ mRNA.
10. The poly A+ mRNA isolation kit we use was originally designed for mammalian RNA. Taking into consideration that total RNA isolated from some plant tissues contains about threefold less poly A+ mRNA than total RNA from mammalian tissues, we suggest starting with 150–200 mg of total plant RNA rather than 75 mg as recommended in the standard protocol.
11. Try to elute the poly A+ mRNA in a minimal volume of buffer to maximise the concentration (10 μL usually gives good results for this kit). For the synthesis of the first-strand cDNA a maximum of 2 μL of RNA solution can be used, containing, at best, ~150–400 ng of poly A+ mRNA. Lower concentrations of poly A+ mRNA samples lead to lower yields of the first-strand cDNA and a greater number of PCR cycles will be required to generate sufficient double-stranded cDNA. The average length and complexity of the cDNA population is reduced, and the abundance of particular cDNAs may be affected.
12. Thirty PCR cycles are recommended in the Matchmaker Kit manual. However, we were never able to get sufficient amounts of double-stranded cDNA with this number of cycles (*see* **Fig. 3.1c**, **d**). Thirty-six cycles yielded enough cDNA for a high-quality library. Any further increase in the number of cycles is not recommended. Another option

is to set up a large number of PCR reactions and pool them after enriching for large-sized cDNAs. Ethanol-precipitate the entire collection for yeast transformation.

13. As our first libraries, prepared according to the Clontech^{TM} manual, contained large numbers of inserts smaller than 600 bp, we tried using CHROMA SPIN+TE-1000 (Clontech^{TM}) columns instead of CHROMA SPIN+TE-400 columns for cDNA size fractionation. According to the CHROMA SPIN Columns User Manual (PT1300-1), a single purification on the TE-1000 column removes about 60% of cDNAs smaller than 1 kb. This change in the protocol substantially increased the average length of cDNAs and thus the frequency of full-length cDNAs in our libraries compared with the original method, while still maintaining a good representation of short cDNAs. There is no need to use CHROMA SPIN+TE-1000 columns if the random primer is used for the cDNA preparation: even small inserts (>400 bp) in this case can be of high value.
14. All inoculations of liquid media and plating operations must be done in a sterile environment to prevent contamination of the cDNA library with other types of yeast or with fungi. Before washing colonies from the plates look carefully at all plates for any colonies of different shape or colour. Discard any suspicious plates to prevent library contamination.
15. Low speed of rotation is very important for successful mating!
16. The optimal concentration of 3-AT can be determined before the screen as described in (6). However, for most of the bait *cis*-elements that we have tested (**Table 3.1**), 5 mM 3-AT is adequate. The exception is DRE from *Arabidopsis*, which gave better results on SD medium containing 10 mM 3-AT.
17. Traditional methods for DNA extraction from yeast are time consuming, labour intensive and require the use of hazardous chemicals that may also affect the efficiency of future applications of the extracted DNA (such as PCR and bacterial transformation). To extract plasmid DNA from yeast, we recommend using the Y-DER Yeast DNA Extraction Reagent Kit (Pierce Chemical, Rockford, IL, USA), which allows the user to extract yeast plasmid DNA directly from colonies, eliminating a liquid culture step characteristic of traditional methods. However, the standard kit protocol for DNA extraction from colonies yields a concentration of DNA that is adequate for PCR but not sufficient for bacterial transformation by electroporation. We modified

the protocol by increasing the amounts of all reagents. The modified protocol described in this chapter yields sufficient plasmid DNA for both PCR and transformation of bacteria by electroporation.

18. Initially, we confirmed protein–DNA interactions before sequencing by re-transformation of the AH109 yeast strain with the isolated plasmid and subsequent mating of the transformed strain with Y187 (as a negative control) and respective bait yeast strain. However, this is a labour-intensive and time-consuming process. Moreover, we found that the number of successful Y1H screenings with known *cis*-elements was much higher than the number of unsuccessful screenings. Thus, it became more practical to proceed directly with sequencing inserts from representatives of each group of clones with the same characteristics (in our case, this was 1–10 plasmids per screen). The identification of obtained sequences using appropriate database searching tools and publically available protein databases (e.g. BlastX program and NCBI databases) is an easy and rapid process. Even novel transcription factors that are absent in the databases may be easily recognised by the obligatory presence of a DNA-binding domain, which is usually highly conserved within each family of transcription factors and across different plant species. Conversely, proteins which non-specifically activate the reporter/selection gene in the Y1H screen (false positives) usually belong to the most abundant cDNAs in the library and hence the majority of them are well known and easy to identify.

Acknowledgements

The authors are grateful to Dr. P.D.F. Ouwerkerk who kindly provided the pINT1-HIS3NB vector, to Dr. J. Hayes for critical reading of the manuscript and to N. Bazanova for technical support with development of the Y1H screening method in our laboratory.

References

1. Lee, T. I. and Young, R. A. (2000) Transcription of eukaryotic protein-coding genes. *Annu. Rev. Genet.* **34**, 77–137.
2. Riechmann, J. L. and Ratcliffe, O. J. (2000) A genomic perspective on plant transcription factors. *Curr. Opin. Plant Biol.* **3**, 423–434.
3. Fields, S. and Song, O. K. (1989) A novel genetic system to detect protein–protein interactions. *Nature* **340**, 245–246.
4. Zhu, Y. Y., Machleder, E. M., Chenchik, A., Li, R., and Siebert, P. D. (2001) Reverse transcriptase template switching: a SMART

(TM) approach for full-length cDNA library construction. *Biotechniques* **30**, 892–897.
5. Meijer, A. H., Ouwerkerk, P. B. F., and Hoge, J. H. C. (1998) Vectors for transcription factor cloning and target site identification by means of genetic selection in yeast. *Yeast* **14**, 1407–1415.
6. Ouwerkerk, P. B. F. and Meijer, A. H. (2001) Yeast one-hybrid screening for DNA-protein interactions, in *Current Protocols in Molecular Biology* (Ausubel, M., ed.), pp. 12.12.1–12.12.22. Wiley, New York, NY.
7. Sieweke, M. (2000) Detection of transcription factor partners with a yeast one hybrid screen. *Methods Mol. Biol.* **130**, 59–77.
8. Meijer, A. H., de Kam, R. J., d'Erfurth, I., Shen, W., and Hoge, J. H. C. (2000) HD-Zip proteins of families I and II from rice: interactions and functional properties. *Mol. Gen. Genet.* **263**, 12–21.
9. Deng, X., Phillips, J., Meijer, A. H., Salamini, F., and Bartels, D. (2002) Characterization of five novel dehydration-responsive homeodomain leucine zipper genes from the resurrection plant *Craterostigma plantagineum*. *Plant Mol. Biol.* **49**, 601–610.
10. Kim, S. Y., Chung, H. J., and Thomas, T. L. (1997) Isolation of a novel class of bZIP transcription factors that interact with ABA-responsive and embryo-specification elements in the *Dc3* promoter using a modified yeast one-hybrid system. *Plant J.* **11**, 1237–1251.
11. Kizis, D. and Pages, M. (2002) Maize DRE-binding proteins DBF1 and DBF2 are involved in *rab17* regulation through the drought-responsive element in an ABA-dependent pathway. *Plant J.* **30**, 679–689.
12. Lopato, S., Bazanova, N., Morran, S., Milligan, A. S., Shirley, N., and Langridge, P. (2006) Isolation of plant transcription factors using a modified yeast one-hybrid system. *Plant Methods* **2**, 3–17.
13. Seth, D., Gorrell, M. D., McGuinness, P. H., Leo, M. A., Lieber, C. S., McCaughan, G. W., and Haber, P. S. (2003) SMART amplification maintains representation of relative gene expression: quantitative validation by real time PCR and application to studies of alcoholic liver disease in primates. *J. Biochem. Biophys. Methods* **55**, 53–66.
14. Wellenreuther, R., Schupp, I., Poustka, A., and Wiemann, S. (2004) SMART amplification combined with cDNA size fractionation in order to obtain large full-length clones. *BMC Genomics* **5**, 36–43.
15. Fusco, C., Guidotti, E., and Zervos, A. S. (1999) In vivo construction of cDNA libraries for use in the yeast two-hybrid system. *Yeast* **15**, 715–720.
16. Sessa, G., Morelli, G., and Ruberti, I. (1993) The Athb-1 and Athb-2 Hd-Zip domains homodimerize forming complexes of different DNA-binding specificities. *EMBO J.* **12**, 3507–3517.
17. Sessa, G., Morelli, G., and Ruberti, I. (1997) DNA-binding specificity of the homeodomain leucine zipper domain. *J. Mol. Biol.* **274**, 303–309.
18. Abe, M., Takahashi, T., and Komeda, Y. (2001) Identification of a *cis*-regulatory element for L1 layer-specific gene expression, which is targeted by an L1-specific homeodomain protein. *Plant J.* **26**, 487–494.
19. Hofer, J., Gourlay, C., Michael, A., and Ellis, T. H. N. (2001) Expression of a class 1 knotted1-like homeobox gene is down-regulated in pea compound leaf primordial. *Plant Mol. Biol.* **45**, 387–398.
20. Chaboute, M. E., Clement, B., and Philipps, G. (2002) S phase and meristem-specific expression of the tobacco RNR1b gene is mediated by an E2F element located in the 5 leader sequence. *J. Biol. Chem.* **277**, 17845–17851.
21. Yamaguchi-Shinozaki, K. and Shinozaki, K. (1994) A novel *cis*-acting element in an *Arabidopsis* gene is involved in responsiveness to drought, low-temperature, or high-salt stress. *Plant Cell* **6**, 251–264.
22. Busk, P. K., Jensen, A. B., and Pages, M. (1997) Regulatory elements in vivo in the promoter of the abscisic acid responsive gene rab17 from maize. *Plant J.* **11**, 1285–1295.
23. Xue, G. P. (2002) An AP2 domain transcription factor HvCBF1 activates expression of cold-responsive genes in barley through interaction with a (G/a)(C/t)CGAC motif. *Biochim. Biophys. Acta* **1577**, 63–72.
24. Zhang, H. W., Huang, Z. J., Xie, B. Y., Chen, Q., Tian, X., Zhang, X. L., Zhang, H. B., Lu, X. Y., Huang, D. F., and Huang, R. F. (2004) The ethylene-, jasmonate-, abscisic acid- and NaCl-responsive tomato transcription factor JERF1 modulates expression of GCC box-containing genes and salt tolerance in tobacco. *Planta* **220**, 262–270.
25. Choi, T. Y., Cho, N. Y., Oh, Y., Yoo, M. A., Matsukage, A., Ryu, Y., Han, K., Yoon, J., and Baek, K. (2000) The DNA replication-related element (DRE)-DRE-binding factor (DREF) system may be involved in the expression of the *Drosophila melanogaster TBP* gene. *FEBS Lett.* **483**, 71–77.
26. Xue, G. P. (2005) A CELD-fusion method for rapid determination of the DNA-binding sequence specificity of novel plant DNA-binding proteins. *Plant J.* **41**, 638–649.

Chapter 4

A Transposon-Based Activation Tagging System for Gene Function Discovery in *Arabidopsis*

Nayelli Marsch-Martínez

Abstract

Activation tagging is a powerful strategy to find new gene functions, especially from genes that are redundant or show lethal knock-out phenotypes. It has been applied using T-DNA or transposons. *En/Spm-I/dSpm* engineered transposons are efficient activation tags in *Arabidopsis*. An immobilized transposase source and an enhancer-bearing non-autonomous element are used in combination with positive and negative selectable markers to generate a population of single- or low-copy, stable insertions. This method describes the steps required for selection of parental lines, generation of a population of stable insertions, and gene identification.

Key words: Activation tagging, transposons, *Arabidopsis*, *En/Spm-I/dSpm* system, dominant mutants.

1. Introduction

1.1. Activation Tagging

For many genes, conventional knock-out insertional mutagenesis does not provide indications of their functions, mainly due to functional redundancy and lethality or because the particular mutant phenotype can only be visualized under specific conditions (1–3). "Activation tagging" consists of using enhancers inside an insertion tag and can overcome some of these problems allowing the study of gene functions that have not been revealed by classical knock-out strategies (4). Enhancers positively influence gene expression even when located at a considerable distance from the target promoter and can increase endogenous gene expression (5). The major advantages of this strategy over simple knock-out strategies are as follows: (a) it produces dominant

L. Yuan, S.E. Perry (eds.), *Plant Transcription Factors*, Methods in Molecular Biology 754,
DOI 10.1007/978-1-61779-154-3_4, © Springer Science+Business Media, LLC 2011

in place of recessive mutations; (b) it may produce viable individuals in cases where gene knock-outs lead to lethal phenotypes; (c) it may produce evident phenotypes for genes with overlapping functions; (d) it is well suited to perform positive selection screens (i.e., resistance or tolerance to chemical, physical, or biological stresses) and for studying metabolic pathways; (e) where the goal is to produce biotechnological applications, this strategy could lead to the direct application of discovered genes.

First proposed for plants in the 1990s (4), the activation tagging approach has been successfully applied since then. Many plant transcription factors belong to large families, and related genes may mask loss-of-function phenotypes of specific ones. In this case, activation tagging strategies can lead to mutant phenotypes, revealing novel transcription factor functions (i.e., (6–20)). In addition to transcription factors, enzymes, components of signaling pathways, and microRNA gene functions have also been discovered, showing that a diversity of genes can be tagged using this strategy (i.e., (21–32)).

In most examples, the CaMV 35S promoter or enhancer has been used (33). In many cases, tags containing the enhancer cause quantitative increases in expression following the original expression pattern of a gene (11, 29, 34). This is an advantage when studying those genes for which true ectopic or constitutive overexpression may affect early development leading to lethal phenotypes.

1.2. Considerations About Transposons vs. T-DNA

The activation tagging populations described to date for *Arabidopsis* and other plants (i.e., poplar, rice, tomato, barley, and petunia) employ T-DNA or transposons containing promoter or enhancer sequences (i.e., (3, 11, 13, 15, 18, 19, 26, 28, 34–42)).

Modified stabilized transposons can be more effective for activation tagging purposes than T-DNA insertions (35). This might be due to a reduction in the frequency of integration configurations that can lead to epigenetic effects affecting the enhancer (43) or hinder the isolation of adjacent sequences.

Moreover, re-introduction of the proper transposase in the plant, which promotes transposon mobilization, allows subsequent reversion and targeted tagging strategies. Reversion, or the excision of the transposable element from its original locus, can be used to prove that a mutant phenotype is caused by the transposable element. With targeted tagging, a transposon can be remobilized from a known position to a new linked position. The frequency in which this occurs varies among transposon families and has been estimated for the two main maize families used in *Arabidopsis* (44). The transposition frequency to linked loci has been estimated to be 30% for the *En/Spm-I/dSpm* system (45) and 68% for the *Ac-Ds* system, with distribution variations depending on the donor site (46).

1.3. Considerations About the Construct

A CaMV 35S enhancer tetramer has been commonly used in activation tag populations (i.e., (4, 34, 35)). However, other enhancers or inducible systems can also be used (47, 48). For the tag, a modified *En/Spm-I/dSpm* system works effectively in *Arabidopsis* (45).

The native transposon system generally consists of

(a) an "autonomous," active element, which has specific terminal inverted repeats and codes for a functional transposase that recognizes the termini of the transposon and cuts and introduces it into a new position;

(b) a "non-autonomous," inactive element, which has specific terminal inverted repeats, but no longer codes for a functional transposase. These elements either can have or lack a mutated transposase, and they require the transposase to be coded by an active element to transpose (5).

The modified system allows stable insertions to be obtained and reduces transposition variability (i.e., frequency and time of transposition). For this, the transposase is "immobilized" by removing the transposon ends and is regulated by the CaMV 35S promoter. On the other hand, the non-autonomous element consists exclusively of transposon terminal inverted repeats, with suitable markers between them in order to follow transposition and/or other markers or regulatory sequences for trapping (i.e., promoter or gene) or activation tagging purposes. Transposon borders should be kept at the minimum length in order to avoid natural methylation sites present before them that can affect transposition (49, 50) or lower enhancer activity (19). The transposon used here contains 200 and 400 bp long ends.

Selectable markers facilitate the high-throughput identification of stable transpositions, especially when the markers allow selection in the greenhouse. The immobilized transposase is linked to a negative selectable marker, and the mobile non-autonomous element bears a positive selectable marker. It is convenient that the negative marker can also be easily recognized in the plant, for example, by producing a visible phenotype.

A useful double selection system (51) employs the *bar* gene (Basta resistance) as the positive marker (52, 53), placed between the transposon ends, and the *SU1* gene as the negative marker. *SU1* converts the otherwise innocuous compound R7402 to sulfonylurea that inhibits or reduces plant growth (DuPont; (54)). Conveniently, *SU1* produces a dwarf, dark green, reduced apical dominance phenotype in *Arabidopsis* plants. This facilitates the identification of progeny plants (T2) containing the *SU1* gene (and hence the transposase) to be used as parentals for the generation of stable insertions.

In principle, the two components (transposase-negative marker and transposon-positive marker) can be placed in separate

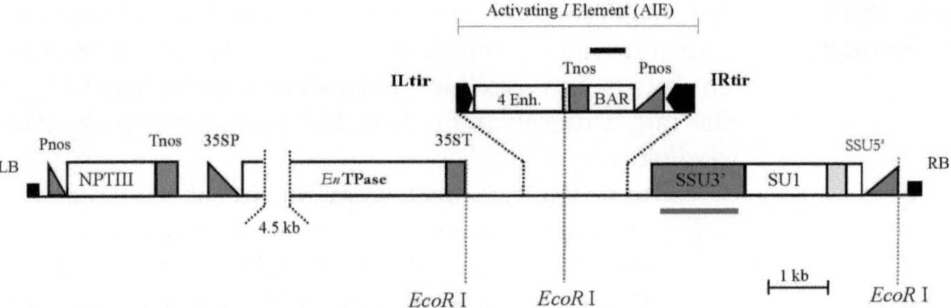

Fig. 4.1. Schematic representation of the WAT construct used for plant transformation. Relevant *Eco*RI sites used for Southern analysis are indicated. LB, Left border; RB, right border; 35SP and 35ST, CaMV 35S promoter and terminator, respectively; *En*TPase, *En* immobile transposase source; ILtir and IRtir, *I* element left and right terminal inverted repeat, respectively; 4 Enh., tetramer of the CaMV 35S enhancer; Pnos and Tnos, promoter and terminator sequences from the nopaline synthase gene, respectively; SSU5′ and SSU3′, promoter and transit signal peptide to the chloroplast and terminator of the small subunit of *Rubisco* gene, respectively. The gene-specific probes (*bar* and SSU3′) used for blot hybridization are indicated as bars above or below the figure. Reproduced with permission from (35). Copyright 2002, American Society of Plant Biologists.

T1
self
STEPS 1 & 2
Double selection
T2 seed
I (BAR) ☑ / En (SU1) ☒
STABLE INSERTIONS
data feedback
I (BAR) ☑
En (SU1) ☑
Plants
Cross x WT
self
F1 seed
T3 seed
STEP 3
I (BAR) ☑
En (SU1) ☒
STABLE INSERTIONS
Double selection
I (BAR) ☑
En (SU1) ☑
Plants
F2 seed
STEP 4

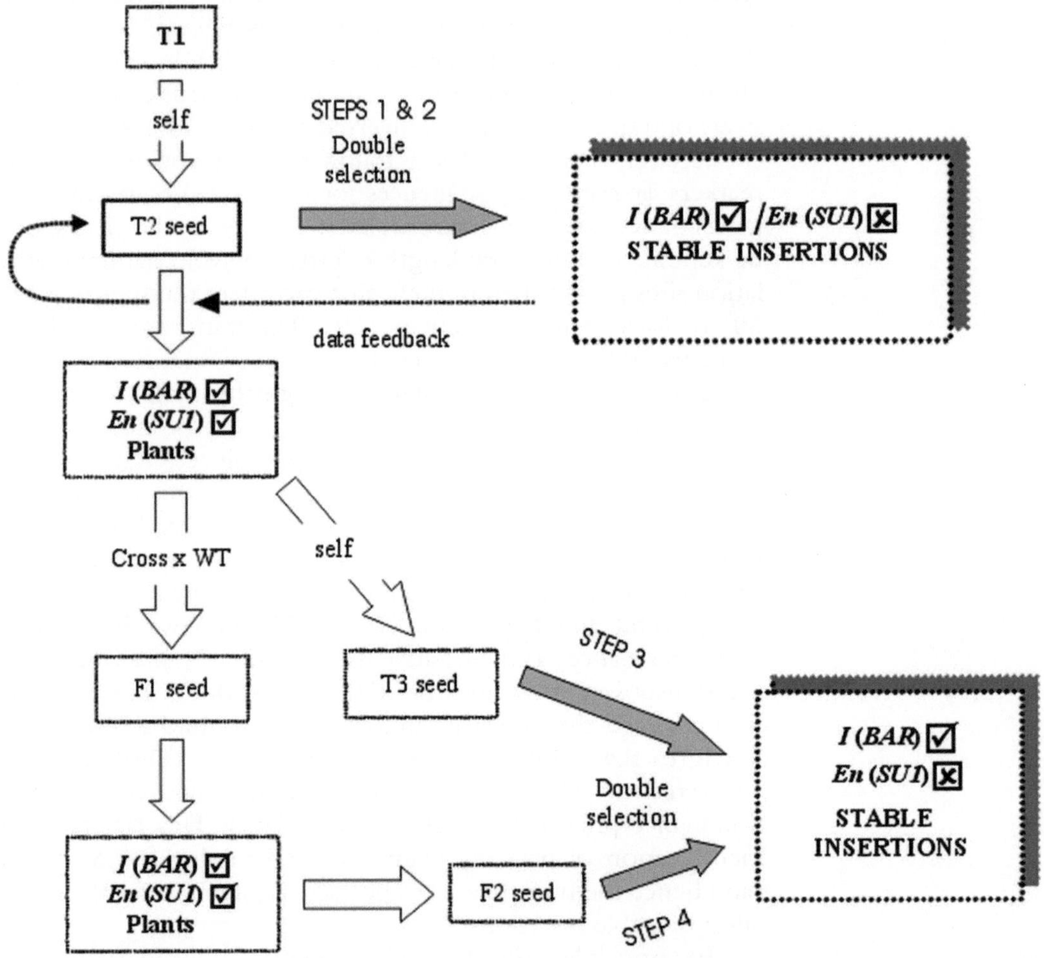

Fig. 4.2. Main steps to be followed to generate a population of stable insertions. The process starts from the first transformants (T_1) containing the construct shown in **Fig. 4.1**. Several rounds of stable insertion selections can be carried out using either self or F_2 seed as parental lines. Reproduced with permission from (35). Copyright 2002, American Society of Plant Biologists.

vectors and plants (*in trans*) or together in a single vector and finally into plants (*in cis*). The "*in cis*" strategy is better suited to obtain large numbers of plants with stable insertions. However, the construct used for the "*in cis*" strategy is large, which may be a disadvantage in plants that are difficult to transform. In that case, and when plants can be crossed, the "*in trans*" strategy may be advantageous, since the constructs are smaller.

The construct used for *Arabidopsis* in this chapter, named WAT, is depicted in **Fig. 4.1** (35). For some other plants or specific strategies, different transposon systems or markers may be better.

1.4. Procedure Outline

This procedure describes the steps required for generating a population of stable insertions to the identification of genes and includes the following:

1. selecting the best genotypes using a double selection assay and calculating stable transposition and independent transposition frequencies;
2. generating collections of stable insertions using T2, T3, or F2 seed (from a cross to WT) – summarized in **Fig. 4.2**;
3. identifying mutants, analyzing mutation segregation, and locating transposon insertions;
4. gene identification, first assessing gene overexpression and then testing using recapitulation assays.

2. Materials

1. First transformants containing the WAT construct.
2. 0.7 mL/L Finale (commercial formulation – Aventis – that contains 150 g/L glufosinate ammonium) (*see* **Note 1**).
3. 100 μg/L R7402 (*see* **Note 2**), with 50 μL/L Silwet L-77.
4. Materials for genomic DNA isolation (*see* **Note 3**).
5. Materials for Southern blot hybridization.
6. *Eco*RI restriction enzyme and buffer (*see* **Note 4**).
7. Agarose.
8. TAE buffer: 40 mM Tris acetate, 1 mM EDTA.
9. Standard gel electrophoresis materials.
10. Hybond N+ membrane.
11. Labeled *bar* fragment (*see* **Note 5**).
12. TAIL-PCR materials (conventional PCR components and TAIL-PCR degenerate oligos (55–57)).

The specific nested primers for the *I* element are as follows:

First TAIL-PCR: **Int2** 5′-CAG GGT AGC TTA CTG ATG TGC G-3′, Second TAIL-PCR: **Irj-201** 5′-CAT AAG AGT GTC GGT TGC TTG TTG-3′, and Third TAIL-PCR: **DSpml** 5′-CTT ATT TCA GTA AGA GTG TGG GGT TTT GG-3′ (51). Sequencing primer: **Itir 3** 5′-CTT ACC TTT TTT CTT GTA GTG-3′ (*see* **Note 6**).

13. Materials for RNA isolation (*see* **Note 7**).
14. DNase I.
15. Materials for RT-PCR.
16. Specific oligos for candidate genes and a control gene (i.e., actin or tubulin).
17. Standard molecular biology materials to make recapitulation constructs.
18. Standard materials for *Escherichia coli* and *Agrobacterium tumefaciens* transformation.
19. Materials for plant transformation and transformant selection.

3. Methods

3.1. Selecting Best Genotypes (Stable Transposition vs. Independent Transposition Frequencies)

3.1.1. Double Selection Assay and Calculation of Stable Transposition

1. Start with 20–30 first transformants (T1) containing the WAT construct. Grow in the best conditions to maximize seed set (*see* **Note 8**).
2. Sow a known quantity of progeny seed (T2 seed) in undivided trays filled with soil. Sow by dispersing the dry seed uniformly over the soil (*see* **Note 9**).
3. Give a three-night cold treatment to the seed before transferring to the greenhouse (*see* **Note 10**). Keep trays covered with a transparent lid or plastic.
4. Transfer to the greenhouse, keep the trays covered for 3 days, and remove the cover afterward.
5. Eight days after transference, spray seedlings with a mix of 0.7 mL/L Finale and 100 μg/L R7402. For the next 8 days, spray R7402 every day and only spray Finale three times during this period (*see* **Note 11**).
6. About 5 to 7 days after the last spray, carefully transfer resistant plants to new soil (pots or trays with individual separations). Some morphological phenotypes can already be identified at this stage (*see* **Note 12**).

7. Count the number of doubly resistant progeny plants from each genotype and calculate the stable transposition frequency (STF) as

$$\text{Stable transposition frequency} = \frac{\text{Number of double-resistant plants}}{\text{Number of sown seed}} \times 100$$

3.1.2. Calculation of Independent Transposition Frequency

1. Isolate genomic DNA from either single or pooled double-resistant plants.
2. Digest at least 500 ng (for single plants) or 1 μg (for up to 10 pooled plants) of genomic DNA with *Eco*RI, run in an electrophoresis gel, and transfer to Hybond N+ membrane (*see* **Note 13**).
3. Perform a Southern hybridization using a *bar* fragment as labeled probe (shown above the *bar* gene in **Fig. 4.1**).
4. The *bar* probe will allow visualization of different *I/dSpm* transpositions in the genome. (**Figure 4.3** shows an example of pooled progeny plants having different insertions with varying frequencies.) Use the number of independent insertions to calculate the frequency of independent insertions from each new transformant (*see* **Notes 20** and **21**). The independent insertion frequency (ITF) is calculated as

$$\text{Independent transposition frequency} = \frac{\text{Number of different visualized insertions}}{\text{Number of progeny plants assayed}} \times 100$$

Lines showing the lowest stable transposition frequency (STF, number of plants that survive the double selection assay), together with the highest independent transposition frequency (ITF, many different insertions in a group of plants), are the best ones with which to build a population. The STF should be less than 20 plants in 1000 seeds. The ITF should be at least 50%, but there are lines that reach more than 100% (meaning as many different insertions as plants in the pool) and even higher when some plants in the pools contain more than one insertion, which is different from any insertion in other plants.

3.2. Generating Large Collections of Stable Insertions (New Double Selection Rounds)

If enough T2 seed from the best genotypes is available, new rounds of selection can be performed as indicated in **Section 3.1.1**. However, to generate large collections of stable insertions using pre-selected best genotypes, it is necessary to use T3 or F2 seed. All parental plants should be checked for the presence of positive and negative selectable markers. Moreover, the

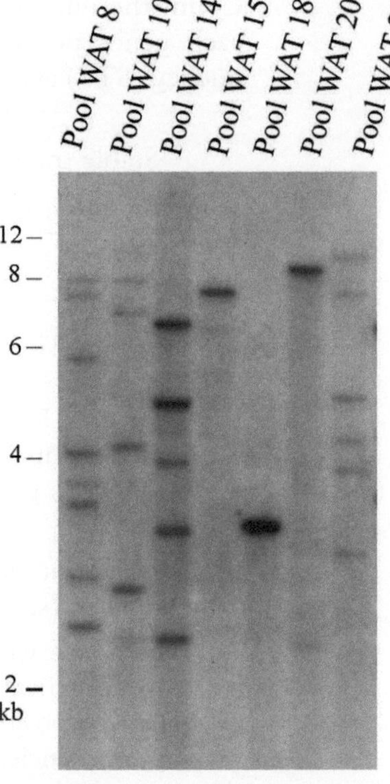

Fig. 4.3. Southern blot hybridization of double-resistant plants revealing stable inserts with a *bar* gene probe. Double-resistant progeny pools from different first transformants (WATs) are displayed. Each band shows an independent insertion. A ladder showing the size in kilobase pairs is indicated on the left side. The numbers of plants per pool are as follows: WAT 8, seven plants; WAT 10, eight plants; WAT 14, six plants; WAT 15, eight plants; WAT 18, ten plants; WAT 20, eight plants; and WAT 21, nine plants. Reproduced with permission from (35). Copyright 2002, American Society of Plant Biologists.

double selection assay will be more efficient if heterozygous *SU1* parentals are pre-selected.

3.2.1. Using T3 Seed Directly (See **Note 14**)

Step 3 of **Fig. 4.2** exemplifies the process of using T3 seed directly.

1. Handle T2 seed as in **Section 3.1.1**, but sow seeds separately (not as a bulk).
2. Spray 0.7 mL/L Finale over the seedlings at days 8 and 11 after transference to the greenhouse.
3. Screen Basta-resistant plants visually for the presence of the *SU1* marker, which confers a dwarf, dark green, reduced apical dominance phenotype.
4. Grow Basta-resistant, *SU1*-positive plants in the best conditions to achieve maximum seed set.
5. To generate new stable transpositions, use T3 seed from single parentals to perform the double selection (Basta and

R7402) treatments as indicated in **Section 3.1.1** (*see* **Notes 15** and **16**). Only transfer double-resistant plants from parentals with low stable transposition frequencies (lower than 5%).

3.2.2. Using F2 Seed

Selfed progeny can harbor "fixed" insertions that occurred early in the transformed lines or initial generations and that will appear in the double selection assay leading to high stable insertion frequencies and masking new transposition events. The use of F2 seed reduces this problem. Moreover, all F1 parentals are heterozygous, avoiding the need to check for *SU1* heterozygosis.

1. Handle as in **Section 3.2.1**, steps 1–3.
2. Cross dwarf, Basta-resistant T2 plants to wild type.
3. Grow F1 seed and select dwarf, Basta-resistant plants again.
4. Grow these plants in the best conditions to achieve maximum seed set.
5. Use F2 seed from single parentals to perform the double selection (Basta and R7402) treatments as indicated in **Section 3.1.1** (exemplified by step 4 in **Fig. 4.2** [*see* **Note 16**]). Only transfer double-resistant plants from parentals with low stable transposition frequencies (lower than 5%).

3.3. Mutant Identification and Establishment of Transposon Position

A variety of mutant phenotypes can be identified at different stages. Many leaf phenotypes can already be identified when transferring double-resistant plantlets to new soil, and set apart from the collection, while other phenotypes (i.e., stem length, flowering time, and floral organ development) can be identified later (*see* **Note 17**). Activation tagging mutations are dominant. Therefore, always have wild-type plants available to cross male or female sterile plants when necessary and to perform genetic analysis. Moreover, collect tissue from plants with compromised or sterile phenotypes that cannot be crossed, so candidate genes causing the phenotype can be identified and assayed with recapitulation constructs (inducible when necessary) to recover the original phenotype.

3.3.1. Genetic Analysis

1. Once a mutant has been identified, backcross it to its wild-type ecotype.
2. Perform genetic analyses to check phenotype segregation using both selfed seed and F1 from crosses to wild-type plants.

3.3.2. Insertion Number Analysis

In most cases mutants have single stable insertions; however, this should be assessed (*see* **Note 18**). To evaluate insertion number

1. collect mutant tissue;
2. isolate DNA and perform a Southern blot assay as in **Section 3.1.2**, steps 1–4.

3.3.3. Using the Isolated DNA to Perform a TAIL-PCR

Use primer Int2 for the first, Irj-201 for the second, and DSpm1 for the third TAIL-PCR as specific nested primers. For sequencing, use a fourth nested primer (Itir3, *see* **Notes 6** and **19–21**).

3.3.4. Finding the Position of the Insert in the Genome

BLAST (58) the sequences to the *Arabidopsis* genome to find the position of the insert and identify adjacent genes (both up and downstream, *see* **Note 22**).

3.4. Gene Identification

3.4.1. Evaluation of Transcript Overaccumulation

1. Isolate total RNA from mutant and wild-type plants (*see* **Note 7**).
2. Perform a DNase I treatment.
3. Perform a reverse transcription reaction.
4. Use the obtained cDNA as a template for PCR to evaluate transcript accumulation of candidate genes in mutant and wild-type backgrounds (*see* **Note 23**).

3.4.2. Candidate Gene Evaluation

In some mutants, more than one gene can be overexpressed, and the phenotype can be the result of the overexpression of a single gene or a combination of two or more genes.

To evaluate this, use the 35S promoter to express the candidate gene(s) in the wild-type background and observe whether it produces a comparable phenotype as the original mutant (recapitulation) (*see* **Note 24**).

4. Notes

1. Finale can be purchased from Bayer (www.bayer.com). It contains 150 g/L glufosinate ammonium as active ingredient. Instead of Finale, other sources of active ingredient can be used after adjusting to a similar final concentration. Pure phosphinothricin (PPT; Duchefa, www.duchefa.com) can also be used at a concentration of 100 mg/L with 50 μL/L Silwet L-77 (Lehle seeds, www.arabidopsis.org) or 0.01% Triton.
2. Dupont (www.dupont.com) owns the rights for the use of the *SU1* gene and should be contacted for use of the gene and to obtain the compound R7402. To make the stock R7402 solution, dissolve the powder in 10 mM KOH (or other dilute base) and then add water.

3. The DNA isolation protocol should produce good quality DNA to be used for Southern blot hybridization. As a DNA isolation method we use the protocol published by Pereira and Aarts (59), a modified version of the method by Liu et al. (55), using either 1.5 mL tubes or 96 tube racks for high-throughput isolation. Good quality DNA can be normally obtained from young flower buds. Green, relatively young, mature leaves in good condition are also a suitable tissue from which to isolate good quality DNA.
4. High-concentration enzyme (i.e., 50 U/μL).
5. The bar fragment can be obtained by PCR in bacteria or plant DNA containing the binary vector or the T-DNA, respectively, using the following primers:
   Bar F1: 5′-ACC ATG AGC CCA GAA CGA CGC-3′
   Bar R1: 5′-CAG GCT GAA GTC CAG CTG CCA G-3′
6. The use of a fourth primer to sequence the product obtained in the third TAIL-PCR increases the reliability of the result.
7. For mRNA transcripts, RNA can be isolated using Trizol, LiCl protocols, or commercially available RNA isolation kits. For microRNAs or small RNAs, special steps should be considered in some of these conventional protocols.
8. The WAT construct is large. For transformation, the aggressive *Agrobacterium* strain AGL0 worked well to obtain the desired number of transformants in the Wassilewskija ecotype, which was used successfully to build a large collection of stable activation tag insertions. It is best to avoid cultivating *Agrobacterium* cultures containing the construct for many generations because the enhancer tetramer can recombine and lose copies. It is better to use direct cultures from transformed *Agrobacterium* or from a glycerol stock (34). Transformed plants can be selected on medium (0.8% agarose, 1/2 MS medium supplemented with 50 mg/L kanamycin or 10 mg/L PPT) or sown in soil and selected by spraying 0.7 mL/L Finale. Some transformants will not be useful due to, i.e., high T-DNA copy number or unfavorable T-DNA location, leading to high early or too low transposition frequencies. Having many initial lines allows one to find suitable starter lines and represent more chances of having starting insertions in different locations in the genome, increasing the likelihood of a better dispersion of inserts (linked transposition events will be eliminated in the double selection rounds). However, in principle, two good starting lines with unlinked original T-DNA insertions are enough to build a population.

9. Seed quantity can be calculated from the weight. Depending on the ecotype, 1 mg contains around 50 seeds. One thousand to three thousand seeds can be sown in a tray of about 35 × 45 × 7 cm (filled with a ~5 cm soil layer). Seeds can also be sown by resuspending in a 0.1% agarose solution and dispersing evenly using a large (i.e., 20 mL) pipette.
10. Depending on seed age, a cold treatment helps to synchronize germination and therefore improves efficiency of the double selection assay. The treatment can be given to seed before sowing, in a regular refrigerator, or to seed sown in trays, placing the trays in a cold room.
11. An alternative to sowing in soil and spraying is to select double-resistant plants in media. Plates should be supplemented with 10 mg/L PPT (Duchefa) and/or 1 μg/L R7402 (Dupont). PPT alone can be supplemented with conventional 1/2 MS, 0.8% w/v agar, and 1% w/v sucrose medium. With R7402, sucrose should be replaced by 100 mg/L myoinositol.
12. Transfer plantlets together with soil using large tweezers to take out and plant directly in the new soil. Resistant plants bear non-autonomous transposon insertions but no T-DNA (and consequently no transposase) and therefore are considered to contain stable insertions.
13. Digest at least 4 h (overnight for best results) with 1 μL of *Eco*RI (50 U/μL) per sample. Electrophorese in a 0.8% w/v agarose gel in 1X TAE buffer (40 mM Tris acetate and 1 mM EDTA). Instead of using Southern blot hybridization, the independent transposition frequency can also be calculated by performing transposon insertion display (51) or another strategy that allows high-throughput visualization or sequencing of transposon insertions.
14. Taking the first transformant as T1, its first progeny seed are T2 and the third generation T3.
15. Alternatively, sow T2 seed as a bulk and transfer Basta-resistant plantlets separately to evaluate the *SU1* phenotype. It is advisable to perform a segregation analysis to identify heterozygous plants for the T-DNA before using their T3 seed for the next stable transposition selection round, since progeny T3 seed from T2 homozygous individuals will not survive the double selection. However, when space is not limiting, F3 seed from all plants can be used directly for the large double selection assay.
16. Avoid mixing T3 or F2 seed from different parentals. Progeny seed from single plants should be used to generate stable insertions. In this way, plants where an insertion

has been fixed in unlinked positions to the T-DNA will not "contaminate" a pool of seed. These plants can be identified as having high stable transposition frequencies and should not be considered or just very few double resistants should be transferred to new pots and used for the population.

17. Stable insertions can also be selected in medium as indicated in **Note 11** when mutant screenings require it.

18. When using the WAT construct in *Arabidopsis*, most plants containing stable insertions had only one insertion (around 70%), and the rest had two (around 20%) or three insertions (around 10%) (35). If the identified mutant has more than one insertion it is necessary to cross it to the wild type to segregate the different insertions to ease analysis. If this is not possible, then obtain the flanking sequences for the different insertions and assay candidate genes using recapitulation constructs.

19. If the third TAIL-PCR reaction produces a single, specific band, purify the reaction and sequence it directly. If the third reaction also produces other unspecific bands, then touch the specific band with a sterile tip or toothpick and introduce it in a tube containing a third TAIL-PCR mix with the corresponding degenerate oligo. This should help to get a single, specific band that can be sequenced. For this re-PCR, the electrophoresis buffer should be 0.5X TBE.

20. Other techniques to isolate flanking sequences can be used, for example, the walk PCR technique (60) or even high-throughput transcriptomic approaches.

21. In many cases, the activation tag does not cause constitutive expression but increased expression of the native expression pattern. Therefore, to be able to detect overexpression, use tissues showing a mutant phenotype.

22. Consider genes located up to 8 kb from the insertion on either side. In most cases, the responsible genes are the nearest to the insertion on either side. Different kinds of genes can be activated, including transcription factors, enzymes, and even microRNA precursors among others. There are documented examples of the enhancing tag activating genes on both sides simultaneously (61). In some cases, the gene identity makes it a better candidate for a specific phenotype.

23. Alternatively, northern blotting can also be used to detect overexpression.

24. There are three main strategies to recapitulate the mutant phenotype in wild-type plants:

1. using the 35S promoter to overexpress candidate genes;
2. cloning the entire genomic region from upstream the enhancer tetramer to the end of the candidate gene;
3. using the candidate gene together with its native promoter and adding the 35S enhancer tetramer.

The use of the 35S promoter to overexpress a gene can have stronger effects than using only the enhancer tetramer in combination with native regulatory regions as in the other two options. Sometimes, a softer effect is more desirable. For example, constitutive overexpression of certain genes will produce lethal or compromised phenotypes that can hinder their study.

References

1. Initiative, T. A. G. (2000) Analysis of the genome sequence of the flowering plant Arabidopsis thaliana. *Nature* **408**, 796–815.
2. Bouché, N. and Bouchez, D. (2001) Arabidopsis gene knockout: phenotypes wanted. *Curr. Opin. Plant Biol.* **4**, 111–117.
3. Kuromori, T., Wada, T., Kamiya, A., Yuguchi, M., Yokouchi, T., Imura, Y., Takabe, H., Sakurai, T., Akiyama, K., Hirayama, T., Okada, K., and Shinozaki, K. (2006) A trial of phenome analysis using 4000 Ds-insertional mutants in gene-coding regions of Arabidopsis. *Plant J.* **47**, 640–651.
4. Walden, R. F. K., Hayashi, H., Miklashevichs, E., Harling, H., and Schell, J. (1994) Activation tagging: a means of isolating genes implicated as playing a role in plant growth and development. *Plant Mol. Biol.* **26**, 1521–1528.
5. Lewin, B. (ed.) (2008) *Genes IX*, Jones and Bartlett Publishers, Sudbury, MA, USA.
6. Borevitz, J. O., Xia, Y., Blount, J., Dixon, R. A., and Lamb, C. (2000) Activation tagging identifies a conserved MYB regulator of phenylpropanoid biosynthesis. *Plant Cell* **12**, 2383–2394.
7. Borghi, L., Bureau, M., and Simon, R. (2007) Arabidopsis JAGGED LATERAL ORGANS is expressed in boundaries and coordinates KNOX and PIN activity. *Plant Cell* **19**, 1795–1808.
8. Kirch, T., Simon, R., Grunewald, M., and Werr, W. (2003) The DORNROSCHEN/ENHANCER OF SHOOT REGENERATION1 gene of Arabidopsis acts in the control of meristem cell fate and lateral organ development. *Plant Cell* **15**, 694–705.
9. Lee, H., Suh, S.-S., Park, E., Cho, E., Ahn, J. H., Kim, S.-G., Lee, J. S., Kwon, Y. M., and Lee, I. (2000) The AGAMOUS-LIKE 20 MADS domain protein integrates floral inductive pathways in Arabidopsis. *Genes Dev.* **14**, 2366–2376.
10. Trigueros, M., Navarrete-Gomez, M., Sato, S., Christensen, S. K., Pelaz, S., Weigel, D., Yanofsky, M. F., and Ferrandiz, C. (2009) The NGATHA genes direct style development in the Arabidopsis gynoecium. *Plant Cell* **21**, 1394–1409.
11. van der Graaff, E., Dulk-Ras, A. D., Hooykaas, P. J., and Keller, B. (2000) Activation tagging of the LEAFY PETIOLE gene affects leaf petiole development in Arabidopsis thaliana. *Development* **127**, 4971–4980.
12. Zuo, J., Niu, Q.-W., Frugis, G., and Chua, N.-H. (2002) The WUSCHEL gene promotes vegetative-to-embryonic transition in Arabidopsis. *Plant J.* **30**, 349–359.
13. Mathews, H., Clendennen, S. K., Caldwell, C. G., Liu, X. L., Connors, K., Matheis, N., Schuster, D. K., Menasco, D. J., Wagoner, W., Lightner, J., and Wagner, D. R. (2003) Activation tagging in tomato identifies a transcriptional regulator of anthocyanin biosynthesis, modification, and transport. *Plant Cell* **15**, 1689–1703.
14. Marsch-Martínez, N., Greco, R., Becker, J. D., Dixit, S., Bergervoet, J. H. W., Karaba, A., de Folter, S., and Pereira, A. (2006) BOLITA, an Arabidopsis AP2/ERF-like transcription factor that affects cell expansion and proliferation/differentiation pathways. *Plant Mol. Biol.* **62**, 825–843.
15. van der Fits, L. and Memelink, J. (2000) ORCA3, a jasmonate-responsive transcriptional regulator of plant primary and secondary metabolism. *Science* **289**, 295–297.
16. Aharoni, A., Dixit, S., Jetter, R., Thoenes, E., van Arkel, G., and Pereira, A. (2004) The SHINE clade of AP2 domain transcription factors activates wax biosynthesis, alters

cuticle properties, and confers drought tolerance when overexpressed in Arabidopsis. *Plant Cell* **16**, 2463–2480.

17. Karaba, A., Dixit, S., Greco, R., Aharoni, A., Trijatmiko, K. R., Marsch-Martinez, N., Krishnan, A., Nataraja, K. N., Udayakumar, M., and Pereira, A. (2007) Improvement of water use efficiency in rice by expression of HARDY, an Arabidopsis drought and salt tolerance gene. *Proc. Natl. Acad. Sci. USA* **104**, 15270–15275.
18. Wilson, K. L. D., Swinburne, K., and Coupland, G. (1996) A dissociation insertion causes a semidominant mutation that increases expression of TINY, an Arabidopsis gene related to APETALA2. *Plant Cell* **8**, 659–671.
19. Schneider, A., Kirch, T., Gigolashvili, T., Mock, H.-P., Sonnewald, U., Simon, R., Flügge, U.-I., and Werr, W. (2005) A transposon-based activation-tagging population in Arabidopsis thaliana (TAMARA) and its application in the identification of dominant developmental and metabolic mutations. *FEBS Lett.* **579**, 4622–4628.
20. Chalfun-Junior, A., Franken, J., Mes, J. J., Marsch-Martinez, N., Pereira, A., and Angenent, G. C. (2005) ASYMMETRIC LEAVES2-LIKE1gene, a member of the AS2/LOB family, controls proximal–distal patterning in Arabidopsis petals. *Plant Mol. Biol.* **57**, 559–575.
21. Huang, S., Cerny, R. E., Bhat, D. S., and Brown, S. M. (2001) Cloning of an Arabidopsis patatin-like gene, STURDY, by activation T-DNA tagging. *Plant Physiol.* **125**, 573–584.
22. Kakimoto, T. (1996) CKI1, a histidine kinase homolog implicated in cytokinin signal transduction. *Science* **274**, 982–985.
23. Ito, T. and Meyerowitz, E. (2000) Overexpression of a gene encoding a cytochrome P450, *CYP78A9*, induces large and seedless fruit in Arabidopsis. *Plant Cell* **12**, 1541–1551.
24. Zhao, Y., Christensen, S. K., Fankhauser, C., Cashman, J. R., Cohen, J. D., Weigel, D., and Chory, J. (2001) A role for flavin monooxygenase-like enzymes in auxin biosynthesis. *Science* **291**, 306–309.
25. Palatnik, J. F., Allen, E., Wu, X., Schommer, C., Schwab, R., Carrington, J. C., and Weigel, D. (2003) Control of leaf morphogenesis by microRNAs. *Nature* **425**, 257–263.
26. Busov, V. B., Meilan, R., Pearce, D. W., Ma, C., Rood, S. B., and Strauss, S. H. (2003) Activation tagging of a dominant gibberellin catabolism gene (GA 2-oxidase) from poplar that regulates tree stature. *Plant Physiol.* **132**, 1283–1291.
27. Kardailsky, I., Shukla, V. K., Ahn, J. H., Dagenais, N., Christensen, S. K., Nguyen, J. T., Chory, J., Harrison, M. J., and Weigel, D. (1999) Activation tagging of the floral inducer FT. *Science* **286**, 1962–1965.
28. Zubko, E., Adams, C. J., Machaekova, I., Malbeck, J., Scollan, C., and Meyer, P. (2002) Activation tagging identifies a gene from Petunia hybrida responsible for the production of active cytokinins in plants. *Plant J.* **29**, 797–808.
29. Neff, M. N. S., Malancharuvil, E. J., Fujioka, S., Noguchi, T., Seto, H., Tsubuki, M., Honda, T., Takatsuto, S., Yoshida, S., and Chory, J. (1999) BAS1: a gene regulating brassinosteroid levels and light responsiveness in Arabidopsis. *Proc. Natl. Acad. Sci. USA* **96**, 15316–15323.
30. He, Z., Wang, Z.-Y., Li, J., Zhu, Q., Lamb, C., Ronald, P., and Chory, J. (2000) Perception of brassinosteroids by the extracellular domain of the receptor kinase BRI1. *Science* **288**, 2360–2363.
31. Li, J., Lease, K. A., Tax, F. E., and Walker, J. C. (2001) BRS1, a serine carboxypeptidase, regulates BRI1 signaling in Arabidopsis thaliana. *Proc. Natl. Acad. Sci. USA* **98**, 5916–5921.
32. Li, J., Wen, J., Lease, K. A., Doke, J. T., Tax, F. E., and Walker, J. C. (2002) BAK1, an Arabidopsis LRR receptor-like protein kinase, interacts with BRI1 and modulates brassinosteroid signaling. *Cell* **110**, 213–222.
33. Odell, J. T., Nagy, F., and Chua, N.-H. (1985) Identification of DNA sequences required for activity of the cauliflower mosaic virus 35S promoter. *Nature* **313**, 810–812.
34. Weigel, D., Ahn, J. H., Blazquez, M. A., Borevitz, J. O., Christensen, S. K., Fankhauser, C., Ferrandiz, C., Kardailsky, I., Malancharuvil, E. J., Neff, M. M., Nguyen, J. T., Sato, S., Wang, Z.-Y., Xia, Y., Dixon, R. A., Harrison, M. J., Lamb, C. J., Yanofsky, M. F., and Chory, J. (2000) Activation tagging in Arabidopsis. *Plant Physiol.* **122**, 1003–1014.
35. Marsch-Martinez, N., Greco, R., Van Arkel, G., Herrera-Estrella, L., and Pereira, A. (2002) Activation tagging using the En-I maize transposon system in Arabidopsis. *Plant Physiol.* **129**, 1544–1556.
36. Ayliffe, M. A. and Pryor, A. J. (2009) Transposon-based activation tagging in cereals. *Funct. Plant Biol.* **36**, 915–921.
37. Ayliffe, M. A., Pallotta, M., Langridge, P., and Pryor, A. J. (2007) A barley activa-

tion tagging system. *Plant Mol. Biol.* **64**, 329–347.

38. Wan, S., Wu, J., Zhang, Z., Sun, X., Lv, Y., Gao, C., Ning, Y., Ma, J., Guo, Y., Zhang, Q., Zheng, X., Zhang, C., Ma, Z., and Lu, T. (2009) Activation tagging, an efficient tool for functional analysis of the rice genome. *Plant Mol. Biol.* **69**, 69–80.
39. Robinson, S., Tang, L., Mooney, B., McKay, S., Clarke, W., Links, M., Karcz, S., Regan, S., Wu, Y.-Y., Gruber, M., Cui, D., Yu, M., and Parkin, I. (2009) An archived activation tagged population of Arabidopsis thaliana to facilitate forward genetics approaches. *BMC Plant Biol.* **9**, 101.
40. Kuromori, T., Takahashi, S., Kondou, Y., Shinozaki, K., and Matsui, M. (2009) Phenome analysis in plant species using loss-of-function and gain-of-function mutants. *Plant Cell Physiol.* **50**, 1215–1231.
41. Jeong, D.-H., An, S., Kang, H.-G., Moon, S., Han, J.-J., Park, S., Lee, H. S., An, K., and An, G. (2002) T-DNA insertional mutagenesis for activation tagging in rice. *Plant Physiol.* **130**, 1636–1644.
42. Qu, S., Desai, A., Wing, R., and Sundaresan, V. (2008) A versatile transposon-based activation tag vector system for functional genomics in cereals and other monocot plants. *Plant Physiol.* **146**, 189–199.
43. Chalfun-Junior, A., Mes, J., Mlynárová, L., Aarts, M., and GC, A. (2003) Low frequency of T-DNA based activation tagging in Arabidopsis is correlated with methylation of CaMV 35S enhancer sequences. *FEBS Lett.* **555**, 459–463.
44. Pereira, A. (2000) A transgenic perspective on plant functional genomics. *Transgenic Res.* **9**, 245–260.
45. Aarts, M. G., Corzaan, P., Stiekema, W. J., and Pereira, A. (1995) A two-element enhancer-inhibitor transposon system in Arabidopsis thaliana. *Mol. Gen. Genet.* **247**, 555–564.
46. Bancroft, I. and Dean, C. (1993) Transposition pattern of the maize element Ds in Arabidopsis thaliana. *Genetics* **134**, 1221–1229.
47. Matsuhara, S., Jingu, F., Takahashi, T., and Komeda, Y. (2000) Heat-shock tagging: a simple method for expression and isolation of plant genome DNA flanked by T-DNA insertions. *Plant J.* **22**, 79–86.
48. Zuo, J., Niu, Q.-W., and Chua, N.-H. (2000) An estrogen receptor-based transactivator XVE mediates highly inducible gene expression in transgenic plants. *Plant J.* **24**, 265–273.
49. Lawson, E. J. R., Scofield, S. R., Sjodin, C., Jones, J. D. G., and Dean, C. (1994) Modification of the 5′ untranslated leader region of the maize Activator element leads to increased activity in Arabidopsis. *Mol. Gen. Genet.* **245**, 608–615.
50. Banks, J. A., Masson, P., and Fedoroff, N. (1988) Molecular mechanisms in the developmental regulation of the maize Suppressor-mutator transposable element. *Genes Dev.* **2**, 1364–1380.
51. Tissier, A. F., Marillonnet, S., Klimyuk, V., Patel, K., Torres, M. A., Murphy, G., and Jones, J. D. (1999) Multiple independent defective suppressor-mutator transposon insertions in Arabidopsis: a tool for functional genomics. *Plant Cell* **11**, 1841–1852.
52. De Block, M., Botterman, J., Vanderwiele, M., Dockx, J., Thoen, C., Gossele, V., Movva, R. N., Thompson, C., Van Montagu, M., and Leemans, J. (1987) Engineering herbicide resistance in plants by expression of a detoxifying enzyme. *EMBO J.* **6**, 2513–2518.
53. Thompson, C. J., Movva, N. R., Tizard, R., Crameri, R., Davies, J. E., Lauwereys, M., and Botterman, J. (1987) Characterization of the herbicide-resistance gene bar from Streptomyces hygroscopicus. *EMBO J.* **6**, 2519–2523.
54. O'Keefe, D. P., Tepperman, J. M., Dean, C., Leto, K. J., Erbes, D. L., and Odell, J. T. (1994) Plant expression of a bacterial cytochrome P450 that catalyzes activation of a sulfonylurea pro-herbicide. *Plant Physiol.* **105**, 473–482.
55. Liu, Y. G., Mitsukawa, N., Oosumi, T., and Whittier, R. F. (1995) Efficient isolation and mapping of Arabidopsis thaliana T-DNA insert junctions by thermal asymmetric interlaced PCR. *Plant J.* **8**, 457–463.
56. Liu, Y. G. and Whittier, R. F. (1995) Thermal asymmetric interlaced PCR: automatable amplification and sequencing of insert end fragments from P1 and YAC clones for chromosome walking. *Genomics* **25**, 674–681.
57. Tsugeki, R., Kochieva, E. Z., and Fedoroff, N. V. (1996) A transposon insertion in the Arabidopsis SSR16 gene causes an embryo-defective lethal mutation. *Plant J.* **10**, 479–489.
58. Altschul, S. F., Madden, T. L., Schaffer, A. A., Zhang, J., Zhang, Z., Miller, W., and Lipman, D. J. (1997) Gapped BLAST and PSI-BLAST: a new generation of protein database search programs. *Nucleic Acids Res.* **25**, 3389–3402.
59. Pereira, A. and Aarts, M. G. M. (1998) Transposon tagging with the En-I system. in "Arabidopsis Protocols" (Martinez-Zapater, J., and Salinas, J., Eds.), Methods

in Molecular Biology, Vol. 82, pp. 329–338, Humana Press Inc., Totowa, NJ.

60. Balzergue, S., Dubreucq, B., Chauvin, S., Le-Clainche, I., Le Boulaire, F., de Rose, R., Samson, F., Biaudet, V., Lecharny, A., Cruaud, C., Weissenbach, J., Caboche, M., and Lepiniec, L. (2001) Improved PCR-walking for large-scale isolation of plant T-DNA borders. *Biotechniques* **30**, 496–498, 502, 504.

61. van der Graaff, E., Hooykaas, P. J. J., and Keller, B. (2002) Activation tagging of the two closely linked genes LEP and VAS independently affects vascular cell number. *Plant J.* **32**, 819–830.

Section III

Functional Verification

Chapter 5

CRES-T, An Effective Gene Silencing System Utilizing Chimeric Repressors

Nobutaka Mitsuda, Kyoko Matsui, Miho Ikeda, Masaru Nakata, Yoshimi Oshima, Yukari Nagatoshi, and Masaru Ohme-Takagi

Abstract

*C*himeric *RE*pressor gene *S*ilencing *T*echnology (CRES-T) is a useful tool for functional analysis of plant transcription factors. In this system, a chimeric repressor that is produced by fusion of a transcription factor to the plant-specific EAR-motif repression domain (SRDX) suppresses target genes of a transcription factor dominantly over the activity of endogenous and functionally redundant transcription factors. As a result, the transgenic plants that express a chimeric repressor exhibit phenotypes similar to loss-of-function of the alleles of the gene encoding the transcription factor. This system is simple and effective and can be used as a powerful tool not only for functional analysis of redundant transcription factors but also for the manipulation of plant traits by active suppression of the gene expression. Strategies for construction of the chimeric repressors and their expression in transgenic plants are described. Transient effector–reporter assays for functional analysis of transcription factors and detection of protein–protein interactions using the trans-repressive activity of SRDX repression domain are also described.

Key words: CRES-T, repressor, transcription factor, plant, gene silencing, *Arabidopsis*, protein–protein interaction.

1. Introduction

Transcriptional regulation plays a major role in the control of gene expression especially in plants (1). In *Arabidopsis thaliana*, approximately 2,000 transcription factors (TFs) regulate the first step of the expression of over 26,000 genes located in the genome (2). Because a transcription factor regulates multiple genes and a number of TFs have been identified to be key regulators for

L. Yuan, S.E. Perry (eds.), *Plant Transcription Factors*, Methods in Molecular Biology 754,
DOI 10.1007/978-1-61779-154-3_5, © Springer Science+Business Media, LLC 2011

various phenotypes, a TF is a fascinating target as a tool for genetic manipulation of plant traits. To utilize TFs as a tool to improve crop traits, identification of the biological functions of each transcription factor is necessary. However, plant genes are frequently duplicated and most plant transcription factors form large families. Such structural and functional redundancy often interferes with efforts to identify the function of TFs in plants.

To overcome such difficulties, a novel gene silencing system called *C*himeric *RE*pressor gene *S*ilencing *T*echnology (CRES-T) was developed, in which a transcription factor that is converted into a strong repressor by fusion with the EAR-motif repression domain (SRDX) suppresses target genes of a TF dominantly over endogenous and functionally redundant transcription factors (3). Transgenic plants that express the chimeric repressor exhibit a dominant-negative phenotype similar to loss-of-function of the alleles of the TF gene. Because the CRES-T system effectively induces various phenotypes, including tolerances to stresses, which are not found in single gene knockout or antisense lines, this system can be used as a tool for the improvement of crop traits, in addition to functional analysis of TFs. Moreover, the SRDX repression domain can convert not only a TF but also a transcriptional complex into a repressor via trans-repression and can be used to detect such protein–protein interactions in plants (4).

Here, the protocol and technical points for applying CRES-T are thoroughly described using *Arabidopsis* TFs as a model. In addition, the protocols of transient expression assay using particle bombardment, which is useful for functional analysis of TFs as well as detection of protein–protein interactions using the trans-repressive activity, are also described. The simplicity and universality of the CRES-T system undoubtedly reinforce the utility of TFs as the target of genetic manipulation.

2. Materials

2.1. Plasmid Construction for a Chimeric Repressor (*See* Note 1)

2.1.1. Plasmid Construction for Stably Transformed Transgenic Plants

1. cDNA clones of TF genes.
2. 5′-Phosphorylated 25-nucleotide primers corresponding to the 5′- and 3′-ends of the coding sequence of a TF gene. The 3′ primer must be designed such that the coding sequence of the TF is in frame with that of the repression domain in the p35SSRDXG or pSRDX_NOSG entry vector, without including the stop codon. 5′-Phosphorylation of primers can be prepared in-house using T4 polynucleotide kinase (*see* **Note 2**).

3. p35SSRDXG or pSRDX_NOSG entry vectors available from authors upon request for construction of the chimeric repressor utilizing the GATEWAY (Invitrogen, Carlsbad, CA, USA) system or its modified vectors (**Fig. 5.1**).
4. pBCKH or pBCKK binary destination vectors available from authors upon request for transformation of plants with chimeric repressor gene constructs (**Fig. 5.1**).
5. High-fidelity DNA polymerase that generates blunt ends (Phusion High-Fidelity DNA Polymerase, New England Biolabs Inc., USA).
6. *Escherichia coli* competent cells (DH5α).
7. *E. coli* DB3.1 (Life Technologies Inc., USA) for GATEWAY system.
8. Bacterial alkaline phosphatase (*E. coli* C75 Alkaline Phosphatase, Takara Bio Inc., Japan).
9. DNA ligase or DNA ligation kit (Takara Bio Inc., Japan).
10. LR clonase II (Life Technologies Inc., USA).
11. Lysogeny broth (LB) liquid medium and LB agar plates containing appropriate antibiotics for selection of transformants. LB is 10 g of tryptone, 5 g of yeast extract, and 10 g of NaCl in 1 L final volume, pH 7. For agar plates, add 15 g of agar per liter. Autoclave.

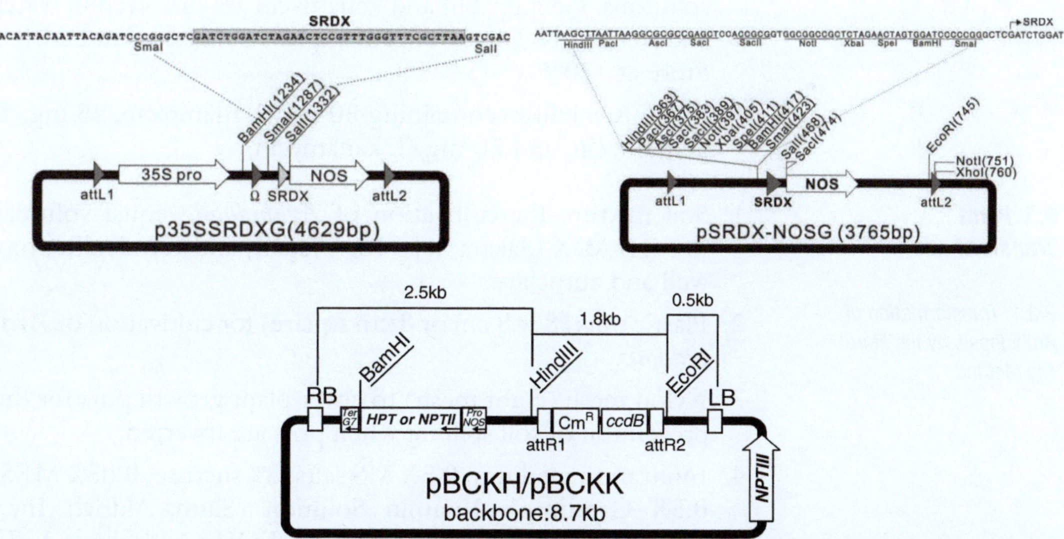

Fig. 5.1. Schematic diagram of plasmid vectors for CRES-T application. p35SSRDXG and pSRDX_NOSG are entry vectors for the construction of chimeric repressors. p35SSRDXG contains CaMV 35S promoter, while pSRDX_NOSG does not contain a promoter for driving the chimeric repressor gene. pBCKH and pBCKK are destination vectors for plant transformation; they contain hygromycin and kanamycin selection markers, respectively. The chimeric repressor gene constructs cloned into p35SSRDXG and pSRDX_NOSG vectors are transferred into pBCKH and pBCKK by GATEWAY LR reaction.

12. Ampicillin, kanamycin, and/or hygromycin depending on vectors used.
13. Conventional plasmid purification tools and sequencing reagents.
14. *Sma*I, *Bam*HI, *Sal*I, *Eco*RI, and *Hin*dIII restriction enzymes.
15. Conventional DNA purification tool, for example, Glassmilk DNA purification kit (BioStar International, Canada).

2.1.2. Plasmid Construction for Transient Expression Reporter–Effector Assay

In addition to many of the items listed in **Section 2.1.1**

1. p430T1.2, p190LUC, p35SSG, p35SSRDXG, and p35SVP16 plasmids available from authors upon request.
2. *Hin*dIII, *Bgl*II, and/or *Sal*I.

2.2. *Agrobacterium* Transformation

1. *Agrobacterium tumefaciens* GV3101 C58C1 Rif^r.
2. 1 mM HEPES, pH 7.0.
3. Solution for preparation of *Agrobacterium* competent cells for electroporation: 1 mM HEPES, pH 7.0, 10% glycerol in distilled water. Store at 4°C after autoclaving.
4. Liquid nitrogen.
5. Electroporation cuvette with 0.1 cm gap and Electroporator (Bio-Rad Inc., USA).
6. 50 mg/mL gentamycin, rifampicin, and kanamycin stock solutions. Gentamycin and kanamycin are dissolved in water and sterilized by filtration. Rifampicin is dissolved in DMSO. Store at –20°C.
7. LB plate medium containing 50 mg/L rifampicin, 25 mg/L gentamycin, and 50 mg/L kanamycin.

2.3. Plant Transformation

2.3.1. Transformation of Arabidopsis by the Floral Dip Method

1. Soil mixture for cultivation of *Arabidopsis*: equal volumes of SOILMIX (Sakata Seed Inc., Japan) and vermiculite, mix well and autoclave.
2. Plastic pots (8 × 8 cm or 8 cm square) for cultivation of *Arabidopsis*.
3. Nylon mesh (eight mesh) to cover plant growth pots for the prevention of soil spilling when pots are inverted.
4. Infiltration medium: 0.5X MS salts, 5% sucrose, 0.05% MES, 0.5X Gamborg's Vitamin Solution (Sigma-Aldrich Inc., USA), adjust to pH 5.7 with 1 M KOH. Add Silwet L-77 and benzylaminopurine to final concentrations of 0.04% and 0.044 μM, respectively.
5. 500 mL plastic cups with diameter smaller than 11 cm.
6. Paper bags, 25 cm × 35 cm or larger.

7. Transparent plastic bags (34 cm × 48 cm) with zipper.
8. Sieve (30 mesh).

2.3.2. Leaf Disk Method for Nicotiana tabacum

1. Tobacco leaves from plants grown in sterile culture.
2. *Agrobacterium* with the desired construct.
3. LB medium with appropriate antibiotics.
4. MS agar medium: 1X MS salts, 100 mg/L myo-inositol, vitamin mix (0.5 mg/L paydoxine-HCl, 0.5 mg/L thiamine-HCl, 2 mg/L glycine, 5 mg/L nicotinic acid, 0.05 mg/L folic acid), 0.8% agar, 3% sucrose, 0.05% MES, adjust to pH 5.7 with 1 M KOH. Autoclave and add appropriate growth regulators and antibiotics. Pour into 90 mm diameter petri plates.
5. MS agar plates with 1.0 mg/L 6-benzylaminopurine (BA), 0.5 g/L claforan, and appropriate antibiotics.
6. MS agar plates with 1.0 mg/L BA, augmentin (2 tablets/L), and appropriate antibiotics.
7. MS agar plates with 1.0 mg/L naphthalene acetic acid (NAA), augmentin (2 tablets/L), and appropriate antibiotics.
8. MS agar plates with augmentin (2 tablets/L) and appropriate antibiotics.
9. Soil to grow regenerated plants.

2.4. Selection of Transgenic Arabidopsis

1. Seed sterilization solution: 2% sodium hypochlorite, 0.02% Triton X-100.
2. 0.1% agarose solution. Store at room temperature after autoclaving.
3. MS selective medium: 1X MS salts, 0.8% agar, 0.5% sucrose, 0.05% MES, adjust to pH 5.7 with 1 M KOH. Add 250 μg/mL vancomycin and 25 μg/mL kanamycin or 30 μg/mL hygromycin after autoclaving and pour into 150 mm diameter petri dishes.
4. Sterile water.
5. Surgical tape to seal plates.

2.5. Transient Reporter–Effector Assay by Particle Bombardment

1. Gold microcarriers (1.0 μm), macrocarrier holders, rupture disks (1,100 psi), and stopping screens (Bio-Rad Inc., USA). Rinse with ethanol and dry.
2. 70 and 100% ethanol.
3. Sterile water.
4. Sterile 50% glycerol.
5. Appropriate vectors shown in **Fig. 5.2** available from authors upon request.

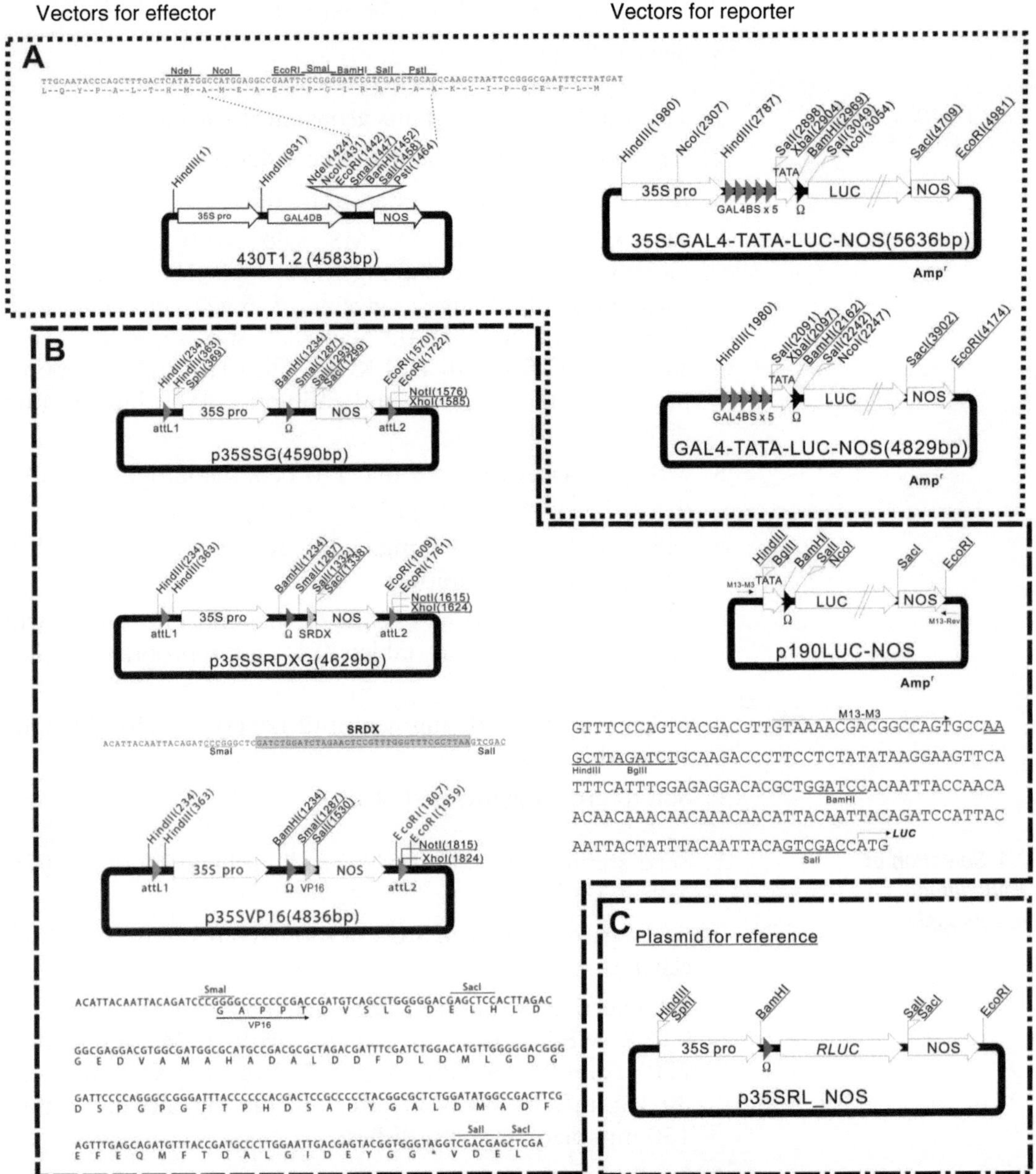

Fig. 5.2. Schematic diagrams of plasmid vectors for transient effector–reporter expression analysis. (**a**) Plasmid vectors designed to analyze transcriptional activation or repressive activity of a TF whose target genes are unknown. 430T1.2 vector for the construction of effector fused with GAL4 DNA binding domain. Plasmid vectors for construction of reporter gene to analyze repressive activity (35S-GAL4-TATA-LUC-NOS) and activation activity (GAL4-TATA-LUC-NOS) of a TF gene using 430T1.2-based effectors. (**b**) Plasmid vectors designed to examine the transcriptional activation or repressive activity of a TF on a promoter sequence of a putative target gene of TF. Three plasmid vectors for effectors of ectopic expression (p35SSG), dominant repressor (p35SSRDXG), and ectopic activation (p35SVP16), respectively. p190LUC-NOS reporter plasmid vector has cloning site for promoter sequence followed by luciferase reporter gene. (**c**) p35SRL_NOS plasmid for reference containing *Renilla* luciferase gene.

6. 2.5 M $CaCl_2$. Store in aliquots at 4°C.
7. 1 M spermidine. Store in aliquots at –20°C.
8. Platform mixer (MicroMixer E-36; Titec Inc., Japan); ultrasonic cell disruptor (Microson XL; Misonix Inc., USA); Biolistic PDX-1000/He particle delivery system (Bio-Rad Inc., USA); pressurized helium gas.
9. Fully expanded *Arabidopsis* rosette leaves in vegetative growth phase.
10. Circular filter papers (70 mm diameter) immersed with distilled water.
11. Lysis buffer: Dilute 5X passive lysis buffer (supplied in the Dual-Luciferase Reporter Assay System, Promega Inc., USA) to 1X with distilled water.
12. Luciferase assay buffer: Resuspend the luciferase assay substrate in 10 mL of luciferase assay buffer II from the Dual-Luciferase Reporter Assay System (Promega Inc., USA). Store in aliquots at –80°C.
13. Stop & Glo reagent: Add 0.2 mL of 50X Stop & Glo substrate to 10 mL of Stop & Glo buffer supplied in the Dual-Luciferase Reporter Assay System (Promega Inc., USA). Store in aliquots at –80°C.
14. Luminometer TD-20/20 (Turner Designs Inc., USA).

3. Methods

The CRES-T system is a powerful tool for the analysis of biological functions of redundant plant transcription factors and their interacting proteins. By fusion with the SRDX repression domain, transcriptional activators or transcriptional complexes can be converted into strong repressors that dominantly suppress the expression of their target genes. To operate the CRES-T system efficiently, the design of gene constructs for chimeric repressors is important. Even though the chimeric repressor acts dominantly over the endogenous transcription factor, higher expression levels of the chimeric repressor result in higher efficiency. Strong and ubiquitous promoters such as the CaMV 35S promoter and, in the case of monocots, the promoter of the maize polyubiquitin gene (*ZmUbi1*) (5) are usually employed. It is often pointed out that phenotypes induced by ectopic expression do not always reflect biological function. In most cases, however, phenotypes induced by the expression of the chimeric repressor are very similar to those of loss-of-function mutants and thus reflect their biological functions, despite being driven by the CaMV 35S promoter.

An advantage of the CRES-T system is that any promoter, including the native promoter of the TF gene, can be used to drive expression of the chimeric repressor. Although a ubiquitous promoter is useful, induction of the chimeric repressor by its own promoter is recommended when the promoter activity of the TF gene of interest is significantly higher than that of the CaMV 35S promoter or if its activity is limited to a specific tissue where the CaMV 35S promoter is not active. Additionally, the native promoter may provide additional insight into the biological function of a TF.

Because the expression of TF genes usually precedes that of non-TF genes, using promoters of non-TF genes to drive the expression of the chimeric repressor is not recommended even if active in desired tissues or conditions. In particular, promoters derived from the target genes of the TF of interest should not be used because the chimeric repressor may then bind the promoter of the chimeric repressor gene and suppress its own expression. Heterologous combinations of a promoter and a chimeric repressor sometimes result in higher efficiency. For example, the NST1 chimeric repressor driven by the promoter of *NST3*, a paralogous gene of *NST1*, resulted in a more dramatic effect than when driven by the promoter of the *NST1* gene itself (6). In addition, when the CRES-T system is applied to plants other than *Arabidopsis*, construction of chimeric repressors using TFs derived from the plant species of interest is recommended. Such chimeric repressors often exhibit higher activities than those constructed using *Arabidopsis* TF genes.

When the SRDX repression domain is fused to the N-terminal end of a TF, the resultant chimeric repressor also acts as well as when fused to a C-terminal region. However, SRDX is usually fused to the C-terminal end of a TF because most endogenous transcriptional repressors contain the EAR-motif at the C-terminus. When a linker sequence, such as that for the GATEWAY system, is inserted in the junction between coding sequences of the TF and the SRDX domain, the resulting chimeric repressors act well but the efficiency to induce dominant-negative phenotypes in transgenic plants is often significantly lower than those without any linker sequence.

In addition, the SRDX domain can convert a transcriptional complex into a repressor via trans-repressive activity that is mediated by protein–protein interactions. This trans-repressive activity can be used to detect protein–protein interactions. Moreover, this activity can be used to detect protein factors that could interact with a TF or be incorporated into a transcriptional complex. If the expression of a non-TF protein gene fused with SRDX induces an aberrant phenotype in transgenic plants, and such phenotype is not induced by the ectopic expression of the gene alone, without

SRDX, the non-TF protein is likely to interact with a TF or a transcriptional complex.

If the TF of interest is a transcriptional activator, the transgenic plants that express the chimeric repressor are expected to have a phenotype similar to loss-of-function mutants. This phenotype would be predicted to be opposite to ectopic expression of the TF lacking the SRDX domain. If the TF of interest acts as a transcriptional repressor, the plants expressing the chimeric repressor would, in some cases, exhibit an enhanced phenotype compared to ectopic/overexpression of the TF lacking the SRDX domain. Aspects of these phenotypes may be opposite to those observed in loss-of-function alleles of a transcriptional repressor. The transient expression analysis using *Arabidopsis* leaves is useful to evaluate molecular function of TFs.

3.1. Plasmid Construction for Chimeric Repressor

3.1.1. Plasmid Construction for Stably Transformed Transgenic Plants

1. For amplification of the promoter sequence to drive a chimeric repressor gene in pSRDX_NOSG vector, design oligonucleotide primers corresponding to 5′ and 3′ regions of the promoter sequence to include suitable restriction enzyme sites for cloning into the multi-cloning sites of pSRDX_NOSG vector (**Fig. 5.1**). Confirm that those restriction enzyme digestion sites are not included in the promoter region. Sequences 1,000–3,000 bp upstream from the first ATG codon (initiation codon) are favorably used (*see* **Note 3**). PCR-amplify the promoter region using high-fidelity polymerase, digest both ends with appropriate restriction enzymes, and ligate into pSRDX_NOSG vector (**Fig. 5.1**).
2. Amplify the coding region of the TF gene with appropriate primers. The polymerase should generate blunt (not A-overhang) ends. Purify amplified fragments using, for example, agarose gel electrophoresis and silica matrix such as available from BioStar (Glassmilk DNA purification kit, Biostar International, Canada).
3. Digest p35SSRDXG or pSRDX_NOSG entry vector with *Sma*I at 25°C for more than 3 h. After digestion, dephosphorylate thoroughly with bacterial alkaline phosphatase at 65°C for 1 h or longer (*see* **Note 4**).
4. Ligate 50–500 ng of purified PCR fragments and 50–200 ng of p35SSRDXG or pSRDX_NOSG entry vector predigested by *Sma*I, at 16°C overnight (**Fig. 5.1**) (*see* **Note 5**).
5. Transform *E. coli* with the chimeric repressor gene construct. After incubation at 37°C for 16 h on LB plate with ampicillin, pick several colonies and perform colony PCR to confirm that the transgene is present.

6. Inoculate several colonies separately into liquid culture, purify plasmids, and digest with *Bam*HI and *Sal*I to check size of the insertion (*see* **Note 6**). When the expected size of insertion is confirmed, perform sequence analysis to check the coding sequence of the TF gene and ensure that it is in frame with that of the repression domain and that an extra ATG codon is not included between the TATA box and the first ATG of the coding region.
7. Using the GATEWAY LR reaction, transfer the chimeric repressor gene constructs cloned in p35SSRDXG or pSRDX_NOSG entry vector to pBCKH or pBCKK destination vector (**Fig. 5.1**). Mix 100–300 ng of p35SSRDXG- or pSRDX_NOSG-based entry clone and 50–100 ng of pBCKH or pBCKK with 1/5 volume of LR clonase II (*see* **Note** 7). Incubate the mixture at 25°C for 1 h or longer. After incubation, treat mixture with proteinase-K according to the manufacturer's instructions.
8. Transform *E. coli* (DH5α) with destination plasmid constructs. After incubation at 37°C for 16 h on LB plate with kanamycin (for pBCKK) or hygromycin (for pBCKH), pick several colonies and culture in liquid medium separately, purify plasmids from individual colonies, and digest with *Eco*RI and *Hin*dIII to check the size of insertion (*see* **Note 8**).

3.1.2. Plasmid Construction for Transient Expression Assay

1. For the construction of effector plasmids as a fused gene with GAL4 DNA binding domain (GAL4DB), design the 5′-end primer for the coding sequence of the TF to be in frame with that of GAL4DB in p430T1.2. The 3′-end primer includes a stop codon (**Fig. 5.2**). Amplify the coding sequence and clone into the *Sma*I site of p430T1.2.
2. The p35SSG, p35SSRDXG, and p35SVP16G vectors are designed for overexpression, dominant repression, and ectopic activation, respectively (**Fig. 5.2**). For the construction of effector plasmids design the 3′-end primer such that the coding sequence of the TF gene is in frame with that of SRDX and VP16 when cloning into p35SSRDXG and p35SVP16G vectors, respectively. For cloning in p35SSG, design the 3′-end primer to include a stop codon. Amplify the TF coding sequence and clone into the *Sma*I site of each vector.
3. To drive the luciferase reporter gene by the promoter of interest, amplify 500–3,000 bp of the promoter region with appropriate primers and clone into the *Hin*dIII (or *Bgl*II)/*Sal*I sites of p190LUC (**Fig. 5.2**). Design primers for the promoter sequence to be in frame with the luciferase

gene if the promoter region includes the initiation codon of the protein coding region of the gene.

3.2. Agrobacterium Transformation

3.2.1. Preparation of Agrobacterium Competent Cells for Electroporation

1. Inoculate a single *Agrobacterium* colony into 200 mL of LB medium containing the appropriate antibiotics and culture at 28°C with vigorous shaking for 12–24 h. Cool the culture medium by placing on ice for 15 min and centrifuge at 2,500×*g* for 15 min at 4°C. Wash the pellet with 50 mL of 1 mM HEPES and centrifuge at 3,900×*g* for 15 min at 4°C. Repeat this step three times.
2. Resuspend the pellet with 45 mL of 10% glycerol and centrifuge at 3,900×*g* for 15 min at 4°C. Discard the supernatant and add 500 μL of 10% glycerol. Aliquot 40 μl of the *Agrobacterium* suspension into 1.5 ml tubes, freeze them by liquid nitrogen, and store at –80°C.

3.2.2. Preparation of Competent Cells for Freezing Method

1. Inoculate a single colony into 500 mL of LB medium and culture at 28°C with vigorous shaking until OD_{650} reaches 1.
2. Centrifuge at 6,000×*g* for 5 min at 4°C. Resuspend the pellet with 100 mL of LB medium, centrifuge at 6,000×*g* for 5 min at 4°C, and resuspend the pellet with 25 mL of LB medium. Aliquot 200 μL of the *Agrobacterium* suspension into 1.5 mL tubes.

3.2.3. Agrobacterium Transformation by Electroporation

1. Thaw competent cells on ice. Add 0.1–1 μg of plasmid DNA. Transfer competent cells into an electroporation cuvette. Keep the cuvette on ice.
2. Wipe moisture off the cuvette and expose to an electric pulse (2.2 kV) using a MicroPulser electroporator (Bio-Rad Inc., USA). Immediately add 1 mL of LB medium. Incubate at 28°C for 1 h without shaking. Spread 50–300 μL of the culture on LB medium containing appropriate antibiotics and incubate at 28°C for 48 h.

3.2.4. Agrobacterium Transformation by Freezing Method

1. Add 5–10 μg of plasmid DNA and 100 μL of LB medium to competent cells. Freeze the mixture in liquid nitrogen and thaw it by incubating at 37°C for 25 min.
2. Add 15 mL of LB medium and incubate at 28°C for 16 h without shaking. Centrifuge the mixture at 1,000×*g* for 10 min at 4°C and resuspend in 200 μL of LB medium. Spread the mixture onto LB medium containing appropriate antibiotics and incubate at 28°C for 48 h.

3.3. Plant Transformation

1. Cover plant growth pots with mesh after filling with SOILMIX/vermiculite mixture. Grow three plants per pot

3.3.1. Transformation of Arabidopsis by the Floral Dip Method

for 4–5 weeks under long-day condition (16 h of light per day). Nine plants (three pots) are typically used for one transformation, from which 1–2 mL of seeds is usually obtained.

2. Remove first inflorescence shoot when it reaches approximately 5 cm to promote the development of secondary inflorescences. Plants are ready for transformation when secondary inflorescences reach 10–15 cm. Cut all flowers and siliques just before transformation.
3. Scratch colonies of *Agrobacterium* grown on LB plates, transfer them into 250 mL of LB medium containing appropriate antibiotics, and culture at 28°C for approximately 48 h with vigorous shaking. The OD_{600} usually reaches 1.5. Pellet by centrifugation at 5,000×*g* for 20 min and resuspend *Agrobacterium* with 400–500 mL of infiltration medium. Transfer the medium into a plastic cup of which the top circumference is smaller than that of plant growth pots such that only inflorescence shoots are immersed.
4. Invert plant pots and dip only inflorescence shoots into the infiltration medium containing the *Agrobacterium* for 2 min.
5. Put plant pots into transparent plastic bags with zippers, close the zippers, and lay them on their sides in a growth chamber or greenhouse for 24 h. Grow them under normal conditions until siliques are completely matured and dried. Stop watering when the terminal flower appears.
6. To harvest seeds, cut inflorescence stems and put them into paper bags. Store bags in a drafty place for more than 1 week. Remove tissue debris using sieves and collect seeds. Store seeds in low humidity (lower than 40%) at 4°C.

3.3.2. Leaf Disk Method for N. tabacum

1. Prepare around 10 mm square pieces of tobacco leaves including veins (*see* **Note 9**) and float them on *Agrobacterium* suspension in LB medium for 3 min. Wipe medium and incubate the leaf pieces on MS plate medium without antibiotics in a culture room (16 h light/8 h dark, 25°C). After 3 days, transfer leaf pieces onto MS plates containing 1.0 mg/L 6-benzylaminopurine (BA), 0.5 g/L claforan, and appropriate antibiotics (*see* **Note 10**) and incubate in a culture room for 2 weeks.
2. Transfer the leaf pieces to MS plate medium containing 1.0 mg/L BA, augmentin (2 tablets/L), and appropriate antibiotics for selection and incubate until regenerated shoots become visible. Every two weeks, transfer the leaf samples on to new MS medium (*see* **Note 11**).

3. Cut regenerated shoots from the explant and transfer them on MS plate medium containing 1.0 mg/L NAA, augmentin (2 tablets/L), and appropriate antibiotics.
4. After plants develop roots, transfer regenerated plants into MS plate medium containing augmentin and appropriate antibiotics (*see* **Note 12**), then grow transgenic plants on soil.

3.4. Selection of Transgenic Arabidopsis

1. For sterilization of seeds, transfer 100 μL of primary transformed (T1) seeds into a 50 mL conical tube (*see* **Note 13**). Add 30 mL of bleach solution. Shake gently for 7 min. Collect surface-sterilized seeds by brief centrifugation at 1,000×*g* for 2 min. Rinse seeds three to five times with 30 mL of sterile distilled water.
2. Suspend seeds with 10 mL of 0.1% agar. Mix gently to allow seeds to be uniformly dispersed in the conical tube. Pour seeds onto MS solid medium containing the appropriate antibiotic selection marker. Swirl dish to allow seeds to distribute evenly. Place dish uncovered on a clean bench for 2 h to dry the top agar. Seal with surgical tape.
3. Incubate plates at 4°C for 2–4 days. Transfer plates and grow under normal growth conditions (22°C, 16 h of light per day). Seedlings that are resistant to the antibiotic selection marker become visible within 3 weeks after sowing. Pick and transfer 10–40 independent plants to soil and grow in a greenhouse or growth chamber (*see* **Note 14**). Collect seeds separately from each plant.

3.5. Transient Reporter–Effector Assay by Particle Bombardment

3.5.1. Preparation of Gold Particles

1. Prepare 60 mg of gold particles in a 1.5 mL tube for use in a Biolistic PDX-1000/He particle delivery system (Bio-Rad Inc., USA).
2. Add 1 mL of 70% EtOH. Vortex vigorously for 3–5 min. Incubate for 15 min. Centrifuge for 5 s and remove supernatant.
3. Add 1 mL of sterile water, vortex vigorously for 1 min, incubate for 1 min, centrifuge briefly, and remove supernatant. Repeat this step three times.
4. Resuspend the particle with 1 mL of sterile 50% glycerol to adjust the final concentration of gold particles to 60 mg/mL. Gold particles in solution can be stored at 4°C for up to 1 month.

3.5.2. DNA Coating

1. Disrupt agglomeration of gold particles by brief sonication and vortex. Dispense 25 μL of particle solution into 1.5 mL tubes. Aliquot this amount for three replicated experiments (*see* **Note 15**).

2. Add 2.4 μg of the reporter gene construct, 1.8 μg of effector genes, and 1.2 μg of p35SRL_NOS plasmid for *Renilla* luciferase as an internal control for the dual-luciferase assay system.
3. Add 25 μL of 2.5 M $CaCl_2$ and 1 μL of 1 M spermidine. Vortex for 3 min, incubate tube for 1 min, centrifuge briefly, and remove supernatant. Rinse particles successively, first with 140 μL of 70% ethanol and then with 100% ethanol.
4. Add 25 μL of 100% EtOH and resuspend particles by brief sonication. Aliquot 8 μL of the particle suspension and load respectively onto the center of three macrocarriers set in metal holders. Do not move the macrocarrier until ethanol is completely evaporated.

3.5.3. Transient Biolistic Transformation of Arabidopsis Leaves

1. Prepare fully expanded leaves from unbolted *Arabidopsis* plants grown in short-day condition (16 h of dark per day). Place one leaf abaxial side up in the center of a pre-wetted filter paper. When *Arabidopsis* leaves grown under long-day condition are used, place four leaves on a filter to form a square shape leaving the central area open (*see* **Note 16**).
2. Set pressure of helium gas to around 200 psi higher than the burst pressure of the selected rupture disk. For example, when a rupture disk of 1,100 psi burst pressure is used, set the gas pressure to 1,300 psi.
3. Place the rupture disk in a retaining cap, stopping screen and macrocarrier onto the macrocarrier launch assembly.
4. Evacuate chamber until it reaches up to 27 in. Hg. Press the "FIRE" button until the rupture disk bursts. Incubate bombarded sample plates at 23°C in the dark for 6–16 h.

3.5.4. Measurement of Luciferase Activity

1. Turn on luminometer TD-20/20 15 min before measurement. Set the program to perform a 2-s pre-measurement delay, followed by a 10-s measurement period for each reporter assay.
2. Wipe moisture off the leaves and grind leaves in liquid nitrogen using a pestle in a 2 mL tube (*see* **Note 17**). Add 250 μL of lysis buffer and homogenize thoroughly using a homogenizer. Centrifuge at 15,000 rpm for 15 min at 4°C in a microcentrifuge. Transfer 20 μL of the supernatant into the tube containing 100 μL of Luciferase Assay Reagent II (LAR II provided in the kit). Mix the solution gently by pipetting two or three times but not by vortexing (*see* **Note 18**).
3. Place the tube into luminometer and measure firefly luciferase activities. Add 100 μL of Stop & Glo reagent and

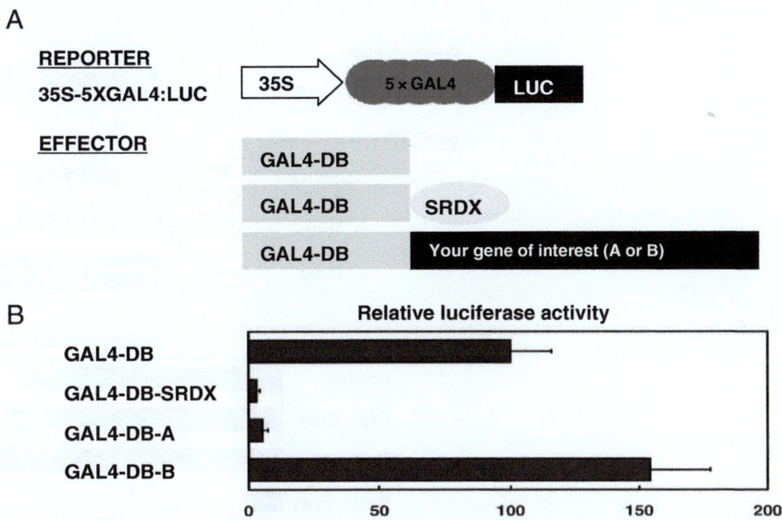

Fig. 5.3. A representation of reporter–effector transient expression assay. (**a**) Schematic representation of reporter and effector constructs. (**b**) Results of transient expression assays for the analysis of molecular function of transcription factors. Results indicate protein "A" and "B" have repressive and activation activity, respectively, as GAL4 DB binding proteins. Bars in the graph indicate averages of relative luciferase activities after co-bombardment of *Arabidopsis* leaves with each combination of reporter and effector genes. SRDX represents a modified repression domain from *Arabidopsis* SUPERMAN (7).

vortex briefly. Measure *Renilla* luciferase activity. Calculate the ratio of firefly and *Renilla* luciferase activities. An example is shown in **Fig. 5.3**.

3.6. Detection of Protein–Protein Interaction

1. For the analyses of interaction of factors by transient expression assay, prepare effector plasmids by insertion of the coding regions of the protein factors into the *Sma*I site of 430T1.2 (bait) and p35SSRDXG (prey), respectively (**Fig. 5.2**).
2. Co-transform bait, prey, and 35S-GAL4-TATA-LUC-NOS reporter gene constructs (**Fig. 5.2**) into *Arabidopsis* leaves. If the two protein factors form a complex, the reporter activity would be suppressed by trans-repression. Confirm that the prey alone does not affect the reporter activity (**Fig. 5.4**).
3. To analyze protein factors that could interact with TFs, prepare transgenic plants that express the transgene of the protein factor of interest fused with SRDX. If an aberrant phenotype is observed, it is possible that the protein "A" interacts with a transcription factor or transcriptional complex. Confirm that the phenotype is observed only with the protein factor–SRDX fusion but not when the protein factor alone is expressed.

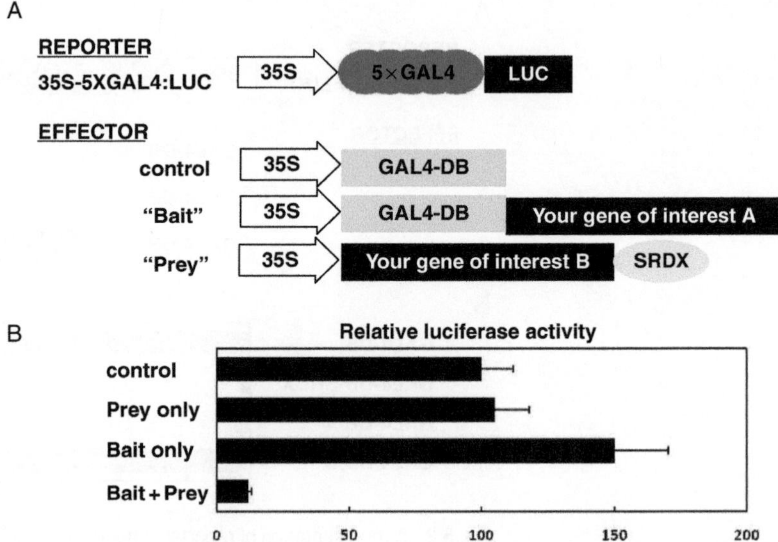

Fig. 5.4. Detection of protein–protein interaction using SRDX repression domain by transient expression assay. (**a**) Schematic representation of reporter and effector constructs. (**b**) Results of transient expression assays for the analysis of protein–protein interaction between "A" and "B." Results indicate protein "A" and "B" interact together to form a complex. Bars in the graph indicate averages of relative luciferase activities after co-bombardment of *Arabidopsis* leaves with each combination of reporter and effector genes.

4. Notes

1. Recommended manufacturers are shown.
2. For 5′-phosphorylation of oligonucleotides by T4 polynucleotide kinase, mix 7 μL of 100 μM oligonucleotide, 2 μL of 10 mM ATP, 2 μL of 10X Protruding end kinase buffer, 2 μL of T4 polynucleotide kinase (Toyobo Inc., Japan), and 7 μL of ultra-pure water and incubate at 37°C for 1 h followed by 5 min incubation at 95°C and cooling. Add 50 μL of ultra-pure water and use as 10 μM oligonucleotide for PCR.
3. It is better for the promoter region to include the ATG initiation codon of the protein coding sequence of the gene to maintain the distance between the TATA box and the initiation codon of the native gene. In this case, design the 3′ primer of the promoter sequence to be in frame with the coding sequence of the TF gene cloned into pSRDX_NOSG vector. Alternatively, perform PCR of the promoter and coding region together and clone it into the pSRDX-NOS entry vector in a manner such that it is in frame with the coding region of SRDX.

4. Note that, in case of digestion of p35SRDXG vector by *Sma*I restriction enzyme, we recommend *Sma*I with high quality such as manufactured by New England Biolabs (NEB Inc.,USA) and NEBuffer 4, because low quality of SmaI frequently causes one or two end nucleotide deletion during digestion, which causes out-of-frame sequences when ligating to SRDX. *Sma*I manufactured by New England Biolabs (NEB Inc., USA) and NEBuffer 4 is recommended. Alkaline Phosphatase (*E. coli* C75) (Takara Bio Inc., Japan) is recommended for dephosphorylation. A thermal cycler is convenient for accurate incubation.
5. DNA ligation kit ver. 1 (Takara Bio Inc., Japan) results in high-efficiency ligation and may shorten reaction time up to 3 h.
6. Confirm that only a single fragment, not multiple fragments, is ligated in p35SSRDXG or pSRDX_NOSG. Double digestion by *Bam*HI and *Sal*I or PCR using primers corresponding to 3′-end region of the CaMV 35S promoter and Nos terminator helps to check the size of inserted fragment. The sequences encoding the SRDX domain will add 42 bp to the TF.
7. The GATEWAY cassette vectors pBCKH and pBCKK contain the *ccdB* lethal gene and therefore can only be amplified in *E. coli* DB3.1 strain (Life Technologies Inc., USA).
8. Check that the entry clone is not contaminated in the destination vector plasmid. Such contamination can be checked by electrophoresis.
9. Leaf samples of sterile plants should be used.
10. Appropriate antibiotics are different between plant species. For tobacco plants, kanamycin is more effective than other antibiotics for selection of transgenic plants.
11. It sometimes takes a longer time to regenerate transgenic shoots when the chimeric repressor gene construct is transformed, probably because the dominant effect of the chimeric repressor often induces severe phenotypes and may prevent shoot regeneration.
12. Because CRES-T transgenic lines exhibit wide variations in their phenotypes, more than 15 independent lines should be observed for characterization of phenotype.
13. If seeds are stored at 4°C, warm them to room temperature before use. From 100 μL of T1 seeds, 10–100 transgenic plants that exhibit tolerance to antibiotics are usually obtained.
14. If no transformant is obtained, the transformed chimeric repressor may induce embryonic lethality or severe growth

retardation. In that case, inducible expression systems, such as those by estrogen, are useful.

15. To analyze slight differences in activities, such as twofold or lower, five to six replicates are recommended to allow statistical confirmation.
16. Use of leaves grown under short-day condition induces higher and more reproducible luciferase activities than those grown in long-day condition. Rosette leaves grown under short-day condition are larger and thicker than those grown in long-day condition and one leaf sample is sufficient for bombardment transient assay.
17. The beads cell disrupter, which is useful in handling multiple samples at once, does not require liquid nitrogen. Put two ceramic balls of 3-mm diameter (YTZ ball; Nikkato Inc., Japan) and 100 μL of lysis buffer into a screw cap tube. Smash the leaves in a beads cell disrupter/microhomogenizing system (Micro Smash MS-100; Tomy Inc., Japan) at 4,000 rpm for 2 min.
18. Do not use a vortex mixer to mix the sample solution and LAR II regent. Vortexing may cause adhesion of microfilm of luminescent solution, which prevents the activity of Stop & Glo reagent.

Acknowledgments

The authors express appreciation to Dr. Masaru Ohta, Dr. Keiichiro Hiratsu, Dr. Tomotsugu Koyama, Dr. Yoshimi Umemura, Dr. Masahito Shikata, and Dr. Akira Iwase who exerted great effort for the establishment and development of the protocols described in this chapter.

References

1. Riechmann, J. L., Heard, J., Martin, G., Reuber, L., Jiang, C.-Z., Keddie, J., Adam, L., Pineda, O., Ratcliffe, O. J., Samaha, R. R., Creelman, R., Pilgrim, M., Broun, P., Zhang, J. Z., Ghandhari, D., Sherman, B. K., and Yu, G.-L. (2000) *Arabidopsis* transcription factors: genome-wide comparative analysis among eukaryotes. *Science* **290**, 2105–2110.
2. Mitsuda, N. and Ohme-Takagi, M. (2009) Functional analysis of transcription factors in *Arabidopsis*. *Plant Cell Physiol.* **50**, 1232–1248.
3. Hiratsu, K., Matsui, K., Koyama, T., and Ohme-Takagi, M. (2003) Dominant repression of target genes by chimeric repressors that include the EAR motif, a repression domain, in *Arabidopsis*. *Plant J.* **34**, 733–739.
4. Matsui, K. and Ohme-Takagi, M. (2010) Detection of protein-protein interactions in plants using the transrepression activity of the

EAR motif repression domain. *Plant J.* **61**, 570–578.

5. Christensen, A. H., Sharrock, R. A., and Quail, P. H. (1992) Maize polyubiquitin genes: structure, thermal perturbation of expression and transcript splicing, and promoter activity following transfer to protoplasts by electroporation. *Plant Mol. Biol.* **18**, 675–689.
6. Mitsuda, N. and Ohme-Takagi, M. (2008) NAC transcription factors NST1 and NST3 regulate pod shattering in a partially redundant manner by promoting secondary wall formation after the establishment of tissue identity. *Plant J.* **56**, 768–778.
7. Hiratsu, K., Ohta, M., Matsui, K., and Ohme-Takagi, M. (2002) The SUPERMAN protein is an active repressor whose carboxy-terminal repression domain is required for the development of normal flowers. *FEBS Lett.* **514**, 351–354.

Chapter 6

Analysis of a Transcription Factor Using Transient Assay in *Arabidopsis* Protoplasts

Yuji Iwata, Mi-Hyun Lee, and Nozomu Koizumi

Abstract

Regulation of gene expression by transcription factors is a fundamental mechanism in essentially all aspects of cellular processes. Transient expression assay of a reporter plasmid containing a reporter gene driven by a promoter of interest and an effector plasmid expressing a transcription factor has been a powerful tool for analyzing transcription factors. Here we present a protocol for polyethylene glycol (PEG)-mediated transformation of *Arabidopsis* protoplasts. It details preparation of protoplasts from *Arabidopsis* suspension cultured cells or leaves of soil-grown *Arabidopsis* plants and subsequent PEG-mediated transformation with reporter and effector plasmids. This protocol can be completed within 24 h from protoplast preparation to reporter assay. As an example, analysis of the membrane-bound transcription factor AtbZIP60 and its target *BiP3* promoter is shown.

Key words: *Arabidopsis thaliana*, luciferase, promoter, protoplast, transcription factor, transient assay.

1. Introduction

Regulation of gene expression by transcription factors is a fundamental mechanism in essentially all aspects of cellular processes in eukaryotes including plants. Upon developmental or environmental cues, a transcription factor activates transcription of its target genes by recognizing and binding to a specific nucleotide sequence in their promoters. Therefore, a promoter is the major determinant for responsiveness of a corresponding gene mediated by a transcription factor. To clarify whether and how a transcription factor of interest activates a specific subset of genes, a feasible

L. Yuan, S.E. Perry (eds.), *Plant Transcription Factors*, Methods in Molecular Biology 754,
DOI 10.1007/978-1-61779-154-3_6, © Springer Science+Business Media, LLC 2011

experimental method that measures the ability of a transcription factor to activate a given promoter is important.

Transient expression systems using reporter genes have been widely used to analyze transcription factor–promoter interactions in both animal and plant cells. A reporter plasmid harboring a reporter gene driven by a promoter of interest and an effector plasmid expressing a transcription factor to be analyzed are co-transformed. To normalize the transformation efficiency that would be expected to vary among each transformation, a reference plasmid harboring another reporter gene driven by a constitutive promoter is co-transformed as well. A diagram of plasmids generally used for transient transformation of *Arabidopsis* is shown in **Fig. 6.1**. For quantitative measurements of gene expression, luciferase (*Luc*) and β-glucuronidase (*GUS*) genes have been commonly used (1, 2). We have routinely used a dual-luciferase reporter assay system that utilizes two *Luc* genes, isolated from firefly (*Photinus pyralis*) and *Renilla* (sea pansy; *Renilla reniformis*), respectively. This system allows us to measure two distinguishable reporter activities within a single tube (3).

There are many different transient expression methods in plants. These include *Agrobacterium* infiltration, particle bombardment, and polyethylene glycol (PEG)-mediated protoplast transformation (4–7). Although *Agrobacterium* infiltration has been widely used for *Nicotiana benthamiana* leaves, some researchers claim that it is not sufficiently efficient and reproducible in *Arabidopsis* (5). We use PEG-mediated protoplast

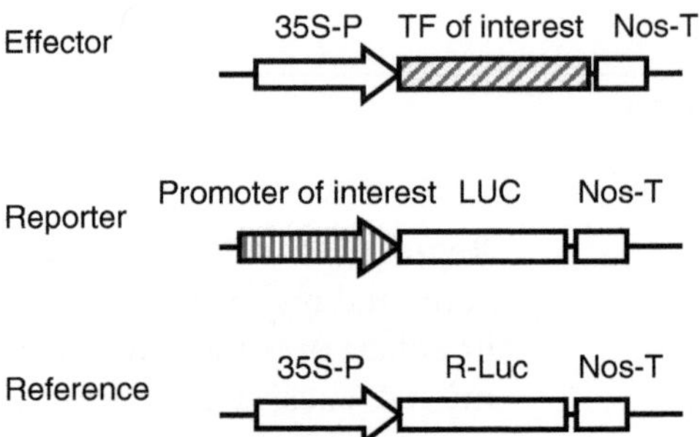

Fig. 6.1. Schematic representation of plasmids generally used in dual-luciferase reporter assays in *Arabidopsis*. To analyze a transcription factor using a transient reporter assay, effector, reporter, and reference plasmids are introduced into protoplasts. An effector plasmid harbors a transcription factor (TF) gene under the control of the constitutive CaMV 35S promoter (35S-P). A reporter plasmid harbors a firefly luciferase (*Luc*) reporter gene under the control of a promoter of interest. To normalize the transformation efficiency that would be expected to vary in each transformation, a third, reference plasmid, harboring *Renilla* luciferase (*R-Luc*) gene under the control of CaMV 35S promoter, is introduced. Nos-T, Nopalin synthase terminator.

transformation because it does not require any special equipment. Here we provide a detailed protocol for PEG-mediated transient transformation of protoplasts prepared from either *Arabidopsis* suspension cultured cells or leaves of soil-grown *Arabidopsis* plants to analyze transcription factor–promoter interactions.

2. Materials

2.1. Preparation of Protoplasts from Arabidopsis Suspension Cells

1. *Arabidopsis* suspension cells: We routinely use the rapidly dividing cell line MM2d, *Arabidopsis thaliana* ecotype L*er* (8) (*see* **Note 1**).
2. Rotary shaker.
3. Modified LS medium: Murashige and Skoog (MS) medium, pH 5.8, supplemented with 3% sucrose. Adjust pH with 1 M KOH. Make 1 L and disperse into 50 mL aliquots in 250-mL flasks. Cover with aluminum foil and autoclave.
4. Filtration set: Buchner funnel, filter paper, and aspirator.
5. Miracloth (Calbiochem).
6. 50-mL conical tube.
7. 1 M mannitol: Sterilize using a 0.45-μm filter.
8. 0.5 M 2-(*N*-morpholino)ethanesulfonic acid (MES), pH 5.7: Sterilize using a 0.45-μm filter.
9. 1 M KCl: Autoclave.
10. 1 M $CaCl_2$: Sterilize using a 0.45-μm filter.
11. 1 M $MgCl_2$: Sterilize using a 0.45-μm filter.
12. Cellulase Onozuka R-10 (Yakult Pharmaceutical Ind. Co., Ltd., Japan).
13. Macerozyme R-10 (Yakult Pharmaceutical Ind. Co., Ltd., Japan).
14. Enzyme solution: 20 mM MES, pH 5.7, 0.4 M mannitol, 20 mM KCl, 10 mM $CaCl_2$, 0.1% bovine serum albumin (BSA), 1.5% Cellulase Onozuka R-10, and 0.4% Macerozyme R-10. To prepare Enzyme solution, warm the solution at 55°C without $CaCl_2$ and BSA, to inactivate nucleases and proteases. Cool to room temperature before adding $CaCl_2$ and BSA. Prepare Enzyme solution the same day protoplast transformation is performed.
15. Wash solution: 2 mM MES, pH 5.7, 150 mM NaCl, 125 mM $CaCl_2$, and 5 mM KCl.

16. MaMg solution: 0.4 M mannitol, 15 mM $MgCl_2$, and 5 mM MES, pH 5.7.
17. Light microscope.
18. Hemocytometer.

2.2. Preparation of Protoplasts from Arabidopsis Leaves

1. *Arabidopsis* plants: We have obtained satisfactory results with *A. thaliana* ecotype Col-0. Sufficient yield of protoplasts for transfection largely depends on growth conditions of plants (*see* **Note 2**).
2. Razor blades.
3. Desiccator with a vacuum.
4. Nylon mesh (75 μm).
5. Petri dishes (100 × 25 mm).
6. Same materials as in steps 6–18 in **Section 2.1**.

2.3. PEG-Mediated Transformation

1. 2-mL round-bottomed tube.
2. 6- or 12-well plate.
3. Plasmids (2 μg/μL): Plasmids should be of high purity (*see* **Note 3**).
4. Polyethylene glycol (PEG) 4000.
5. PEG solution: 0.4 M mannitol, 40% (w/v) PEG 4000, and 0.1 M $CaCl_2$. PEG solution needs to be warmed up to 37°C to completely dissolve PEG 4000. Prepare PEG solution the same day that protoplast transformation is performed (*see* **Note 4**).
6. Wash solution: same as step 15 in **Section 2.1**.
7. Protoplast culture solution: 0.5 M mannitol, 20 mM KCl, and 4 mM MES, pH 5.7.

2.4. Dual-Luciferase Reporter Assay

1. Luminometer.
2. Dual-luciferase reporter assay system (Promega).
3. Prepare 1X Passive Lysis Buffer and Stop & Glo reagent as described in the manufacturer's instructions. These are best when prepared just prior to use.

3. Methods

Protoplasts can be easily prepared from suspension cultured cells by collecting cells by filtration and incubating them in an enzyme solution that digests cell walls. Protoplasts can also be obtained from leaves of soil-grown plants. Although somewhat laborious and time consuming, an advantage of using leaves as a source of protoplasts is that one can take advantage of the available collection of *Arabidopsis* mutants defective in a gene of interest. We provide protocols using both suspension cells (**Section 3.1**) and leaves (**Section 3.2**) as a source of protoplasts. The same protocols for PEG-mediated transformation (**Section 3.3**) and reporter assay (**Section 3.4**) can be utilized regardless of the source of protoplasts.

3.1. Preparation of Protoplasts from Arabidopsis Suspension Cells

3.1.1. Maintenance of Arabidopsis Suspension Cells

1. Incubate the suspension cell line MM2d at 25°C in the dark with shaking at a speed of approximately 130 rpm.
2. Subculture suspension cells once a week by adding 2.5 mL of 7-day-old cells into 50 mL of modified LS medium in 250 mL flasks (*see* **Note 5**).
3. Suspension cells at 4 days or longer following subculture are suitable for protoplast preparation/transformation (*see* **Note 6**).

3.1.2. Protoplast Preparation

4. Collect suspension cells by filtration using a funnel and filter paper with an aspirator (*see* **Note 7**).
5. Add suspension cells to 25 mL of Enzyme solution in a 50-mL conical tube using a spatula. Up to 2 g of suspension cells can be used (*see* **Note 8**).
6. Incubate suspension cells in Enzyme solution at room temperature for 2 h with gentle shaking (*see* **Note 4**). The temperature can be increased up to 30°C to facilitate enzyme digestion.
7. Check round protoplasts under the light microscope (**Fig. 6.2**). Incubate an additional 30–60 min if enzyme treatment is not sufficient.
8. Pass protoplasts through two layers of Miracloth.

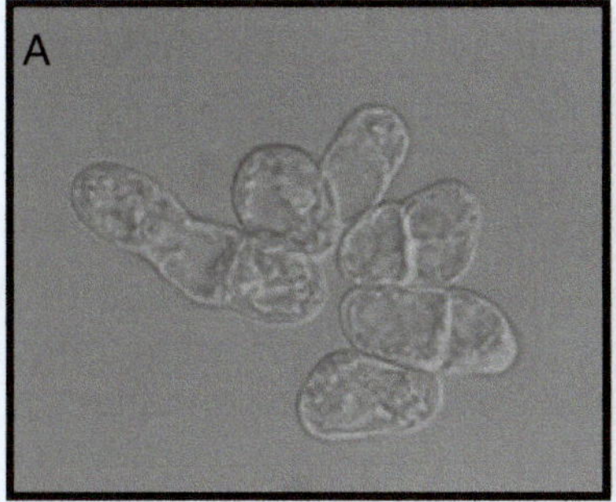

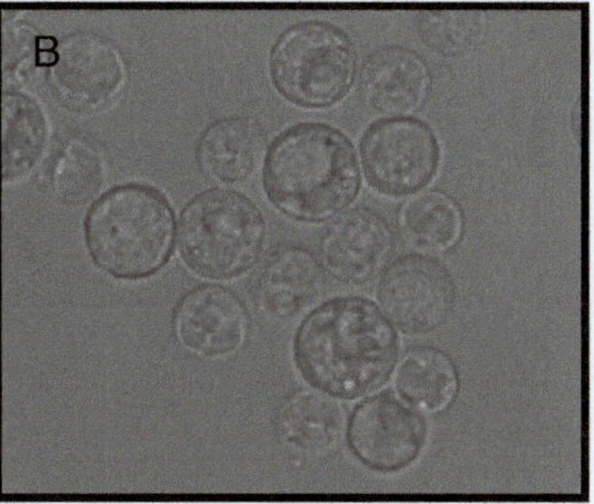

Fig. 6.2. Protoplasts prepared from the *Arabidopsis* cell line MM2d. Shown are MM2d cells before (**a**) and after (**b**) treatment with Enzyme solution.

9. Centrifuge at 100×*g* for 5 min with a swing rotor (*see* **Note 9**).
10. Discard the supernatant using a pipette.
11. Gently add 25 mL of Wash solution and resuspend protoplast pellets. Gentle inverting should be enough to resuspend the protoplast pellets.
12. Centrifuge at 100×*g* for 5 min.
13. Discard the supernatant using a pipette.
14. Repeat steps 11–13.
15. Resuspend protoplasts with MaMg solution to make a density of 2–5 × 10^5 cells/mL (*see* **Note 10**). Keep on ice. Proceed to **Section 3.3**.

3.2. Preparation of Protoplasts from Arabidopsis Leaves

1. Grow *Arabidopsis* plants for 3–4 weeks in sterile vermiculite or appropriate soil in a growth room or chamber (*see* **Note 2**).
2. Collect well-expanded leaves from plants before flowering (*see* **Note 11**).
3. Cut 0.5-mm leaf strips from the middle part of a leaf using a razor blade on clean sheets of Xerox copy paper and submerge in Enzyme solution in a petri dish (*see* **Note 12**). Razor blades may get blunt and should be replaced every 10–15 leaves.
4. Cover the petri dish and apply vacuum in the dark using a desiccator for 30 min.
5. Incubate leaf strips in Enzyme solution at room temperature in the dark for 2.5–3 h without vacuum or shaking (*see* **Note 13**).
6. Release the protoplasts by gently shaking the petri dish.
7. Check for round protoplasts under the light microscope.
8. Add an equal volume of Wash solution and mix gently.
9. Transfer the Enzyme solution containing protoplasts from the petri dish to a 50-mL conical tube through a 75-μm nylon mesh to remove undigested leaves and cell debris.
10. Centrifuge at 100×*g* for 2 min to pellet protoplasts (*see* **Note 9**). Remove the supernatant using a pipette.
11. Resuspend the protoplasts with Wash solution to a density of 2 × 10^5 cells/mL after counting cells under the microscope using a hemocytometer.
12. Keep on ice for 30 min.

13. Remove Wash solution with a pipette and resuspend protoplasts (that will have settled during the 30-min incubation on ice) with MaMg solution kept at room temperature (*see* **Note 14**). Proceed to **Section 3.3**.

3.3. PEG-Mediated Transformation

1. Place 10 μg each (2 μg/μL, 5 μL) of an effector and reporter plasmid and 2 μg (2 μg/μL, 1 μL) of a reference plasmid in a 2-mL tube (*see* **Note 15**).
2. Add 100 μL of protoplast solution, and then add 100 μL of PEG solution. Mix by gently tapping tube several times (*see* **Note 16**).
3. Keep on ice for 10–20 min.
4. Add 400 μL of Wash solution to the protoplast–plasmid–PEG solution mixture. Mix by gently inverting several times (*see* **Note 17**).
5. Centrifuge at 100×*g* for 5 min.
6. Discard supernatant. Resuspend protoplasts gently with Protoplast culture solution by pipetting (*see* **Note 18**).
7. Transfer protoplasts to a 6- or 12-well plate. Incubate at 22–25°C for 12–20 h.

3.4. Dual-Luciferase Reporter Assay

1. Transfer protoplasts to a 2-mL tube using a pipette and centrifuge at 100×*g* for 10 min to collect protoplasts.
2. Discard the supernatant.
3. Add 100 μL of 1X Passive Lysis Buffer (supplied by dual-luciferase reporter assay system) to the pellet.
4. Vortex for 10 s to disrupt protoplasts (*see* **Note 19**).
5. Centrifuge at 10,000×*g* for 2 min.
6. Transfer supernatant to a new tube. Keep on ice until dual-luciferase reporter assay measurement (*see* **Note 20**).
7. Perform dual-luciferase reporter assay according to the manufacturer's instructions. Briefly, dispense 100 μL of Luciferase Assay Reagent II (LAR II), which contains firefly Luc substrate, into the appropriate number of luminometer tubes. Add up to 20 μL of cell lysate to a tube containing LAR II, mix by pipetting two or three times, and initiate reading. Add 100 μL of Stop & Glo reagent, which quenches the firefly Luc reaction and contains R-Luc substrate, vortex briefly, and initiate reading again. Proceed to the next measurement. As an example, analysis of the *Arabidopsis* transcription factor AtbZIP60 and its target *BiP3* promoter is shown in **Fig. 6.3**.

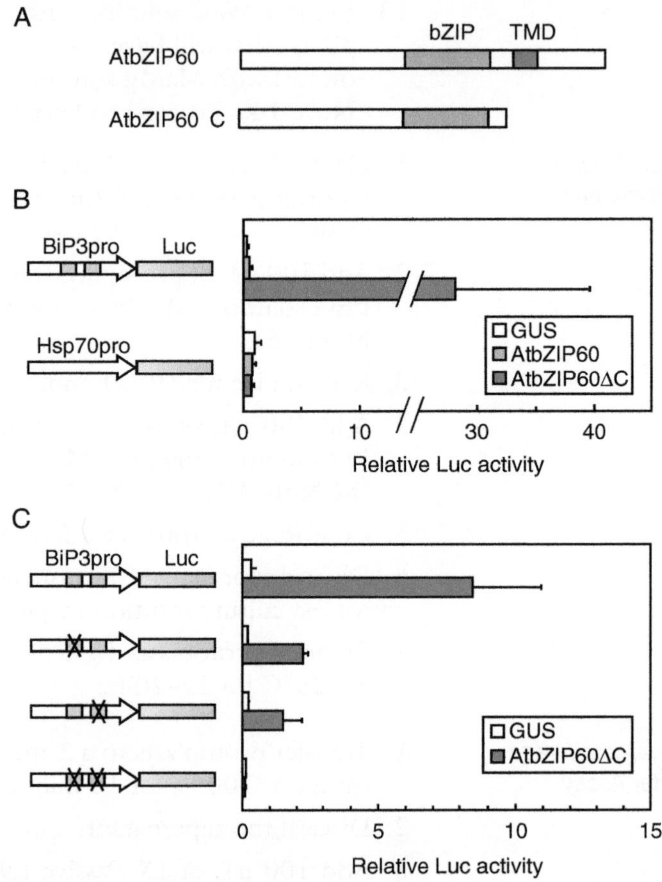

Fig. 6.3. Activation of the *BiP3* promoter by the *Arabidopsis* transcription factor AtbZIP60. (**a**) Schematic representation of the domain structure of AtbZIP60, an *Arabidopsis* transcription factor used for effector constructs in the experiment. AtbZIP60 is synthesized as a membrane-bound form with its transmembrane domain (TMD). AtbZIP60 is activated by possibly proteolysis near or within the TMD, allowing the cytosolic domain that contains a basic leucine zipper (bZIP) domain to translocate into the nucleus where it activates transcription of a number of genes including *BiP3* (9, 10). (**b**) Effector plasmids expressing either a precursor or active forms of AtbZIP60 (AtbZIP60 or AtbZIP60△C) were co-transfected with the *BiP3* promoter–*Luc* reporter plasmid and a *R-Luc* reference plasmid. The results show that the active form of AtbZIP60 (AtbZIP60△C) activated the *BiP3* promoter whereas the full-length AtbZIP60 did not. *Hsp70* promoter and GUS were used as a control reporter and effector plasmid, respectively. Relative Luc activity represents activity relative to that of constructs with the *BiP3* promoter and *GUS* (modified from (9) with Copyright (2005) National Academy of Sciences, USA). (**c**) Effect of mutation in the *BiP3* promoter on activation of the *BiP3* promoter by AtbZIP60△C. Mutations were introduced into two copies of a *cis*-acting element present in the *BiP3* promoter. PEG-mediated protoplast transformation and dual-luciferase assay were carried out as in (**b**). Mutation in either one of two elements reduced the activation of the *BiP3* promoter by AtbZIP60△C, and mutation in both abolished the activation completely (modified from (11) with permission from Japan Society for Bioscience, Biotechnology, and Agrochemistry).

4. Notes

1. *Arabidopsis* suspension cells other than MM2d can be used. In that case, one may need to change the subculture protocol accordingly.
2. We usually grow *Arabidopsis* in 16 h light/8 h dark conditions at 22–23°C. The growth condition of plants is important for experimental reproducibility.
3. Plasmids should be of good quality. We have obtained transfection-grade plasmids by using Wizard SV Midiprep Kit (Promega) or Plasmid Midiprep Kit (Qiagen).
4. We usually make PEG solution while suspension cells or leaf strips are incubating in Enzyme solution because it takes approximately 1 h to dissolve the PEG. We recommend that PEG solution be warmed up to 37°C to dissolve PEG completely.
5. Although the original protocol maintains MM2d cells with different culture medium at 27°C (8), the protocol described here also works. If a 25°C rotary shaker is not available, suspension cells can be grown at a lower temperature (e.g., 22–23°C) commonly used for growing *Arabidopsis* plants. In that case, culture conditions may need to be optimized (i.e., change the subculture volume or subculture period).
6. We do not recommend using younger cultures (i.e., 1–3 days after subculture) because they tend to be fragile during the protoplast isolation/transformation procedure.
7. Because the reporter assay can be completed within 24 h, it is not necessary to handle protoplasts under sterile conditions.
8. Using too large an amount of suspension cells decreases the efficiency of enzyme digestion and therefore results in low protoplast recovery. In our experience, 50 mL of 4-day-old MM2d culture contains 2–3 g of MM2d cells. One gram of fresh weight of MM2d cells produces $0.5–2 \times 10^7$ protoplasts, which is enough for at least 100 transformations.
9. Use "no brake" or "minimum deceleration" function of a centrifuge to avoid the protoplast pellets being disturbed.
10. A rough estimation is that 0.1 mL packed volume of protoplast pellets corresponds to 5×10^6 cells. First dissolve protoplast pellets in a smaller volume of MaMg solution than the volume that would be expected to give the desired density of protoplasts. Count cells using a hemocytometer, and then add the appropriate volume of MaMg solution to bring to the desired density.

11. Using healthy leaves at the proper developmental stage is very important for protoplast preparation. Well-expanded third and fourth pairs of leaves are recommended for the protoplast isolation. Avoid using curled leaves that may be too old or dried to obtain competent protoplasts. We have used 10 mL of Enzyme solution for digesting 10–20 leaves, which could yield approximately 0.5–1 $\times$ 10^6 protoplasts. Experiments can be scaled up or down as long as the ratio of the volume of Enzyme solution to the number of leaves is kept constant.
12. Immediate dipping and submerging of both sides of strips is critical for protoplast yield.
13. Enzyme solution will turn green after incubation. This indicates the release of protoplasts. The digestion time may vary depending on plant materials.
14. Add the same volume of MaMg solution that is used at step 11 of **Section 3.2** to maintain a density of 2 $\times$ 10^5 cells/mL.
15. It is useful to include a "no DNA" control (use water in place of plasmid solution) to check background for luciferase activity.
16. It is critical for the plasmid–protoplast–PEG solution to be mixed completely; otherwise transformation efficiency will be dramatically decreased.
17. Complete mixing of the protoplast mixture with Wash solution is critical for efficient transformation of protoplasts.
18. Use 1 mL (for 6-well plates) or 0.5 mL (for 12-well plates) of Protoplast culture solution.
19. Vortexing is sufficient to disrupt protoplasts.
20. Western blot analysis of cell extracts can be used to check the expression of a transcription factor encoded by an effector plasmid. If an antibody specific to the transcription factor of interest is not available, the transcription factor can be tagged with an epitope tag (e.g., FLAG, HA, and c-Myc) and detected using the corresponding antibody.

References

1. Gallagher, S. R. (1992) GUS Protocols: Using the GUS Gene as a Reporter of Gene Expression. Academic, Boston, MA.
2. Ow, D. W., DE Wet, J. R., Helinski, D. R., Howell, S. H., Wood, K. V., and Deluca, M. (1986) Transient and stable expression of the firefly luciferase gene in plant cells and transgenic plants. *Science* **234**, 856–859.
3. Grentzmann, G., Ingram, J. A., Kelly, P. J., Gesteland, R. F., and Atkins, J. F. (1998) A dual-luciferase reporter system for studying recoding signals. *RNA* **4**, 479–486.

4. Yang, Y., Li, R., and Qi, M. (2000) In vivo analysis of plant promoters and transcription factors by agroinfiltration of tobacco leaves. *Plant J.* **22**, 543–551.
5. Ueki, S., Lacroix, B., Krichevsky, A., Lazarowitz, S. G., and Citovsky, V. (2009) Functional transient genetic transformation of Arabidopsis leaves by biolistic bombardment. *Nat. Protoc.* **4**, 71–77.
6. Abel, S. and Theologis, A. (1994) Transient transformation of Arabidopsis leaf protoplasts: a versatile experimental system to study gene expression. *Plant J.* **5**, 421–427.
7. Yoo, S. D., Cho, Y. H., and Sheen, J. (2007) Arabidopsis mesophyll protoplasts: a versatile cell system for transient gene expression analysis. *Nat. Protoc.* **2**, 1565–1572.
8. Menges, M. and Murray, J. A. (2002) Synchronous Arabidopsis suspension cultures for analysis of cell-cycle gene activity. *Plant J.* **30**, 203–212.
9. Iwata, Y. and Koizumi, N. (2005) An Arabidopsis transcription factor, AtbZIP60, regulates the endoplasmic reticulum stress response in a manner unique to plants. *Proc. Natl. Acad. Sci. USA* **102**, 5280–5285.
10. Iwata, Y., Fedoroff, N. V., and Koizumi, N. (2008) Arabidopsis bZIP60 is a proteolysis-activated transcription factor involved in the endoplasmic reticulum stress response. *Plant Cell* **20**, 3107–3121.
11. Iwata, Y., Yoneda, M., Yanagawa, Y., and Koizumi, N. (2009) Characteristics of the nuclear form of the Arabidopsis transcription factor AtbZIP60 during the endoplasmic reticulum stress response. *Biosci. Biotechnol. Biochem.* **73**, 865–869.

Chapter 7

Microarray-Based Identification of Transcription Factor Target Genes

Maartje Gorte, Anneke Horstman, Robert B. Page, Renze Heidstra, Arnold Stromberg, and Kim Boutilier

Abstract

Microarray analysis is widely used to identify transcriptional changes associated with genetic perturbation or signaling events. Here we describe its application in the identification of plant transcription factor target genes with emphasis on the design of suitable DNA constructs for controlling TF activity, the experimental setup, the statistical analysis of the microarray data, and the validation of target genes.

Key words: Microarray, transcription factor, target gene, cycloheximide, inducible gene expression systems, relative quantification, quantitative real-time RT-PCR, endogenous reference gene, amplification efficiency.

1. Introduction

Elucidating the signal transduction cascades activated by transcription factors (TFs) is an essential step toward understanding TF function. Analysis of loss- and gain-of-function mutants with altered phenotypes often provides the first clues to a TF's function; however, additional approaches are required to identify the specific gene expression cascades that lead to the observed phenotypes. One way to obtain insight into these signaling cascades is through transcriptional profiling. Transcription profiling can be applied to stable loss- and gain-of-function TF mutants to identify global expression changes that are associated with the mutant phenotype, thereby facilitating placement of the TF in a developmental pathway or process. Ultimately, one would also like to

L. Yuan, S.E. Perry (eds.), *Plant Transcription Factors*, Methods in Molecular Biology 754,
DOI 10.1007/978-1-61779-154-3_7, © Springer Science+Business Media, LLC 2011

know the direct targets activated by TF binding to be able to distinguish between primary targets and secondary downstream signaling events. A number of techniques can be used to identify the primary targets of TF binding, including microarrays (1), high-throughput transcriptome sequencing (2), and chromatin immunoprecipitation (ChIP) (3), each with its associated advantages and disadvantages. While microarrays and mRNA sequencing provide information about direct and indirect transcriptional changes that take place upon TF binding, ChIP identifies TF-binding sites and does not provide any information about whether a target gene is expressed. Ultimately, a combined approach is required to identify gene expression changes and DNA-binding sites (4). Here we focus on microarray-based transcriptional profiling as an initial method to identify TF target genes due to its high-throughput nature, low cost, technical ease, and comparatively well-established framework for data analysis.

The chapter that follows is not a detailed step-by-step protocol for identification of TF target genes using microarray technology, but rather a compendium of points that need to be addressed when setting up the experiments and analyzing the data.

2. Materials

2.1. Construct Development

Measurement of the direct transcriptional response of a transcription factor (TF) requires control over its function in both time and space. Temporal regulation can be achieved by using a chemically inducible form of the TF, while spatial regulation can be achieved by using cell- or tissue-specific promoters. Construct design is an essential first step in the identification of TF targets, as the choice of both the promoter and the inducible system will influence the results and should therefore be adapted to answer the research question. Here we discuss several one- and two-component chemical induction systems that can be used to identify TF target genes, as well as approaches to ensure sufficiently high expression and/or activity of the TF of interest (TOI) for microarray analysis.

2.1.1. One-Component Chemical Induction Systems

The frequently used one-component inducible expression systems make use of a TOI fused in frame to a tag that ensures chemical inducibility. These systems commonly use (parts of) animal nuclear receptors (NRs), which are TFs that are translocated to the nucleus upon binding of hormones or other molecules [reviewed in (5)]. These NRs have a modular structure containing a DNA-binding domain (DBD) and a ligand-binding domain (LBD). In the absence of its ligand, the NR is bound by heat-shock proteins (HSPs) in the cytosol. However, when the ligand is present, the NR dissociates from the HSP complex and moves to the nucleus, where it binds to DNA and regulates gene transcription. This mechanism has been exploited to create chemically inducible LBD-TF chimeras in many different organisms.

A disadvantage of any one-component system is that fusion of additional protein sequence, such as an LBD, to a TF may negatively influence its activity or reduce stability. The functionality of the fusion protein must therefore be evaluated in advance, using N- and C-terminal fusions. A fusion protein can be considered functional if it either rescues a mutant phenotype or, in the absence of a mutant phenotype, generates the expected overexpression phenotype.

2.1.1.1. Glucocorticoid Receptor

The rat glucocorticoid receptor (GR) is an example of an NR that is widely used to create inducible TFs. The synthetic glucocorticoid dexamethasone (DEX) is commonly used as the inducing agent. The GR system has been used to identify targets of AGAMOUS and LEAFY during flower development (6, 7), and BABY BOOM during somatic embryo development (8). In this approach, the TOI is fused to the GR LBD. In the absence of DEX, the TOI:GR fusion protein is sequestered by HSPs in the cytosol (9), thereby preventing TOI activity. When DEX is added, the interaction with HSPs is disrupted, allowing the TOI to enter the nucleus and activate gene expression. DEX treatment does not appear to affect plant development or gene expression (10, 11).

The manner in which DEX is applied, the concentration of DEX used, and the amount of time required for TF target gene expression after DEX induction are all points that need to be considered when adjusting the system to a specific TOI. DEX can be applied by spraying or watering. When plants are watered with DEX, it is taken up via the roots and transported rapidly to the shoot (12). The spraying method requires the addition of soap to the medium to lower surface tension (12). Alternatively, when using seedlings, these can be grown in vitro on a nylon mesh on agar-containing Petri dishes, which facilitates the transfer of the seedlings to dishes containing induction medium supplemented with DEX (8, 13). The concentration of DEX that is used varies greatly between studies, but 5–20 μM DEX is commonly used to achieve full TF induction. One way to identify the optimal DEX concentration in your experimental system is to test the dose–response of the TOI-induced phenotype or complementation (14). The amount of time needed to induce nuclear localization and transcriptional activation by TFs has not been rigorously studied in plants and must therefore be assessed on a case-by-case basis. In one study, an APETALA3 (AP3):GR fusion protein was used to identify AP3 target genes in inflorescences, with downregulation of its target *APETALA1* being observed within

2 h of DEX induction (15). In another example, activation of a SHORT ROOT (SHR):GR fusion in the root led to upregulation of SHR target genes within 6 hours (13). Nuclear localization can also be confirmed using antibodies to the TOI and/or the GR LBD.

2.1.1.2. Estradiol Receptor

A similar approach that is also widely used in plants employs the LBD of the human estradiol receptor (ER) fused to the TOI. In this system, the chimeric TF is translocated to the nucleus in response to estradiol (18). The ER system has been used to identify the target genes of a number of plant TFs, including ENHANCER OF SHOOT REGENERATION 2 (ESR2) (16) and ARR2 (17).

As in the GR system, the manner in which estradiol is applied, the dosage, and the induction period need to be considered. Estradiol is efficiently taken up by seedlings, but its accumulation appears to be slower (18), as it is relatively immobile, i.e., it does not move from treated leaves to newly developing leaves, as is the case for DEX (19). Nevertheless, the system can be used if induction of the fusion protein gives the expected phenotype and generates a transcriptional response within several hours.

Estradiol treatment does not appear to affect plant development (18); however, it has been suggested that the ER system is unsuitable for plants species, such as soybean, that produce large quantities of phytoestrogens as these can activate the ER LBD (20). The estradiol concentration used to induce TF expression is in the range used for the GR-based system. In the study on ESR2 target genes, ESR2:ER was activated for 1 h with 10 μM estradiol before RNA was isolated from root explants for expression profiling (16).

2.1.2. Two-Component Chemical Induction Systems

A chemically inducible two-component system can be used to control the activity of a TF in cases where the tagged TOI is not functional. The first component in these systems is usually an artificial TF, consisting of an activation domain and a DNA-binding domain, which is made inducible through the addition of an NR LBD. Expression of the artificial transcription factor can be placed under control of any suitable promoter. The second component in these systems is a promoter that is specifically bound by the artificial TF and that is placed upstream of the gene encoding the TOI. Upon chemical induction, the artificial TF binds to the promoter and activates transcription of the TOI. This system is therefore similar to the one-component system, but with one major difference: in one-component systems, the TOI is present, but sequestered in an inactive form in the cytoplasm awaiting chemical induction, while in two-component systems, the TOI is transcribed and translated only after chemical induction of, and activation by, the artificial TF. This fundamental difference has implications for the identification of direct target genes in combination with the translational inhibitor cycloheximide (described in **Section 3.1.1**). A variety of two-component systems exist, including ones that use alcohol, tetracycline, or copper to control gene expression (21). Here we focus on NR-based systems, as only these have been extensively used to identify TF gene targets.

One example of an NR-based, two-component induction system that has been used to identify TF target genes is the GVG system. The first component of this system is a chimeric TF (GVG), consisting of the DNA-binding domain of the yeast *GAL4* TF, the herpes virus *VP16* transactivation domain, and a *GR* LBD (12). Expression of this artificial TF can be controlled by any promoter of choice. The second component is the TOI, which is transcribed from a promoter containing multiple copies of the GAL4 upstream activating sequence (UAS). Induction by DEX results in translocation of GVG to the nucleus where it activates transcription of the TOI. The kinetics of this system was studied using a luciferase (Luc)-based construct (12). *Luc* mRNA was first detected 1 h after DEX induction and reached a maximum level after 4 h.

The GVG system was used to identify targets of MYB46 and REPRESSOR OF ga1-3 (RGA), which are involved in secondary wall biosynthesis and GA signaling, respectively (22, 23). MYB46 mRNA was detected 30 min after spraying with 10 μM DEX and its targets were upregulated within 3–6 h of induction. The same treatment induced RGA protein synthesis within 1 h, while RGA target transcripts accumulated after 2–4 h.

It has been reported that activation of GVG itself can cause developmental growth defects in *Arabidopsis*, rice, and lotus (24–26). In *Arabidopsis*, high-level GVG expression was correlated with severe phenotypes (24). This effect appears to be due to the presence of a GR domain in the GVG protein, as overexpression of the GAL4–VP16 (GV) protein lacking the GR domain does not cause growth defects (27). For experimental purposes it may be sufficient to select lines that do not show growth defects; however, such a selection process may itself influence the inferences that are drawn.

An additional drawback of the GVG system is that expression of the transgene may be variable from generation to generation (27), or even lost over generations (28, 29). In tobacco and *Arabidopsis*, GVG-based expression could be restored by treatment with the cytosine methylation inhibitor 5-azacytidine, suggesting that the loss of expression is caused by methylation of the UAS sequence.

The second example of an inducible two-component system is the XVE system. It is comparable to the GVG system but uses a different artificial TF that is composed of the DNA-binding domain of the bacterial repressor LexA and the viral *VP16* transactivation domain and is fused to the ER LBD (18). Once estradiol is added, XVE activates the second component, consisting of multiple LexA operator sequences fused to the 35S minimal promoter, which is placed upstream of the gene encoding the TOI.

Zuo et al. (18) used a GFP reporter to evaluate the XVE system. GFP transcripts were first detected 30 min after estradiol treatment and reached a maximum after 24 h, suggesting that the XVE system is slower than the GVG system. The minimal concentration of estradiol required to induce the system was 8 nM and the response was saturated at 5 μM.

The XVE system was used to identify target genes of AGAMOUS-LIKE 24 (AGL24) (30), ENHANCER OF SHOOT REGENERATION 1 (ESR1) (31), OCTADECANOIDRESPONSIVE ARABIDOPSIS AP2/ERF 59 (ORA59) (32), and the NAC transcription factor ANAC092 (33). The expression of *AGL24*, *ESR1*, and *ANAC092* was induced using 10 μM estradiol, whereas the expression of *ORA59* was induced using 2 μM estradiol. The induction period ranged from 5 h for *ANAC092* to 24 h for *ESR1*.

2.1.3. Enhancing Transcription Factor Activity

An important point to consider when designing the construct is whether the TOI will induce gene expression changes at a sufficiently high level to be detected on a microarray. There are two aspects to consider here: the expression level of the TOI and the transcriptional activation or repression strength of the TOI.

When the endogenous promoter is too weakly expressed to allow microarray analysis, the TOI can be expressed at a higher level by placing it under the control of a semi-constitutive promoter, such as the cauliflower mosaic virus 35S promoter or the maize *ubiquitin* promoter (*ubi*). In fact, most studies make use of the 35S promoter, probably because its high expression level in a wide range of tissues provokes a strong transcriptional response. It is important to remember, however, that a TF may display ectopic functions upon ubiquitous expression compared to the wild-type situation, probably leading to identification of a different set of target genes. The choice of promoter will depend therefore on the TF that is being studied and on the biological question. Another way in which the expression level of the TOI can be increased is to add a translation enhancer to the DNA construct, although again, care must be taken to avoid generating overexpression phenotypes. Commonly used translation enhancers include the 5′ non-coding leader sequences of the tobacco mosaic virus genomic RNA, alfalfa mosaic virus RNA 4, and tobacco etch virus [reviewed in (34)]. Translation enhancers can also be used to compensate for reduced protein stability caused by protein tags (such as the fusions described in **Section 2.1.1**).

The native promoter was used to identify SHR targets (13). In this experiment, a *pSHR::SHR:GR* construct was used that also included a translational enhancer element from tobacco etch virus to counterbalance a possible decrease in protein function due to fusion of the GR domain. The experiment was performed in an *shr* mutant background to avoid generating overexpression phenotypes.

3. Methods

3.1. Microarray Experimental Setup

3.1.1. Direct Versus Indirect Target Genes

Analysis of the gene regulatory networks downstream of TFs requires identification of direct targets. However, induction of a TF will result in the activation of many downstream genes, some of which will be TFs that will induce expression of secondary targets. Microarray-based transcriptional profiling of TF-activated cells or tissues identifies gene expression changes but, unlike ChIP, does not differentiate between direct and indirect targets. Two ways are commonly used to enrich for direct targets in microarray analyses: a time course analysis and application of the translation inhibitor cycloheximide.

3.1.1.1. Time Course Analysis

A time course analysis gives insight into gene expression patterns, allows researchers to study and model gene dynamics and regulatory networks, and enables researchers to correlate early and late genes with their possible function. Generally, sampling of earlier time points will yield a set of genes that are enriched in direct targets, while later time points will identify gene sets that are both direct and indirect targets. It is important to decide not only on the duration of the time course but also on the developmental stage at which the analysis will take place. An additional consideration when complementing mutant lines is that the induction times for the TF:NR need to take place within the time frame of full complementation or ectopic phenotype formation.

An example of an extensive time course microarray analysis is the study by Wagner et al. (7), who treated *35S::LFY:GR* root explants with DEX for 0, 2, 4, 6, 8, 10, 12, and 22 h. Sixty-one genes were differentially expressed, most of which were upregulated. There was no full overlap between the targets identified in this experiment and previous studies, which the authors attribute to the difference in LFY activity in the different target tissues that were used. While the experimental setup does not differentiate between direct and indirect targets, the use of a time course allowed the authors to make regulation profiles for the target genes, retrieve slow-responding known targets, and identify novel quick-responding TFs as targets (7).

Another example is the study by Ko et al. on MYB46. This TF was previously shown to induce biosynthesis and deposition of cellulose, xylan, and lignin (35), important components of secondary cell wall thickening. The authors studied DEX-induced MYB46 expression in 2-week-old leaves at 0, 1, 3, and 6 h after DEX application. Of the 282 genes that were more than threefold upregulated after 6 h, 42 were TFs. Thirty-two of these TFs were already upregulated after 3 h. Because the authors used a two-component system, they determined putative direct targets based on the speed of regulation and later confirmed in vitro binding of MYB46 to their target promoters using EMSA (22).

3.1.1.2. Cycloheximide

In plant biology, cycloheximide (CHX) is often used together with an inducible TF:NR fusion to discriminate between direct and indirect target genes. CHX is a drug that inhibits protein synthesis by preventing the transfer of amino acids from aminoacyl-tRNA to nascent proteins (36). Transcriptional activation by the TF will occur in the presence of CHX, but subsequent translation of the primary target genes will be blocked, thereby preventing the activation of secondary or indirect, target genes. It is not possible to use cycloheximide with two-component systems that first require translation of the TOI (*see* **Section 2.1.2** above).

Two important issues to consider when using CHX are the concentration and the length of application. On one hand, failure to completely block protein synthesis will contaminate the candidate direct target gene set with secondarily-induced genes, while completely blocking protein synthesis for too long, while other cellular processes continue, will eventually cause cell death. In general, a concentration of 10 μM is used in liquid culture systems [e.g., (7)]. However Lee et al. (37) used concentrations up to 50 μM to treat plants grown on agar plates because of the relatively small contact surface between the tissue and the medium. CHX pre-incubation times of 30 min prior to DEX induction have been used (38), as have induction times of up to 8 h in the presence of DEX (8). Reversibility of the CHX treatment can be used to test for cell-lethal effects of CHX during the sampling period. Cell cultures or seedlings can be grown in the presence of CHX for the desired time and then transferred to CHX-free medium and assayed for recovery (8).

Even a non-toxic CHX dose will affect the transcriptional profile of a cell. We have observed that 10 μM CHX alters the mRNA levels of roughly one-third of the *Arabidopsis* genome. Many short-lived proteins that repress gene expression, like the PS-IAA4/5-related *Arabidopsis* early response Aux/IAAs, need to be constantly supplied through de novo protein synthesis (39, 40). CHX is also known to stabilize certain short-lived mRNAs (41). Nevertheless, even though CHX causes genome-wide effects on gene expression, our observations indicate that array data from CHX treatments cluster together. CHX-treated samples are therefore different from non-CHX-treated samples in a specific, non-random way that enables researchers to discriminate between genes regulated only by CHX and genes regulated by both CHX and the TOI. Given that CHX treatment induces significant changes in gene expression, it is essential to include control samples in which CHX is applied, but the TOI is not induced.

To summarize, when using an inducible TF:NR fusion in combination with CHX and time course experiments, the ideal situation would be to perform the fully replicated analysis, with and without CHX. This allows for separation of early and late direct targets, and early and late indirect targets, and therefore aids in generating hypotheses concerning downstream effects (42).

3.1.2. Enrichment

Identification of TF target genes becomes more difficult when the TOI is expressed in a limited number of cells (dilution problem) or when it is expressed in many different cell types (specificity problem). A number of methods exist for identification of TF targets in specific cell types, provided the cells of interest can be isolated or marked. An excellent method for purifying fluorescently marked cells or tissues is fluorescence-activated cell sorting (FACS). FACS is a type of flow cytometry, where cells are sorted into two or more containers based on the amount and type of laser scattering and fluorescence (43).

Plant lines that express a fluorescent protein (FP) can be used for sorting but first require tissue dissociation into single cells (protoplasts) by enzymatic digestion of their cell walls (44). The power of this approach is exemplified by the transcriptional profiling performed for individual tissues in both the Arabidopsis root and shoot (45–47). The transcriptional identity of protoplasts appears rather stable but can be monitored as a quality control (48). In addition, because of the limited amount of cells, amplification of the RNA population is required before analysis on microarrays.

Cells can be marked either by a promoter::FP reporter or directly by the expression of the TOI translationally fused to FP. Plants carrying the TOI:FP fusion protein can be used both to sort the specific cells and to study the transcriptional consequences of expressing the protein in these cells, while the corresponding promoter::FP reporter can be used as a control to sort wild-type cells. For example, Levesque et al. (13) sorted the epidermis and lateral root cap cell types ectopically expressing the SHR:GFP fusion protein from the *WEREWOLF* (*WER*) promoter. A *WER::GFP* line was used to sort the same cell types as a control sample. This approach allowed the authors to focus specifically on the transcriptome of tissues that were previously shown to respond to ectopic SHR activation.

In a slightly different setup, this method can be used to identify direct TF targets. A TF:NR fusion protein can be expressed in a specific tissue, while a promoter::FP transcriptional fusion can be used simultaneously to mark the cells of interest. Application of the NR inducer and CHX followed by cell sorting and microarray hybridization of labeled transcripts will identify regulated genes when compared to the mock-treated control sample. Although technically feasible, this particular approach has not yet been used to identify plant TF direct targets.

3.2. Data Validation

The result of a microarray data analysis is a long list of candidate target genes that are significantly up- or downregulated to different extents under the chosen experimental conditions. The next challenge is to verify the differential expression of the identified genes using an independent technique, a step referred to as validation. Here we discuss quantitative real-time RT-PCR (qRT-PCR) as a first step to validating the expression profiles of the significantly regulated genes observed in the microarray analysis. A brief description of ChIP-based techniques, which can be used after initial validation by qRT-PCR to verify TF binding to the promoter of the identified target genes, is also included.

3.2.1. Quantitative Real-Time RT-PCR

qRT-PCR is a method of quantifying transcript abundances that is often used to confirm microarray results (49–51). The primary strengths of qRT-PCR are its unprecedented sensitivity and large dynamic range, which make it particularly well suited for quantifying low-abundance transcripts and transcripts that vary widely in abundance between groups of interest (52–54). However, despite these attractive features, there are a number of difficulties associated with generating high-quality qRT-PCR data. A number of previous reviews have described in detail the issues surrounding generating sound qRT-PCR data including the following: ensuring that RNA is of a sufficient quality and purity (55, 56), the pros and cons of the various approaches to generating cDNA via reverse transcription (55, 57), the importance of proper sample storage (56), the need for careful primer design and assay validation (56), and the advantages and drawbacks of different detection chemistries (55, 58).

With respect to biological materials, there are three general points to consider: the samples used to validate the microarray data, the endogenous reference gene (ERG) used to normalize the data (*see* also **Section 3.3.2** for a detailed discussion), and the primer design. Strictly speaking, for validation purposes, qRT-PCR only needs to be performed on the same RNA samples that were used for the microarray analysis. Nonetheless, qRT-PCR analysis of candidate target genes using a new set of biological replicates is an important step that can be used to determine the generality of the results (50). For greatest accuracy, the expression level of the ERG should be similar to that of the gene of interest (GOI) so that distortions that occur at different stages in the PCR are comparable (52). The ERG should also be stably expressed in the tissues and conditions under study. So-called housekeeping genes defined in the pre-genomic era, e.g., *ACTIN* and *UBIQUITIN*, were thought to be ubiquitous or invariant in their expression. However, recent studies have shown that the expression of classic plant ERGs is too variable to be used in qRT-PCR experiments, and as a result, more stable ERGs have been identified (59, 60). Finally, qRT-PCR primers should be designed to be gene-specific and their standard curves and efficiency should be determined. Mispriming, primer–dimer formation (52), and amplification bias (50) introduce inaccuracies in the reaction that will be amplified exponentially (61).

Even when qRT-PCR experiments are robustly designed, the results may not corroborate the microarray data. Whereas the qualitative validation (i.e., up- or downregulation) usually correlates well, quantitative correlation (i.e., similar fold change estimates between the two methods) may be lacking. These differences are often due to inherent differences between the qRT-PCR and microarray methodologies. For example, candidate GOIs showing very low levels of expression are more difficult to validate with qRT-PCR (61, 62). Nonetheless, many potentially interesting GOIs will show low fold changes. The researcher should consider whether low fold changes can be explained by other factors (e.g., restricted expression of the candidate gene in the tissue sample) and whether they can be handled appropriately in subsequent steps of experimentation.

3.2.2. Chromatin Immunoprecipitation

ChIP allows researchers to determine TF-binding sites by pulling down only those DNA fragments that are bound by the TOI. In the plant research community, complementing or overexpressing transgenic lines (63), and antibodies against a protein tag, such as FP or GR (13, 64) are commonly used, in contrast to a specific antibody against the TOI, which is the standard in other fields (65).

An important caveat of ChIP is that it only validates TF binding, not TF functionality – binding does not imply that the expression of the gene is regulated by this TF in this particular assay or time frame (2). A combined approach using microarrays and ChIP can be used to identify a TF's direct target genes with confidence (4). Within the context of microarrays, ChIP is used to validate whether a certain target gene is directly or indirectly regulated. ChIP can be combined with PCR of selected target genes, tiling microarrays or next-generation parallel sequencing to determine TF binding at a genome-wide level (66). For details on the procedure, controls, and considerations for ChIP, see **Chapters 16** and **17**.

3.3. Statistical Analysis

3.3.1. Microarray Data

In the plant sciences community, the most commonly used microarray platforms for TF target gene identification are single-channel, whole-genome printed oligonucleotide arrays and these will be the focus of our discussion. However, most of what is said here also applies to two-channel spotted arrays carrying oligonucleotides, cDNAs, or PCR fragments. This section summarizes part of Fu et al. (67).

In a non-microarray experiment, sample size is typically determined by estimating variability in the data, determining the difference you want to detect, and deciding on alpha, the probability of a false positive and beta, and the probability of a false negative. In a microarray experiment, which is really many thousands of individual experiments all done at the same time, the variability in expression will be different for each gene, and a traditional power analysis would result in thousands of different sample sizes. Genes that have similar, but not identical, expression in two groups would require very large sample sizes to detect the minor difference, while genes with dramatic differences can be detected with very small sample sizes. That said, here are some general guidelines. One array per group does not typically allow any assessment of variability and is therefore rarely useful. Two arrays per group are typically used for pilot data to see which genes change the most. Although some studies have been published using two arrays per group, in our experience, this is rare. Using three arrays per group is generally the minimum for publication and is reasonable for samples with similar genetic backgrounds where between-subject variability is small. Studies where variability is larger benefit from at least five to ten subjects per group. Obviously, a larger sample size is always better.

The use of replicates in microarray experiments is under constant debate. Technical replicates use the same mRNA sample on multiple chips. They are useful for establishing the reliability of the platform, but they cannot be used to increase the sample size for statistical calculations. Biological replicates use different mRNA samples on each chip and contribute to the overall statistical sample size for the experiment. In general, for professionally produced microarrays, technical replicates are not useful, as the reliability of the platform has already been well established.

A related issue is the use of pooling, which means putting more than one mRNA sample on each microarray. Pooling reduces individual variability and thus increases power, but at the price of not being able to use individual covariates in the statistical model. When pooling, it is essential to extract RNA from every sample and then combine equal amounts of RNA from each sample to go on each chip. Pooling is discussed extensively by Peng et al. (68).

Once the RNA has been appropriately extracted, hybridized to chips, and scanned, it is necessary to normalize the chips so that data between chips can be "fairly" compared. Although plenty of options are available for chip normalization, the most common are MAS 5 from Affymetrix and gcRMA from Bioconductor (www.bioconductor.org). MAS 5 is the easiest to use and will be the focus of this discussion. gcRMA will typically yield similar results, unless gene expression is very low. In that case, gcRMA is likely better than MAS 5.

Using MAS 5 results in an output file for each chip that contains the probe set ID, the probe set expression, the presence/absence call, and the presence/absence *p* value. The presence/absence *p* value is used to declare each probe set either "P" for present ($p < 0.05$), "A" for absent ($p > 0.065$), or "M" for marginal ($0.05 < p < 0.065$). The *p* value cutoffs for each label are adjustable. Technically, the assumptions of independence that the statistical test makes are not satisfied, but the P/A call is still useful as we will see below.

The next step in the statistical analysis is data reduction. If there are probe sets that are not of interest, then statistical calculations should not be done on those probe sets. On most, if not all, Affymetrix chips, the first approximately 50 probe sets are quality control probe sets used by the MAS 5 software. Typically, there is no need to do statistics on these probe sets. Another group of probe sets that are typically removed is probe sets that are labeled as absent (A) on all the chips in the experiment. If the P/A call determines that the probe set is not expressed on any chip in the experiment, there is no reason to do statistical analysis on that probe set. At this stage, researchers should also identify any subsets of the probe sets that are of particular interest. These subgroups can be statistically analyzed separately as well as together with the rest of the probe sets.

The final step prior to statistical analysis of the data is to decide whether or not to take the log of the data. For most microarray datasets, the probe sets with larger expression levels benefit from a log transformation, but the smaller expression levels should not be logged. Most researchers choose to log their data, but many do not. Typically, for one color microarrays, there is not much difference in gene lists with or without logging the data.

The end result of the statistical analysis will be one or more *p* values for each probe set. The overall *p* value for each row tests whether there are any statistical differences between the rows. A histogram of these *p* values provides useful information. If that histogram looks like that of a uniform distribution (a rectangle), then there may be little, if any differences, between the treatment groups. On the other hand, a histogram with a large peak for low p values indicates that large differences exist between the treatment groups. Histograms with a low or moderate peak for small p values indicate that more chips would likely result in smaller p values for probe sets that are actually differentially expressed.

The next decision is how to determine the list of probe sets that have changed. Traditionally, a p value less than 0.05 rejects the null hypothesis of no change. In a microarray experiment in which say 10,000 tests are done, using a p value cutoff of 0.05 could mean as many as $0.05 \times 10,000 = 500$ false positives. The false discovery rate (FDR) of Benjamini and Hochberg (69) chooses the cutoff by a user-specified expected proportion of false positives on this list. Ten or twenty percent are common choices. For experimental conditions that cause differential expression, but no large changes, the FDR method may not find any genes that change. As an alternative to using $p = 0.05$ or the FDR method, many researchers simply use $p = 0.01$ as the cutoff.

Once the overall p values are used to determine the list of genes that changed, many researchers attempt to use cluster analysis to determine genes that are responding similarly to the experimental conditions. To avoid excess noise in the gene clusters, be sure to cluster only genes that are determined by statistical methods to be differentially expressed. Many different types of cluster analyses are possible, and they often yield results that are hard to interpret. Statistical pattern matching [e.g. (70)] is an alternative that can be used to divide the list into sub-lists of genes that changed similarly. For example, if two sample t-tests were used to generate the overall p values, then the list should be sorted into upregulated and downregulated genes. For more complicated experimental designs, the patterns will be more complicated. Consult with a statistician to determine appropriate patterns.

The biological interpretation of the resulting list(s) is done by first annotating the gene lists using the manufacturer's Web site. There may be obvious biological conclusions that can be made at this point. A more statistical approach is to provide the list plus a larger list, say the entire chip, to a statistical software package that statistically determines gene ontology categories that are over-represented on the smaller list compared to the larger list.

3.3.2. Quantitative Real-Time RT-PCR Data

Compared to the technical issues discussed in **Section 3.2.1**, little effort has been made reviewing the issues associated with the statistical analysis of qRT-PCR data. Nevertheless, it is now clear that applying objective statistical methods to qRT-PCR data poses special challenges (53, 54, 71–73), and over the past decade, a number of approaches to analyzing qRT-PCR data have been suggested in the literature. We briefly describe the nature of the numerical data generated via most relative expression qRT-PCR experiments and then discuss issues that researchers who seek to draw valid statistical inferences from such data are likely to face.

3.3.2.1. The Quantification Cycle: The Central Value of qRT-PCR

The strategy underlying qRT-PCR is to record the accumulation of fluorescent dyes that label a specific nucleic acid product throughout the course of a PCR. The amount of product yielded by a PCR approximates a logistic (S-shaped) curve when it is plotted as a function of the number of reaction cycles completed. Thus, setting a threshold within the exponential phase of the amplification curve and recording the number of fractional cycles required to eclipse this threshold provide a correlate to the initial amount of template, known as the quantification cycle (C_q; lower C_q values correspond to more starting template). However, while C_q is the value of interest in the majority of qRT-PCR experiments, its determination requires exclusion of ground phase cycles (also known as the background or the baseline) as well as determination of where along the *y*-axis, within the exponential phase, the threshold should be placed. Determination of the baseline and the threshold is usually handled by proprietary software that comes with real-time PCR hardware. Because this section focuses on how to analyze C_q values rather than how to ensure that they are valid, we do not discuss baseline and threshold determination further and refer interested readers to (55) for discussions of when to adjust the baseline and threshold manually.

3.3.2.2. The Relative Expression Quantification Strategy

There are two general approaches to conducting qRT-PCR experiments. The first, known as absolute quantification, is based on calibration to a standard curve generated from a known external source (e.g., recombinant DNA) that enables one to express data in terms of transcripts per biological unit (e.g., copies per micrograms of tissue). The second, known as relative quantification, seeks to describe expression in arbitrary units that are based on comparisons to a calibrator sample or a series of calibrator samples (e.g., RNA isolated from control or un-manipulated sources). Because the relative quantification approach makes fewer assumptions, is less labor intensive, and is sufficient for most applications [*see* (53)], it is the method most frequently used in basic research.

The traditional, and still broadly used, approach to relative expression qRT-PCR is to enter the relevant C_q values, or their averages, into one of a number of mathematical models that generate relative expression ratios (R_E) describing expression in non-calibrator samples in terms of fold change relative to calibrator samples (53). Usually, R_E is normalized to one or more ERGs (53) because, in principle, this approach enables one to correct for variations in the amount and/or the quality of starting template that are introduced during upstream phases of the workflow. The simplest and most widely used approach to calculating R_E

was proposed by Livak and Schmittgen (74) and is known as the $2^{-\Delta\Delta Ct}$ method.

Although the $2^{-\Delta\Delta Ct}$ method is popular due to its simplicity and ease of calculation, it is valid only if a number of implicit assumptions are met. First and foremost, the expression level of the ERG must be invariant across the groups being considered. Second, there must be a doubling of the reaction product following every cycle. Finally, the reaction efficiencies (E; varies between 1 and 2) must be equal among all reactions that go into the calculation of R_E. In practice, any combination of these assumptions can be violated with the end result being an inaccurate estimate of R_E, spurious statistical significance, or both (53, 60, 73).

3.3.2.3. Addressing Assumptions That Are Likely to be Violated

Perhaps the most critical assumption of all relative expression qRT-PCR analyses is that the ERG used for normalization is invariant across all of the treatments/categories being considered. The reason for this is that variation in the ERG will bias estimates of R_E and may give misleading results (53, 60). As mentioned in **Section 3.2.1**, it is unlikely that any genes are universally suitable as ERGs across all tissues and research paradigms, and it is therefore important to verify that ERGs are invariant each time one wishes to investigate a new experimental system or tissue (57, 58, 60). However, ERG validation is often challenging and can become a circular problem [*see* (75, 76)].

Several software packages are available for assessing ERG stability. A widely used approach to normalization put forth by Vandesompele et al. (75), and implemented in the geNorm software package, is to assume that candidate ERGs are not co-regulated and compute metrics of stability for each of these candidates based on pair-wise calculations between the candidates and a set of RNA samples corresponding to the groups to be compared. These metrics are in turn used to arrive at a subset of candidate ERGs from which a normalization factor based on the geometric mean of this subset is calculated. Another strategy put forth by Andersen et al. (76) and implemented by the NormFinder software package uses a model-based approach to arrive at estimates of intragroup and intergroup variations in gene expression for candidate ERGs. In turn, these estimates are used to derive "stability values" (i.e., metrics) for each candidate that enable researchers to identify the candidate ERGs with the lowest intergroup and intragroup variations.

Another assumption of the $2^{-\Delta\Delta Ct}$ model that is likely to be violated is the assumption of 100% reaction efficiency [i.e., $E = 2$; *see* (73)]. Implicit to this assumption is the additional assumption that the E values that go into the calculation of R_E are all equal. Thus, while situations in which E differs from two but is more or less equal among the reactions used to calculate R_E will result in inaccurate estimates of R_E, they are not likely to result in erroneous inferences about differences between groups. However, cases in which the E values of the reactions used to calculate R_E are qualitatively different will lead to poor estimates of R_E and may lead to erroneous inferences about differences between groups. Models that incorporate the concept of E into the calculation of R_E have been put forward by Pfaffl (77) and Hellemans et al. (78). However, it is important to note that while these models account for differences in E between the GOI and the ERG (i.e., allow for gene-specific efficiencies), they assume that the efficiencies of the GOI and ERG(s) do not vary between calibrator and non-calibrator samples (i.e., they do not allow for group/treatment-specific efficiencies). Nevertheless, it is advisable to investigate the validity of this assumption, and methods for doing so have been presented by Burns et al. (79) and Yuan et al. (72, 73).

The introduction of E into relative quantification models means that E must be empirically estimated. The most commonly used approach for doing this is to estimate the average E from a series of reactions that were set up using a variety of cDNA template concentrations (i.e., a dilution series) [*see* (53, 72)]. However, despite being the most commonly used method, estimating E from dilution series data has several drawbacks. Most obviously, the dilution series method requires considerable amounts of RNA and is laborious. Hence, for large experiments, it may not be feasible to estimate E for every sample and gene combination to be investigated. In addition, the dilution series approach does not estimate reaction-specific efficiencies, but rather the average E across several reactions (see below). Finally, the dilution series method occasionally yields estimates of E that are > 2, suggesting that it is prone to overestimating E (53).

The second method for estimating E is to use the cycle-by-cycle fluorescence data that are collected during the course of a real-time PCR. This approach has the advantage of being able to yield an estimate of E for every reaction. Furthermore, unlike the dilution series approach, it does not require additional labor as fluorescence data are acquired during the course of conducting an experiment. A number of strategies for estimating E from fluorescence data have been suggested in the literature; however, the most straightforward approaches involve identifying the exponential phase of the amplification curve (80, 81) and regressing the resulting $\log_{10}$ (80, 81) or $\log_2$ (73) transformed subset of fluorescence values against cycle number.

While estimating E from fluorescence data has several advantages over the dilution series method (see above), there are also some drawbacks. An obvious concern is that for large experiments, involving thousands of reactions, using florescence-based approaches creates a considerable informatics problem. Another concern that arises when using fluorescence-based methods that rely on linear regression is that estimates of E will be based on small sample sizes due to the exclusion of a large number of reaction cycles from the analysis. Further, it is not clear that using the estimates of E generated for every reaction in a dataset is the most appropriate use of this information as analyzing data based on reaction-specific E values may introduce considerable noise into a dataset (80, 82). Peirson et al. (80) and Cikos et al. (82) have suggested that analyses based on averaged efficiencies provide more robust results and that reaction-specific E values should be used primarily to exclude reactions that have outlying Es.

3.3.2.4. Statistical Inference

By far the most common way in which qRT-PCR data are analyzed is via the use of standard parametric statistical tests (i.e., *t*-test, ANOVA, etc.). As described in the previous sections, there are a number of situations that can render such analyses invalid. Nevertheless, if care is taken to ensure that the assumptions that are essential to calculate unbiased estimates of R_E are met, the application of objective statistical methods to relative expression qRT-PCR data is acceptable. However, meeting the assumptions of relative expression models is an altogether distinct issue from meeting the assumptions of conventional parametric statistical tests (e.g., constant and normally distributed error variance, independence of observations, and symmetrically scaled continuous response variable). One particularly egregious error that occurs far too often during the analysis of qRT-PCR experiments is the use of R_E values as a response variable in standard parametric tests. The reasons for this being inappropriate are simple. First, R_E is not symmetrically scaled, as upregulated values of R_E lie on one scale ($1 < R_E < \infty$) and downregulated values lie on another scale ($1 > R_E > 0$). Second, R_E is rarely, if ever, normally distributed and is unlikely to be described adequately by linear functions. For these reasons, it is necessary to logarithmically transform R_E before conducting standard parametric analyses. We suggest analyzing $\log_2(R_E)$ as transformation to this base results in a straightforward interpretation that has a long history of use in the microarray literature. It is important to note that analyzing $\Delta\Delta C_q$ values (as opposed to $2^{-\Delta\Delta C_q}$) is essentially the same as analyzing $\log_2(R_E)$. Because $\log_2(R_E)$ is symmetrically scaled and is usually approximately normally distributed, it has a far greater chance of meeting the assumptions of conventional parametric tests than R_E.

3.3.2.5. Summary of Data Processing and Error Propagation

It has been repeatedly demonstrated that the way in which qRT-PCR data are processed and analyzed can strongly influence the biological conclusions drawn from the data (71). Although a large number of processing procedures have been described in the literature, there is currently no consensus on which methods are the most robust. The quality control and processing procedures that are essential to relative expression qRT-PCR experiments consist of several steps including the following: baseline determination, threshold determination, control gene validation, estimation of E, removal of reactions with outlying C_q (79) and/or E values, calculation of R_E or some other measure of expression, and a statistical analysis of these calculated expression measures. One of the most important issues associated with data processing is the issue of error propagation [*see* (78)]. Of particular concern within the context of the approaches to data analysis discussed in this section is that all of the components that go into the calculation of R_E such as estimates of E and C_q are themselves measured with uncertainty. Thus, there is a pressing need to develop processing procedures that account for this uncertainty, as well as user-friendly software implementations of these procedures that make them readily accessible to biologists.

Acknowledgments

M.G. was supported by a Netherlands Genomic Initiative Horizon grant. A.H. was supported by a Technology Top Institute Green Genetics grant. A.S. was supported by NIH P20 RR16481 and NSF EPS-0814194.

References

1. Gregory, B. D., and Belostotsky, D. A. (2009) Whole-genome microarrays: applications and technical issues. *Methods Mol. Biol.* **553**, 39–56.
2. Wang, Z., Gerstein, M., and Snyder, M. (2009) RNA-Seq: a revolutionary tool for transcriptomics. *Nat. Rev. Genet.* **10**, 57–63.
3. Collas, P. (2010) The current state of chromatin immunoprecipitation. *Mol. Biotechnol.* **45**, 87–100.
4. Kirmizis, A., and Farnham, P. J. (2004) Genomic approaches that aid in the identification of transcription factor target genes. *Exp. Biol. Med. (Maywood).* **229**, 705–721.
5. Aranda, A., and Pascual, A. (2001) Nuclear hormone receptors and gene expression. *Physiol. Rev.* **81**, 1269–1304.
6. Gomez-Mena, C., de Folter, S., Costa, M. M. R., Angenent, G. C., and Sablowski, R. (2005) Transcriptional program controlled by the floral homeotic gene AGAMOUS during early organogenesis. *Development* **132**, 429–438.
7. Wagner, D., Wellmer, F., Dilks, K., William, D., Smith, M. R., Kumar, P. P., Riechmann, J. L., Greenland, A. J., and Meyerowitz, E. M. (2004) Floral induction in tissue culture: a system for the analysis of LEAFY-dependent gene regulation. *Plant J.* **39**, 273–282.
8. Passarinho, P., Ketelaar, T., Xing, M., van Arkel, J., Maliepaard, C., Hendriks, M. W., Joosen, R., Lammers, M., Herdies, L., den Boer, B., van der Geest, L., and Boutilier, K. (2008) BABY BOOM target genes provide diverse entry points into cell proliferation and cell growth pathways. *Plant Mol. Biol.* **68**, 225–237.
9. Rohila, J. S., Chen, M., Cerny, R., and Fromm, M. E. (2004) Improved tandem affinity purification tag and methods for isolation of protein heterocomplexes from plants. *Plant J.* **38**, 172–181.
10. Craft, J., Samalova, M., Baroux, C., Townley, H., Martinez, A., Jepson, I., Tsiantis, M., and Moore, I. (2005) New pOp/LhG4 vectors for stringent glucocorticoid-dependent transgene expression in *Arabidopsis. Plant J.* **41**, 899–918.
11. Hanson, J., Hanssen, M., Wiese, A., Hendriks, M. M., and Smeekens, S. (2008)

The sucrose regulated transcription factor bZIP11 affects amino acid metabolism by regulating the expression of ASPARAGINE SYNTHETASE1 and PROLINE DEHYDROGENASE2. *Plant J.* **53**, 935–949.
12. Aoyama, T., and Chua, N. H. (1997) A glucocorticoid-mediated transcriptional induction system in transgenic plants. *Plant J.* **11**, 605–612.
13. Levesque, M. P., Vernoux, T., Busch, W., Cui, H., Wang, J. Y., Blilou, I., Hassan, H., Nakajima, K., Matsumoto, N., Lohmann, J. U., Scheres, B., and Benfey, P. N. (2006) Whole-genome analysis of the SHORT-ROOT developmental pathway in *Arabidopsis. PLoS Biol.* **4**, e143.
14. Hay, A., Jackson, D., Ori, N., and Hake, S. (2003) Analysis of the competence to respond to KNOTTED1 activity in *Arabidopsis* leaves using a steroid induction system. *Plant Physiol.* **131**, 1671–1680.
15. Sundstrom, J. F., Nakayama, N., Glimelius, K., and Irish, V. F. (2006) Direct regulation of the floral homeotic APETALA1 gene by APETALA3 and PISTILLATA in *Arabidopsis. Plant J.* **46**, 593–600.
16. Ikeda, Y., Banno, H., Niu, Q. W., Howell, S. H., and Chua, N. H. (2006) The ENHANCER OF SHOOT REGENERATION 2 gene in *Arabidopsis* regulates CUP-SHAPED COTYLEDON 1 at the transcriptional level and controls cotyledon development. *Plant Cell Physiol.* **47**, 1443–1456.
17. Che, P., Lall, S., and Howell, S. H. (2008) Acquiring competence for shoot development in *Arabidopsis*: ARR2 directly targets A-type ARR genes that are differentially activated by CIM preincubation. *Plant Signal. Behav.* **3**, 99–101.
18. Zuo, J., Niu, Q. W., and Chua, N. H. (2000) Technical advance: an estrogen receptor-based transactivator XVE mediates highly inducible gene expression in transgenic plants. *Plant J.* **24**, 265–273.
19. Tornero, P., Chao, R. A., Luthin, W. N., Goff, S. A., and Dangl, J. L. (2002) Large-scale structure–function analysis of the *Arabidopsis* RPM1 disease resistance protein. *Plant Cell* **14**, 435–450.
20. Zuo, J., and Chua, N. H. (2000) Chemical-inducible systems for regulated expression of plant genes. *Curr. Opin. Biotechnol.* **11**, 146–151.
21. Gatz, C., and Lenk, I. (1998) Promoters that respond to chemical inducers. *Trends Plant Sci.* **3**, 352–358.
22. Ko, J. H., Kim, W. C., and Han, K. H. (2009) Ectopic expression of MYB46 identifies transcriptional regulatory genes involved in secondary wall biosynthesis in *Arabidopsis. Plant J.* **60**, 649–665.
23. Zentella, R., Zhang, Z. L., Park, M., Thomas, S. G., Endo, A., Murase, K., Fleet, C. M., Jikumaru, Y., Nambara, E., Kamiya, Y., and Sun, T. P. (2007) Global analysis of della direct targets in early gibberellin signaling in *Arabidopsis. Plant Cell* **19**, 3037–3057.
24. Kang, H. G., Fang, Y., and Singh, K. B. (1999) A glucocorticoid-inducible transcription system causes severe growth defects in *Arabidopsis* and induces defense-related genes. *Plant J.* **20**, 127–133.
25. Ouwerkerk, P. B., de Kam, R. J., Hoge, J. H., and Meijer, A. H. (2001) Glucocorticoid-inducible gene expression in rice. *Planta* **213**, 370–378.
26. Andersen, S. U., Cvitanich, C., Hougaard, B. K., Roussis, A., Gronlund, M., Jensen, D. B., Frokjaer, L. A., and Jensen, E. O. (2003) The glucocorticoid-inducible GVG system causes severe growth defects in both root and shoot of the model legume *Lotus japonicus. Mol. Plant Microbe Interact.* **16**, 1069–1076.
27. Weijers, D., Van Hamburg, J. P., Van Rijn, E., Hooykaas, P. J., and Offringa, R. (2003) Diphtheria toxin-mediated cell ablation reveals interregional communication during *Arabidopsis* seed development. *Plant Physiol.* **133**, 1882–1892.
28. Galweiler, L., Conlan, R. S., Mader, P., Palme, K., and Moore, I. (2000) Technical advance: the DNA-binding activity of gal4 is inhibited by methylation of the gal4 binding site in plant chromatin. *Plant J.* **23**, 143–157.
29. Engineer, C. B., Fitzsimmons, K. C., Schmuke, J. J., Dotson, S. B., and Kranz, R. G. (2005) Development and evaluation of a Gal4-mediated LUC/GFP/GUS enhancer trap system in *Arabidopsis. BMC Plant Biol.* **5**, 9.
30. Liu, C., Chen, H., Er, H. L., Soo, H. M., Kumar, P. P., Han, J. H., Liou, Y. C., and Yu, H. (2008) Direct interaction of AGL24 and SOC1 integrates flowering signals in *Arabidopsis. Development* **135**, 1481–1491.
31. Matsuo, N., Mase, H., Makino, M., Takahashi, H., and Banno, H. (2009) Identification of ENHANCER OF SHOOT REGENERATION 1-upregulated genes during in vitro shoot regeneration. *Plant Biotechnol.* **26**, 385–393.
32. Pre, M., Atallah, M., Champion, A., De Vos, M., Pieterse, C. M. J., and Memelink, J. (2008) The AP2/ERF domain transcription factor ORA59 integrates jasmonic acid and

ethylene signals in plant defense. *Plant Physiol.* **147**, 1347–1357.

33. Balazadeh, S., Siddiqui, H., Allu, A. D., Matallana-Ramirez, L. P., Caldana, C., Mehrnia, M., Zanor, M. I., Kohler, B., and Mueller-Roeber, B. (2010) A gene regulatory network controlled by the NAC transcription factor ANAC092/AtNAC2/ORE1 during salt-promoted senescence. *Plant J.* **62**, 250–264.
34. Turner, R., and Foster, G. D. (1995) The potential exploitation of plant viral translational enhancers in biotechnology for increased gene-expression. *Mol. Biotechnol.* **3**, 225–236.
35. Zhong, R., Richardson, E. A., and Ye, Z. H. (2007) The MYB46 transcription factor is a direct target of SND1 and regulates secondary wall biosynthesis in *Arabidopsis. Plant Cell* **19**, 2776–2792.
36. McKeehan, W., and Hardesty, B. (1969) The mechanism of cycloheximide inhibition of protein synthesis in rabbit reticulocytes. *Biochem. Biophys. Res. Commun.* **36**, 625–630.
37. Lee, D. J., Park, J. W., Lee, H. W., and Kim, J. (2009) Genome-wide analysis of the auxin-responsive transcriptome downstream of iaa1 and its expression analysis reveal the diversity and complexity of auxin-regulated gene expression. *J. Exp. Bot.* **60**, 3935–3957.
38. Ravni, A., Eiden, L. E., Vaudry, H., Gonzalez, B. J., and Vaudry, D. (2006) Cycloheximide treatment to identify components of the transitional transcriptome in PACAP-induced PC12 cell differentiation. *J. Neurochem.* **98**, 1229–1241.
39. Abel, S., Nguyen, M. D., and Theologis, A. (1995) The PS-IAA4/5-like family of early auxin-inducible mRNAs in Arabidopsis thaliana. *J. Mol. Biol.* **251**, 533–549.
40. Theologis, A., Huynh, T. V., and Davis, R. W. (1985) Rapid induction of specific mRNAs by auxin in pea epicotyl tissue. *J. Mol. Biol.* **183**, 53–68.
41. Herrick, D., Parker, R., and Jacobson, A. (1990) Identification and comparison of stable and unstable mRNAs in *Saccharomyces cerevisiae. Mol. Cell Biol.* **10**, 2269–2284.
42. Kiddle, S. J., Windram, O. P., McHattie, S., Mead, A., Beynon, J., Buchanan-Wollaston, V., Denby, K. J., and Mukherjee, S. (2010) Temporal clustering by affinity propagation reveals transcriptional modules in *Arabidopsis thaliana. Bioinformatics* **26**, 355–362.
43. Herzenberg, L. A., Sweet, R. G., and Herzenberg, L. A. (1976) Fluorescence-activated cell sorting. *Sci. Am.* **234**, 108–117.
44. Bargmann, B. O., and Birnbaum, K. D. (2009) Fluorescence activated cell sorting of plant protoplasts. JoVE. 36. http://www.jove.com/index/Details.stp?ID=1673. doi: 10.3791/1673.
45. Birnbaum, K., Jung, J. W., Wang, J. Y., Lambert, G. M., Hirst, J. A., Galbraith, D. W., and Benfey, P. N. (2005) Cell type-specific expression profiling in plants via cell sorting of protoplasts from fluorescent reporter lines. *Nat. Methods* **2**, 615–619.
46. Birnbaum, K., Shasha, D. E., Wang, J. Y., Jung, J. W., Lambert, G. M., Galbraith, D. W., and Benfey, P. N. (2003) A gene expression map of the *Arabidopsis* root. *Science* **302**, 1956–1960.
47. Yadav, R. K., Girke, T., Pasala, S., Xie, M., and Reddy, G. V. (2009) Gene expression map of the *Arabidopsis* shoot apical meristem stem cell niche. *Proc. Natl. Acad. Sci. USA* **106**, 4941–4946.
48. Sheen, J. (2001) Signal transduction in maize and *Arabidopsis* mesophyll protoplasts. *Plant Physiol.* **127**, 1466–1475.
49. Rajeevan, M. S., Ranamukhaarachchi, D. G., Vernon, S. D., and Unger, E. R. (2001) Use of real-time quantitative PCR to validate the results of cDNA array and differential display PCR technologies. *Methods* **25**, 443–451.
50. Chuaqui, R. F., Bonner, R. F., Best, C. J., Gillespie, J. W., Flaig, M. J., Hewitt, S. M., Phillips, J. L., Krizman, D. B., Tangrea, M. A., Ahram, M., Linehan, W. M., Knezevic, V., and Emmert-Buck, M. R. (2002) Post-analysis follow-up and validation of microarray experiments. *Nat. Genet.* **32** *Suppl.*, 509–514.
51. Canales, R. D., Luo, Y., Willey, J. C., Austermiller, B., Barbacioru, C. C., Boysen, C., Hunkapiller, K., Jensen, R. V., Knight, C. R., Lee, K. Y., Ma, Y., Maqsodi, B., Papallo, A., Peters, E. H., Poulter, K., Ruppel, P. L., Samaha, R. R., Shi, L., Yang, W., Zhang, L., and Goodsaid, F. M. (2006) Evaluation of DNA microarray results with quantitative gene expression platforms. *Nat. Biotechnol.* **24**, 1115–1122.
52. Bustin, S. A. (2002) Quantification of mRNA using real-time reverse transcription PCR (RT-PCR): trends and problems. *J. Mol. Endocrinol.* **29**, 23–39.
53. Pfaffl, M. A. (2006) Relative quantification, in *Real-time PCR* (Dorak, M. T., Ed.), pp 63–82. Taylor and Francis, New York, NY.
54. Karlen, Y., McNair, A., Perseguers, S., Mazza, C., and Mermod, N. (2007) Statistical significance of quantitative PCR. *BMC Bioinformatics* **8**, 131.

55. Bustin, S. A., and Nolan, T. (2004) Pitfalls of quantitative real-time reverse-transcription polymerase chain reaction. *J. Biomol. Tech.* **15**, 155–166.
56. Bustin, S. A., Benes, V., Garson, J. A., Hellemans, J., Huggett, J., Kubista, M., Mueller, R., Nolan, T., Pfaffl, M. W., Shipley, G. L., Vandesompele, J., and Wittwer, C. T. (2009) The MIQE guidelines: minimum information for publication of quantitative real-time PCR experiments. *Clin. Chem.* **55**, 611–622.
57. Bustin, S. A., Benes, V., Nolan, T., and Pfaffl, M. W. (2005) Quantitative real-time RT-PCR—a perspective. *J. Mol. Endocrinol.* **34**, 597–601.
58. Wong, M. L., and Medrano, J. F. (2005) Real-time PCR for mRNA quantitation. *Biotechniques* **39**, 75–85.
59. Czechowski, T., Stitt, M., Altmann, T., Udvardi, M. K., and Scheible, W. R. (2005) Genome-wide identification and testing of superior reference genes for transcript normalization in *Arabidopsis. Plant Physiol.* **139**, 5–17.
60. Gutierrez, L., Mauriat, M., Guenin, S., Pelloux, J., Lefebvre, J. F., Louvet, R., Rusterucci, C., Moritz, T., Guerineau, F., Bellini, C., and Van Wuytswinkel, O. (2008) The lack of a systematic validation of reference genes: a serious pitfall undervalued in reverse transcription-polymerase chain reaction (RT-PCR) analysis in plants. *Plant Biotechnol. J.* **6**, 609–618.
61. Freeman, W. M., Walker, S. J., and Vrana, K. E. (1999) Quantitative RT-PCR: pitfalls and potential. *Biotechniques* **26**, 112–125.
62. Beckman, K. B., Lee, K. Y., Golden, T., and Melov, S. (2004) Gene expression profiling in mitochondrial disease: assessment of microarray accuracy by high-throughput Q-PCR. *Mitochondrion* **4**, 453–470.
63. Leibfried, A., To, J. P., Busch, W., Stehling, S., Kehle, A., Demar, M., Kieber, J. J., and Lohmann, J. U. (2005) WUSCHEL controls meristem function by direct regulation of cytokinin-inducible response regulators. *Nature* **438**, 1172–1175.
64. Schlereth, A., Moller, B., Liu, W., Kientz, M., Flipse, J., Rademacher, E. H., Schmid, M., Jurgens, G., and Weijers, D. (2010) MONOPTEROS controls embryonic root initiation by regulating a mobile transcription factor. *Nature* **464**, 913–916.
65. Ye, Q., Zhu, W., Li, L., Zhang, S., Yin, Y., Ma, H., and Wang, X. (2010) Brassinosteroids control male fertility by regulating the expression of key genes involved in Arabidopsis anther and pollen development. *Proc. Natl. Acad. Sci. USA* **107**, 6100–6105.
66. Kaufmann, K., Muino, J. M., Osteras, M., Farinelli, L., Krajewski, P., and Angenent, G. C. (2010) Chromatin immunoprecipitation (ChIP) of plant transcription factors followed by sequencing (ChIP-SEQ) or hybridization to whole genome arrays (ChIP-CHIP). *Nat. Protoc.* **5**, 457–472.
67. Fu, W. J., Stromberg, A. J., Viele, K., Carroll, R. J., and Wu, G. (2010) Statistics and bioinformatics in nutritional sciences: analysis of complex data in the era of systems biology. *J. Nutr. Biochem.* **21**, 561–572.
68. Peng, X., Wood, C. L., Blalock, E. M., Chen, K. C., Landfield, P. W., and Stromberg, A. J. (2003) Statistical implications of pooling RNA samples for microarray experiments. *BMC Bioinformatics* **4**, 26.
69. Benjamini, Y., and Hochberg, Y. (1995) Controlling the false discovery rate – a practical and powerful approach to multiple testing. *J. R. Stat. Soc. Ser. B Methodol.* **57**, 289–300.
70. Liu, H., Tarima, S., Borders, A. S., Getchell, T. V., Getchell, M. L., and Stromberg, A. J. (2005) Quadratic regression analysis for gene discovery and pattern recognition for non-cyclic short time-course microarray experiments. *BMC Bioinformatics* **6**, 106.
71. Skern, R., Frost, P., and Nilsen, F. (2005) Relative transcript quantification by quantitative PCR: roughly right or precisely wrong? *BMC Mol. Biol.* **6**, 10.
72. Yuan, J. S., Reed, A., Chen, F., and Stewart, C. N., Jr. (2006) Statistical analysis of real-time PCR data. *BMC Bioinformatics* **7**, 85.
73. Yuan, J. S., Wang, D., and Stewart, C. N., Jr. (2008) Statistical methods for efficiency adjusted real-time PCR quantification. *Biotechnol. J.* **3**, 112–123.
74. Livak, K. J., and Schmittgen, T. D. (2001) Analysis of relative gene expression data using real-time quantitative PCR and the 2(-Delta Delta C(T)) method. *Methods* **25**, 402–408.
75. Vandesompele, J., De Preter, K., Pattyn, F., Poppe, B., Van Roy, N., De Paepe, A., and Speleman, F. (2002) Accurate normalization of real-time quantitative RT-PCR data by geometric averaging of multiple internal control genes. *Genome Biol.* **3**, RESEARCH0034.
76. Andersen, C. L., Jensen, J. L., and Orntoft, T. F. (2004) Normalization of real-time quantitative reverse transcription-PCR data: a model-based variance estimation approach to identify genes suited for normalization, applied to bladder and colon cancer data sets. *Cancer Res.* **64**, 5245–5250.

77. Pfaffl, M. W. (2001) A new mathematical model for relative quantification in real-time RT-PCR. *Nucleic Acids Res.* **29**, e45.
78. Hellemans, J., Mortier, G., De Paepe, A., Speleman, F., and Vandesompele, J. (2007) qBase relative quantification framework and software for management and automated analysis of real-time quantitative PCR data. *Genome Biol.* **8**, R19.
79. Burns, M. J., Nixon, G. J., Foy, C. A., and Harris, N. (2005) Standardisation of data from real-time quantitative PCR methods – evaluation of outliers and comparison of calibration curves. *BMC Biotechnol.* **5**, 31.
80. Peirson, S. N., Butler, J. N., and Foster, R. G. (2003) Experimental validation of novel and conventional approaches to quantitative real-time PCR data analysis. *Nucleic Acids Res.* **31**, e73.
81. Ramakers, C., Ruijter, J. M., Deprez, R. H., and Moorman, A. F. (2003) Assumption-free analysis of quantitative real-time polymerase chain reaction (PCR) data. *Neurosci. Lett.* **339**, 62–66.
82. Cikos, S., Bukovska, A., and Koppel, J. (2007) Relative quantification of mRNA: comparison of methods currently used for real-time PCR data analysis. *BMC Mol. Biol.* **8**, 113.

Section IV

Protein–Protein Interaction and Other Means for Activity Control

Chapter 8

Yeast Protein–Protein Interaction Assays and Screens

Stefan de Folter and Richard G.H. Immink

Abstract

Most transcription factors fulfill their role in protein complexes. As a consequence, information about their interaction capacity sheds light on a protein's function and the molecular mechanism underlying this activity. The yeast two-hybrid GAL4 (Y2H) assay is a powerful method to unravel and identify the composition of protein complexes. This in vivo based system makes use of two functional protein domains of the GAL4 transcription factor, each fused to a protein of interest. Upon interaction between the two proteins under study, a transcriptional activator gets reconstituted and reporter genes get activated, allowing the yeast to grow on selective medium. In this chapter protocols are given for Y2H library screening, directed Y2H screening, Y2H matrix screening, and Y*n*H screening involving more than two proteins.

Key words: Protein–protein interactions, yeast two-hybrid, Y2H, Y*n*H, GAL4 system, complexes, transcription factors.

1. Introduction

Many regulatory proteins, such as transcription factors, function in complexes. Knowing which other proteins are present in a complex is informative for understanding how a certain transcription factor can fulfill its function. Twenty years ago a method was developed that has had an enormous impact on the available protein–protein interaction information and which has led to the elucidation of large-scale protein–protein interaction networks (e.g., 1–8). This method is called the yeast two-hybrid system (Y2H) and here we will focus on the yeast two-hybrid GAL4 system (9, 10). The basic concept of this technology is that when an interaction occurs between two fusion proteins

L. Yuan, S.E. Perry (eds.), *Plant Transcription Factors*, Methods in Molecular Biology 754,
DOI 10.1007/978-1-61779-154-3_8, © Springer Science+Business Media, LLC 2011

(hybrids) a transcriptional activator gets reconstituted resulting in activation of reporter genes. In more detail, the yeast GAL4 transcription factor that is used for this purpose consists of two functional domains, a DNA binding domain and a transcriptional activation domain. One domain gets fused to protein X and the other domain gets fused to protein Y. Both fusion proteins are expressed in a suitable yeast strain and upon interaction, a GAL4 responsive reporter gene gets expressed, which is often an auxotrophic marker enabling the yeast to grow on selective media (*see* **Figs. 8.1**, **8.2**, and **8.3**). Besides Y2H assays various other methods are available to study protein interactions, like affinity chromatography, co-immunoprecipitation, pull-downs, and protein arrays (e.g., 11). Each method has its own advantages and drawbacks and Y2H has proven itself as a powerful tool that is sensitive, relatively easy to perform, and inexpensive. Nowadays, there are many variations on the Y2H method (12–14), but still the yeast GAL4 system is used often, which is also the focus of this chapter.

To support or validate Y2H interaction data various approaches are available. Most common but time consuming are genetic studies. An alternative is to accomplish co-expression analyses for the genes encoding interacting proteins (e.g., using available microarray expression data) (e.g., 15, 16). Furthermore, a search for interologs (conserved interactions among species) can be performed when protein–protein interaction data from other related species are available (3). Various interaction data are collected in databases (11, 17), which can be used for this purpose. Other possibilities that provide direct confirmation of physical interaction are more sophisticated, such as bimolecular fluorescence complementation (BiFC) (18), fluorescence resonance energy transfer (FRET) (19), co-IP, or pull-downs (17). Finally, interaction data can be visualized with specific software programs (e.g., 20–24), which allow better interpretation of large datasets.

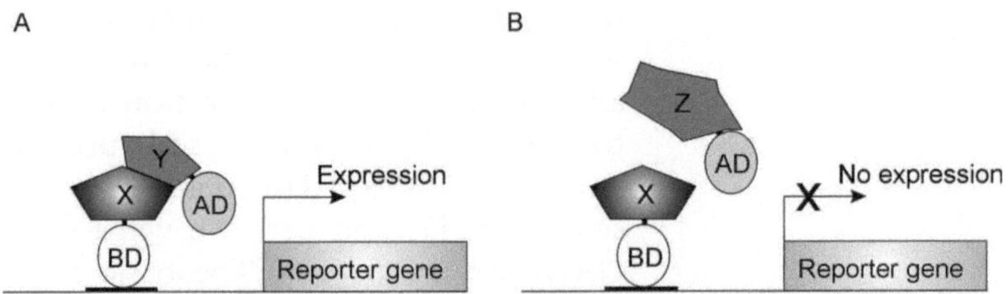

Fig. 8.1. Outline of the yeast two-hybrid (Y2H) GAL4 system. (**a**) The basic concept of this technology is that when an interaction occurs between two fusion proteins (hybrids) the GAL4 transcriptional activator gets reconstituted resulting in activation of reporter genes. (**b**) Reporter genes do not get activated when no protein–protein interaction occurs. BD, binding domain; AD, activation domain; X, protein X; Y, protein Y; Z, protein Z.

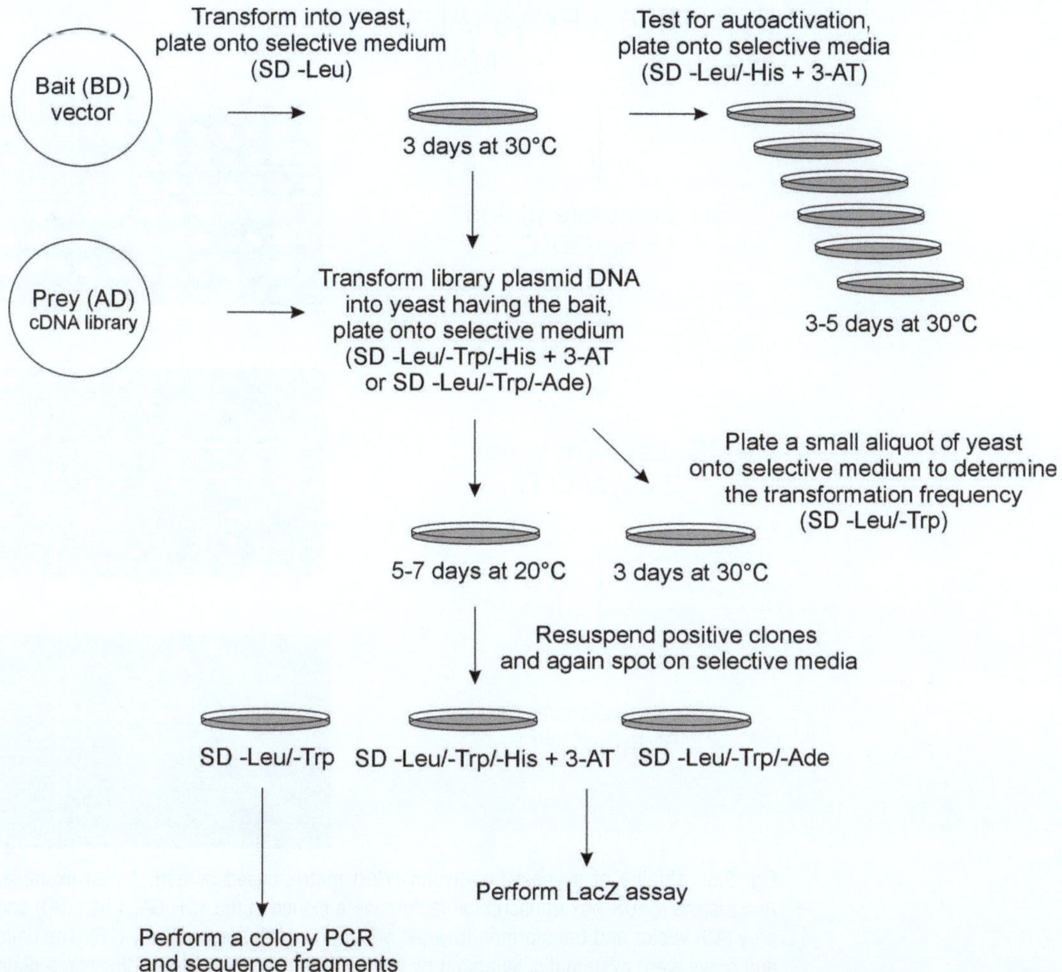

Fig. 8.2. Outline of a typical yeast two-hybrid (Y2H) library screen. The bait vector (BD) first gets transformed to yeast, followed by a test for autoactivation. Then the cDNA expression library in the prey vector (AD) gets transformed into the yeast, which contains already the BD vector, followed by plating on appropriate selective medium. After 5–7 days of incubation positive interactors will be obtained that should be resuspended in sterile MQ water and plated again on selective media to verify and test for activity of the other reporter genes. Finally, a colony PCR is performed on positive interactors and fragments get sequenced to identify the encoding genes.

In this chapter, protocols are given for a Y2H library screen, a directed Y2H screen, a Y2H matrix screen, and a Y*n*H screen involving more than two proteins. Examples of implementation of the protocols described here can be found in our various publications focusing on MADS domain transcription factors (e.g., 25–30). Additional useful information about yeast two-hybrids can be found in the following references (12, 13, 31–33).

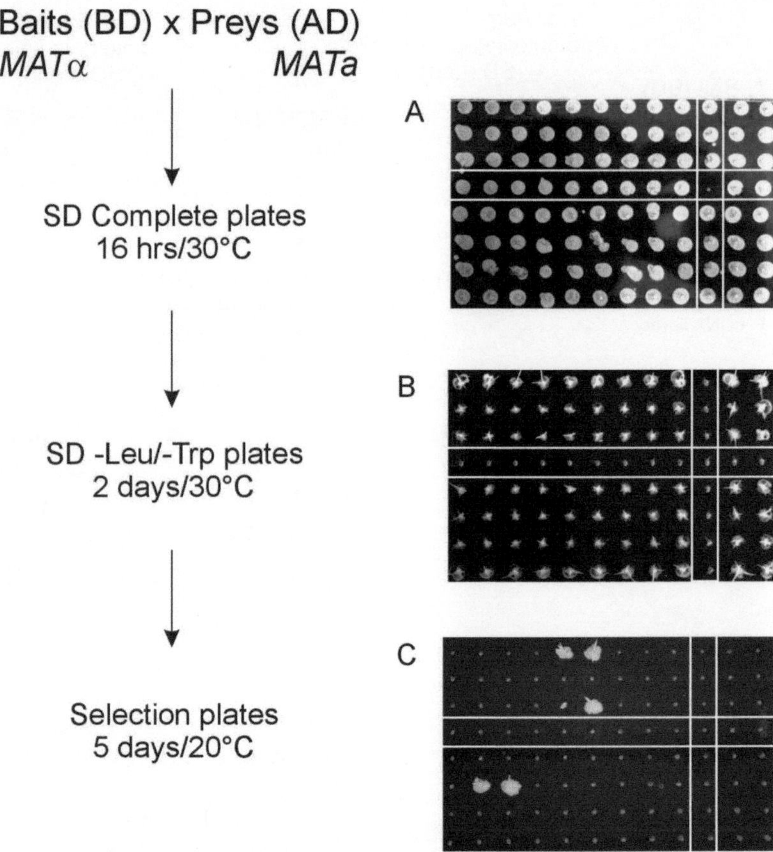

Fig. 8.3. Outline of a yeast two-hybrid (Y2H) matrix-based screen. As an example, *Arabidopsis* MADS-box transcription factors were cloned in the Y2H GAL4 bait (BD) and prey (AD) vector and transformed to yeast *MATα* and *MATa*, respectively (25). The baits and preys were systematically mated by spotting them on top of each other on a plate with non-selective medium (**a**). After overnight incubation, the spots were transferred to a plate selecting for diploid yeast containing the two plasmids (**b**). The plates were incubated for 2 days at 30°C and then the yeast was transferred to two separate selection plates ((**c**), SD –Leu/–Trp/–Ade and SD –Leu/–Trp/–His/+5 mM 3-AT), incubated at 20°C, and after 5 days scored for protein–protein interactions. In the row and column marked with lines water was spotted instead of a prey or bait culture, respectively.

2. Materials

2.1. Vector System

1. pENTR/D-TOPO cloning vector (Invitrogen, Carlsbad, CA, USA): Kanamycin selection for bacteria.
2. pDEST32 (bait vector, BD; Invitrogen): Leu selection marker gene for yeast and gentamycin for bacteria.
3. pDEST22 (prey vector, AD; Invitrogen): Trp selection marker gene for yeast and ampicillin for bacteria.

4. pARC352 (Gateway-compatible version of pTFT1; for Y3H (30)): Ade2 selection marker gene for yeast and ampicillin for bacteria.
5. Yeast strain PJ69-4A (*MAT*a; trp1-901 leu2-3,112 ura3-52 his3-200 gal4 (deleted) gal80 (deleted) LYS2::GAL1-HIS3 GAL2-ADE2 met2::GAL7-lacZ) (34).
6. Yeast strain PJ69-4α (*MAT*α; trp1-901 leu2-3,112 ura3-52 his3-200 gal4 (deleted) gal80 (deleted) LYS2::GAL1-HIS3 GAL2-ADE2 met2::GAL7-lacZ) (34).

2.2. Media for Yeast Growth and Selection

1. SD glu medium: 6.7 g yeast nitrogen base without amino acids, 20 g dextrose (D+glucose). Dissolve contents and add MQ water up to 900 mL. For solid SD glu medium add 17 g/L bacto agar (Becton, Dickinson and Company, NY, USA). Autoclave and before use add 1/10 volume of the appropriate 10X Dropout solution.
2. 10X Dropout solution: For growth of yeast strains without additional plasmids use 10X "Complete." Add all components as indicated in **Table 8.1** and fill to 1 L with MQ water. Note that some amino acids do not dissolve in water, but will dissolve during autoclaving. 10X Dropout stock solution can be stored for at least 1 year at 4°C after autoclaving for 15 min at 120°C (1 atm). Some amino acids are heat instable, and therefore, heating should be kept to a minimum. Use a Dropout solution lacking the appropriate component(s) in order to select for plasmids or protein–protein interaction events.
3. 3-AT (3-amino-1,2,4-triazole) solution: Prepare a 2 M stock in MQ water and filter sterilize. Chemical should be stored at –20°C and stock solution can be stored at 4°C for at least 2 months.

2.3. Solutions for Yeast Transformation

1. 100 mM LiAc, pH 7.5 (sterile).
2. 1 M LiAc, pH 7.5 (sterile).
3. PEG 50% (w/v) solution (polyethylene glycol, M_r 3350) (sterile).
4. Salmon sperm DNA (10 mg/mL) (Invitrogen, Carlsbad, CA, USA).
5. DMSO.
6. 1X TE buffer: 10 mM Tris–HCl, pH 7.5, 1 mM EDTA, pH 7.5.
7. 10X TE buffer: 100 mM Tris–HCl, pH 7.5, 10 mM EDTA, pH 7.5.
8. MQ water (sterile).

Table 8.1
Chemical components necessary for a 10X Dropout (Complete) solution

Components	Weight (mg/L)	Sigma catalog
L-Isoleucine	300	I2752
L-Valine	1500	V0500
L-Adenine hemisulfate salt	200	A9126
L-Arginine HCl	200	A5131
L-Histidine HCl monohydrate	200	H8125
L-Leucine	1000	L8000
L-Lysine HCl	300	L5626
L-Methionine	200	M9625
L-Phenylalanine	500	P2126
L-Threonine	2000	T8625
L-Tryptophan	200	T0254
L-Tyrosine	300	T3754
L-Uracil	200	U0750

9. 1X TE/100 mM LiAc (fresh): prepare from 10X TE buffer and 1 M LiAc solution.
10. PEG/LiAc solution (fresh): 24 mL of PEG 50%, 3 mL of 10X TE, and 3 mL of 1 M LiAc (sufficient for two large-scale yeast transformations).

2.4. LacZ Assay

1. Chloroform.
2. X-GAL (20 mg/mL in DMSO) (store at –20°C, protect from light).
3. RoBlue medium: 1% low-melting agarose, 100 mM KPO_4, pH 7.

2.5. Yeast Colony PCR

1. Taq polymerase.
2. dNTPs (10 mM each).
3. 10X PCR buffer: 100 mM Tris–HCl, pH 8.3, 15 mM $MgCl_2$, 500 mM KCl.
4. 0.1 M $MgCl_2$.
5. FW Oligo AD vector (10 pmol/μL): 5′-AGGGATGTTTAATACCACTAC-3′.
6. RV Oligo AD vector (10 pmol/μL): 5′-CAAACCTCTGGCGAAGA-3′.

2.6. Yeast Plasmid DNA Extraction

1. Lysis buffer: 2% (v/v) Triton X-100, 1% (v/v) SDS, 100 mM NaCl, 10 mM Tris–HCl, pH 8, 1 mM EDTA, pH 8.
2. Phenol/chloroform (50/50; v/v).
3. Acid-washed glass beads (425–600 μm).
4. 100% ethanol.
5. 70% ethanol.
6. 3 M NaOAc, pH 5.2 (sterile).
7. 1X TE buffer: 10 mM Tris–HCl, pH 7.5, 1 mM EDTA, pH 7.5 (sterile).

2.7. Yeast Glycerol Stocks

1. Glycerol (sterile, 99% pure).
2. SD freezing medium (for 96-well plates): 3.35 g of yeast nitrogen base w/o amino acids, 10 g of dextrose, 112.5 mL of glycerol (=22.5% v/v) in 450 mL. Autoclave and afterward mix with appropriate 10X Dropout solution to obtain a total volume of 500 mL.

3. Methods

3.1. Vector System

This chapter is focused on the yeast two-hybrid GAL4 system using Gateway vectors; however, other vector systems may be used (*see* **Note 1**). Most of the protocols described here are based on Clontech and Stratagene manuals.

Sub-clone the coding region of a cDNA encoding for the bait protein of interest in a Gateway recombination-compatible high-copy plasmid, e.g., pENTR/D-TOPO, sequence verify the integrity of the clone, followed by recombining the cDNA of interest in the bait vector pDEST32. The coding region of the gene of interest will be fused in frame at its 5′-end with the coding region of the GAL4 binding domain (BD) (*see* **Note 2**). Depending on the type of interaction screen to be performed, specific cDNA clones of interest or a complete cDNA library may be used (*see* **Note 3**), cloned in frame in the prey vector pDEST22 containing the GAL4 activation domain coding region (AD). For higher-order interaction screens, Y3H or Y4H, one or two extra yeast protein expression vectors have to be used, respectively (*see* **Section 3.9** and **Note 4**).

3.2. Yeast Transformation (Small Scale)

- Use the yeast strain PJ69-4α for bait vectors (BD; pDEST32) (*see* **Note 5**); the selection after transformation is on medium SD glu –Leu.

- Use the yeast strain PJ69-4A for prey vectors (AD; pDEST22) (*see* **Note 5**); the selection after transformation is on medium SD glu –Trp.

1. Prepare an overnight culture of the appropriate yeast strain in SD glu complete medium. Inoculate 2–3 colonies in 50 mL of culture and incubate for 16–18 h at 30°C with shaking at 300 rpm.
2. Use 2 mL of culture per transformation (*see* **Note 6**) in a 2 mL Eppendorf tube.
3. Centrifuge in a microcentrifuge for 10 s at maximum speed (13,000 rpm).
4. Resuspend the pellet in 1 mL of 100 mM LiAc.
5. Incubate for 1 min at room temperature.
6. Centrifuge in a microcentrifuge for 10 s at maximum speed.
7. Remove the supernatant.
8. Add to the pellet (*see* **Note 7**) the following:
   240 μL of 50% PEG (resuspend pellet in the PEG)
   36 μL of 1 M LiAc
   5 μL of salmon sperm DNA (10 mg/mL) (*see* **Note 8**)
   3 μL of plasmid DNA (~0.5–1.5 μg) (*see* **Note 9**)
   67 μL of MQ water
9. Vortex for 1 min.
10. Incubate for 20–30 min at 42°C in a water bath (*see* **Note 10**).
11. Centrifuge in a microcentrifuge for 10 s at maximum speed (13,000 rpm).
12. Resuspend the pellet in 100 μL of MQ water.
13. Plate the yeast on the appropriate selection medium (50 μL of yeast solution per plate ($d = 10$ cm)).
14. Incubate for ~3 days at 30°C.

3.3. Yeast Transformation (Large Scale, for Libraries)

1. Inoculate 2–3 yeast colonies into a flask containing 50 mL of the appropriate medium (selection for the BD plasmid, pDEST32) (*see* **Note 11**).
2. Incubate for 16–18 h at 30°C with shaking at 300 rpm to reach the stationary phase ($OD_{600} > 1.5$).
3. Transfer part of this overnight culture to a 1 L flask containing 300 mL of SD glu medium supplemented with the appropriate 10X Dropout solution so that the OD_{600} will be about 0.1 (*see* **Note 12**).

4. Incubate for 1.5–3 h at 30°C with shaking at 300 rpm to reach an OD_{600} between 0.4 and 0.6.
5. Transfer the cells to 50 mL tubes (six tubes in total) and centrifuge for 5 min at 1000×*g* at room temperature.
6. Discard the supernatant and divide 50 mL of sterile 1X TE buffer over the six tubes.
7. Resuspend the cell pellets by vortexing vigorously.
8. Pool the cells (50 mL in total) in one tube and centrifuge for 5 min at 1000×*g* at room temperature.
9. Discard the supernatant and resuspend the cell pellet in freshly prepared 1.5 mL sterile 1X TE/100 mM LiAc.
10. Add 5–10 μL of cDNA library in pDEST22 (5–10 μg), 200 μL of salmon sperm DNA (*see* **Note 8**), and 12 mL of sterile PEG/LiAc solution.
11. Vortex for 1 min at high speed.
12. Incubate for 10 min at 30°C with shaking at 200 rpm.
13. Add drop by drop 1.4 mL of DMSO and mix well by gentle shaking and inversion of the tube (*see* **Note 13**).
14. Perform a heat shock transformation for 20–30 min at 42°C in a water bath (*see* **Note 10**).
15. Chill the cells for 1–2 min on ice.
16. Centrifuge the cells for 1.5 min at 1000×*g* at room temperature.
17. Discard the supernatant and resuspend the cell pellet in 12 mL of 1X TE buffer.
18. Dissolve 5 μL of this yeast suspension in 45 μL of MQ water and plate on SD glu medium selecting for both of the plasmids (–Leu/–Trp) to determine the transformation frequency (*see* **Note 14**). Plate the remaining suspension (max 500 μL/plate, $d = 15$ cm) on the appropriate medium for protein–protein interaction identification (ADE, or HIS selection + 3-AT) (*see* **Note 15**).
19. Incubate the plate for determination of transformation frequency for 3 days at 30°C and all other plates for 5–7 days at 20°C (*see* **Section 3.5**).
20. Resuspend all identified positive clones in 50 μL of MQ water and spot as 5 μL droplets on a master plate (SD medium –Leu/–Trp) and on selection plates for interaction (SD –Leu/–Trp/–His + a concentration range of 3-AT and SD –Leu/–Trp/–Ade, to select for expression of the second reporter) (*see* **Fig. 8.2**).

3.4. Autoactivation Test

The bait construct (BD) has to be tested for autoactivation before doing any interaction screen. This determines whether the protein encoded by the cloned cDNA has an intrinsic activation domain, which will cause the yeast to grow on selection medium without any protein interaction. When autoactivation occurs there are a few approaches to solve this problem and allow continuation with the screening (1): When the aim is to use the protein of interest in a directed Y2H screen, use the autoactivating clone only in fusion with the GAL4 AD domain and perform the protein–protein interaction screening in the other orientation. Recombine for this purpose the cDNA of interest in the prey vector (AD) (2). Perform the screening using a higher concentration of 3-AT in the selection medium (*see* **Note 16**; 3). Make truncations or point mutations in your cDNA of interest and repeat the autoactivation test. Be aware that even a single amino acid mutation can have a large effect on folding, stability, and functionality of a protein.

1. Transform PJ69-4α (mating type α) with the bait construct (BD) (small-scale yeast transformation, *see* **Section 3.2**).
2. Test the bait for autoactivation by spotting of yeast on the following selection media (resuspend a single yeast colony in 100 μL of sterile water and spot 5 μL droplets):
   (a) –Leu
   (b) –Leu/–Ade
   (c) –Leu/–His
   (d) –Leu/–His + 1 mM 3-AT
   (e) –Leu/–His + 5 mM 3-AT
   (f) –Leu/–His + 10 mM 3-AT
3. Test for growth after 3–5 days incubation at 30°C (*see* **Note 17**).

3.5. Y2H Library Screen

1. Transform the yeast that already contains the bait plasmid (BD) with the cDNA expression library in the prey vector (AD) following the large-scale yeast transformation protocol (*see* **Section 3.3**).
2. Select for positives on medium –Leu/–Trp/–His + 1 mM 3-AT (when there is no autoactivation; *see* **Section 3.4**). Incubate the plates at 20°C and take positive clones after 5–7 days of incubation. In the case of autoactivation, do the final selection either on –Leu/–Trp/–Ade plates or alternatively on –Leu/–Trp/–His plates supplemented with a 3-AT concentration that is at least 3 mM higher than the concentration at which growth due to autoactivation stops (*see* **Section 3.4**).

3. In the case of HIS selection, re-streak positives on ADE selection and/or perform a blue-white assay (*see* **Section 3.10** and **Note 18**).
4. Perform a colony PCR on the positive clones (positive for at least two selection markers) (*see* **Section 3.11**). Amplified PCR products can be sequenced directly when just one product gets amplified (*see* **Fig. 8.2**).

3.6. Yeast Mating

Diploid yeast can be obtained by mating the haploid yeast A strain (PJ69-4A; *MAT*a) with the haploid yeast α strain (PJ69-4α; *MAT*α). Yeast mating can be applied for a directed Y2H analysis or for matrix-based screening (*see* **Sections 3.7** and **3.8**). One yeast strain contains the bait construct (BD) and the other yeast strain contains the prey construct (AD). After mating the resulting diploid yeast will have both constructs.

1. Make an overnight culture of each strain (incubate for 16–18 h at 30°C with shaking at 300 rpm) (*see* **Note 19**) or alternatively resuspend a fresh yeast colony in 50 μL of sterile MQ water.
2. Spot a 5 μL droplet of the A strain overnight culture/suspension on a plate with SD glu complete medium.
3. Spot a 5 μL droplet of the α strain overnight culture/suspension on top of the other droplet of the A strain (*see* **Note 20**).
4. Incubate overnight at 30°C.
5. Transfer a bit of yeast (*see* **Note 21**) by a sterile pipette tip or 96 pin replicator to selection medium for both plasmids (–Leu/–Trp) and incubate for 2–3 days at 30°C to obtain diploid yeast.
6. Transfer a bit of yeast to final selection medium by a 96 pin replicator or as 5 μL droplets after resuspending the diploid yeast in 50 μL of MQ water to test for protein–protein interaction events (*see* **Section 3.5**).

3.7. Y2H Directed Screen

This approach allows testing a cDNA of interest in the bait vector (BD) against a collection of prey vectors (AD) with known cDNA inserts. A directed screen like this gives the advantage of assessing possible interactions between proteins of interest and avoids sequencing, which makes it a rapid method of screening. Furthermore, besides information about positive combinations, it also provides information about proteins that are not able to interact. This information cannot be obtained from a library screen, because the lack of isolation of a protein from a cDNA expression library does not automatically mean that the proteins do not interact.

1. Transform the yeast α strain with the bait vector (BD) containing the cDNA of interest (*see* **Section 3.2**) and check first for autoactivation (*see* **Section 3.4**).
2. The collection of prey vectors (AD) with the cDNAs of interest should be independently transformed to the yeast A strain (*see* **Section 3.2** and **Note 22**).
3. Make diploid yeast by mating (*see* **Section 3.6**).
4. Test for protein–protein interaction events (*see* **Section 3.5**).

3.8. Y2H Matrix-Based Screen

A Y2H matrix-based screen is recommended when the aim is to study all possible protein–protein interactions within a family of, e.g., transcription factors in order to generate a comprehensive protein interaction network.

1. For each gene of the family the cDNA has to be recombined in both the bait vector (BD) and prey vector (AD). Whereas pDEST32 (BD) and pDEST22 (AD) have different selection markers for bacteria, both can be added to one Gateway LR reaction. Plate half of the transformed *Escherichia coli* culture on gentamycin (BD) selective medium and half on ampicillin selective medium (AD) after recombination and transformation.
2. All the generated bait vectors (BD) have to be transformed independently in the yeast α strain.
3. All the obtained prey vectors (AD) have to be transformed independently in the yeast A strain.
4. Make diploid yeast by mating (*see* **Section 3.6**), producing all possible combinations between the bait vectors (BD) and the prey vectors (AD) in a matrix setup (*see* **Note 23**).
5. Test for protein–protein interactions (*see* **Section 3.5**; **Note 24**; **Fig. 8.3**).

3.9. YnH Screen

In addition to testing for binary protein–protein interactions it is also possible to test for interactions between more than two proteins to investigate whether higher-order complexes can be formed. For various transcription factors multimerization and/or interactions with co-factors, such as transcriptional repressors or chromatin modifiers, are important for their function. Hence, generating information about higher-order complex formation can be helpful to decipher how particular transcription factors act at the molecular level. To test for interactions between three proteins (Y3H) a third vector is added to the two Y2H vectors (BD and AD), and in order to study complex formation between four proteins (Y4H) a fourth vector is added. The principle is the same as for the Y2H, just with addition of expression vectors encoding for proteins fused to a nuclear localization signal only and containing different selection markers. An important prerequisite for higher-order complex formation screenings is that the bait protein fused to the GAL4-BD domain is not able to interact with the prey protein expressed as fusion with the GAL4-AD domain from the pDEST22 vector. Therefore, it is always essential to first test all binary interactions between the proteins of interest. Based on this information the appropriate cloning strategy can be defined. The idea is that the additional protein(s) act as bridging molecules that bring bait and prey together in a large protein complex. In case of a Y*n*H library screen, binary interaction capacity for an identified putative third protein complex factor should be analyzed afterward. A second point to be taken into account is that third and fourth proteins should not contain an intrinsic transcriptional activation domain. This can result in autoactivation of the reporters independent of multimerization for all expressed individual proteins. Furthermore, it is possible that neither the bait nor the third protein interacting with the bait contains an intrinsic transcriptional activation domain, but that the binary protein–protein interaction results in a conformational change generating a transcriptional activation domain. Also this possibility should be ruled out by performing autoactivation screenings. The protocol described below is for a Y3H screen.

1. Recombine one of the three cDNAs of interest in the bait vector (BD), one in the prey vector (AD), and one in the Gateway-compatible pTFT1 vector (pARC352).
2. Co-transform yeast (*see* **Section 3.2** and **Note 25**).
3. Select for positives on medium –Leu/–Trp/–Ade/–His + 1 mM 3-AT (when there is no autoactivation; *see* **Section 3.4** and **Note 16**). Incubate the plates at 20°C and check for growth after 5–7 days of incubation.
4. A LacZ assay should be performed to test for activation of the second reporter gene (*see* **Section 3.10**) and rule out false positives due to reversions toward a functional gene at the mutant *his3* locus.

3.10. LacZ Assay

Yeast colonies that grow on selection medium can be further tested for being true positives by blue-white screening. All positives identified in an initial screening based on the HIS or ADE marker should give a blue stain, indicating activation of the LacZ reporter gene, in cases of a true protein–protein interaction event. The protocol described below is based on Duttweiler (35).

1. Make a replicate plate (*see* **Note 26**) by growing the yeast clones on a plate with selection for the presence of the vectors and a very mild selection for protein–protein interactions (*see* **Note 27**).
2. Grow this plate for 2–4 days at 20°C until full-grown yeast spots.

3. Put the plate without the lid upside down in a fume hood on top of a droplet of chloroform and incubate for at least 10 min (*see* **Note 28**).
4. Overlay the plate with a thin layer of RoBlue medium supplemented with 0.5 mg/mL X-GAL (*see* **Note 29**).
5. When the medium has hardened, invert the plate and incubate at 30°C.
6. Check for blue coloring every hour (*see* **Note 30**).

3.11. Colony PCR on Yeast

A colony PCR for yeast works essentially the same as for *E. coli*. The oligonucleotides used are specific for the prey vector (AD) (*see* **Note 31**).

1. Prepare the following PCR mix on ice (total volume of 50 μL):
   dNTPs (10 mM each) 1 μL
   Taq (3–5 U/μL) 0.5 μL
   PCR buffer (10X) 5 μL
   FW AD Oligo (10 pmol/μL) 1 μL
   RV AD Oligo (10 pmol/μL) 1 μL
   $MgCl_2$ (0.1 M) 0.8 μL
   MQ water 40.2 μL
2. Add directly to each tube a bit of yeast with, e.g., a toothpick.
3. Run the PCR reaction with the following program:
   1. 5 min 94°C
   2. 30 s 94°C
   3. 30 s 55°C
   4. 1.5 min 72°C (time depending on expected insert size and used Taq polymerase)
   5. Repeat steps 2–4 for a total of 30 cycles
   6. 5 min 72°C
4. Run 10 μL of each reaction on a 0.8% agarose gel.
5. When a single amplified product is visible on gel, the other 40 μL of the reaction can be purified with a PCR purification kit or by precipitation.
6. The product can be sequenced using the FW AD oligonucleotide.

3.12. Plasmid DNA Extraction from Yeast

Plasmid DNA can be isolated from yeast, but for the relatively fast protocol given below the quality of the DNA is insufficient for direct further analysis. Therefore, isolated DNA should be transformed to *E. coli*, followed by plasmid DNA isolation according to a standard miniprep protocol before further analyses, such as restriction digestion and sequencing, are performed.

1. Make an overnight culture of the yeast in SD medium with the appropriate selection (incubate for 16–18 h at 30°C with shaking at 300 rpm).
2. Transfer the yeast culture to a 1.5 mL Eppendorf tube and spin in a microcentrifuge for 10 s at maximum speed (13,000 rpm).
3. Resuspend the pellet in 0.2 mL of lysis buffer by vortexing.
4. Add 0.2 mL of phenol/chloroform and add 0.3 g of acid-washed glass beads (425–600 μm) and vortex for 2 min.
5. Centrifuge in a microcentrifuge for 5 min at maximum speed (13,000 rpm).
6. Transfer top aqueous phase to a new Eppendorf tube and precipitate the DNA with 2.5 volumes of ethanol (absolute) and 1/10 volume of 3 M NaOAc, pH 5.2.
7. Centrifuge in a microcentrifuge for 10 min at maximum speed (13,000 rpm).
8. Wash the DNA pellet with 70% (v/v) ethanol and centrifuge again in a microcentrifuge for 10 min at maximum speed (13,000 rpm).
9. Dissolve the pellet in 20 μL of TE or sterile water.
10. Use 2 μL of plasmid DNA to transform *E. coli*.

3.13. Preparation of Yeast Glycerol Stocks

1. Make an overnight culture of the yeast in SD medium with the appropriate selection (incubate for 16–18 h at 30°C with shaking at 300 rpm).
2. Transfer 500 μL of yeast culture to an Eppendorf tube or cryo tube, add 125 μL of glycerol, and mix very well (*see* **Note 32**).
3. Store the tube directly in a –80°C freezer for long-term storage (*see* **Note 33**).

4. Notes

1. When using other vector systems, pay attention to the selection markers. BD and AD vectors should have different selection markers and in general Gateway-compatible vectors are not compatible with restriction enzyme-based cloning vectors. Also various yeast strains are available that are suitable for yeast two-hybrid screenings. We prefer the PJ69-4 strains because of their strong growth capacity (34).

2. Follow instructions for standard Gateway cloning (Invitrogen, Carlsbad, CA, USA), which will finally result in an expression clone that is in frame with the BD domain.
3. AD vector, cDNA library, or specific cDNA clones are used depending on the purpose of the experiment.
4. It is important that selection markers for the extra vectors for Y3H and Y4H be different from the other vectors. It is also important to consider the combinations to be made. Make sure that no interaction occurs directly between the protein fused to the BD domain and the protein fused to the AD domain. The idea is to see if an interaction occurs when a third (Y3H) or fourth protein (Y4H) is added.
5. We advise to consequently use bait vectors (BD; pDEST32) in the yeast strain PJ69-4α and prey vectors (AD; pDEST22) in the strain PJ69-4a, facilitating mating. However, the vectors also work in the other yeast strain.
6. Diluting the overnight culture 1 to 10 and growing until the OD_{600} is between 0.4 and 0.6 will result in strongly increased transformation frequencies. However, the protocol gets more laborious and for most direct transformations with a single plasmid this is not essential.
7. If a large number of individual transformations are to be performed, a fresh pre-mix can be made from everything except for the plasmid DNA.
8. Before use, boil the carrier DNA for 5 min and afterward put directly on ice to maintain single-stranded state.
9. The plasmid DNA to be used for yeast transformation may be isolated by using a standard miniprep kit or by any other manual protocol. Purity of DNA is only an issue for library transformations when high transformation frequencies are required.
10. Incubation times can be up to 1 hour; longer times will give higher transformation frequency. However, note that this is yeast strain dependent.
11. For large-scale yeast transformation, use the yeast that already contains the BD vector and transform it with the AD vector containing a cDNA library. The yeast with the BD vector should have been tested for autoactivation before using it in the protein–protein interaction screen.
12. Determine which dilution has to be made to obtain an OD_{600} of 0.1.
13. Do not vortex.
14. The transformation frequency can be calculated as follows: (number of counted colonies) $\times 2.4 \times 10^3$. This gives you

the total number of clones screened in the assay. In order to have a good coverage of all different clones present in the library, a number of transformants need to be analyzed that reflect three to four times the titer of the used AD-cDNA expression library. The protocol should yield 10^5–10^6 colonies per individual screen.

15. The number of positives that can be expected is strongly dependent on the bait proteins used. A typical yeast two-hybrid library screen will yield between 20 and 200 positives. We advise not to combine selection for both markers (ADE and HIS), because this results in a very strong selection pressure demanding too much from the yeast. Double selection will seriously affect transformation efficiency.
16. 3-Amino-1,2,4-triazole (3-AT) is a competitive inhibitor of the product of the *HIS3* gene and can be used to repress autoactivation. Use a concentration range from 1 to 50 mM. Using much higher concentrations of 3-AT is not helpful, because when autoactivation is that strong, an interaction with a second protein fused to the GAL4-AD domain will not result in a stronger expression of the reporter and true interactors cannot be distinguished from background growth.
17. Preferably no autoactivation should occur. When this is the case use 1 mM 3-AT in the final screening to repress the leakiness of the HIS3 reporter construct that is present in the PJ69-4 strains. In case of autoactivation, perform the screen for protein–protein interaction at a concentration of at least 3 mM 3-AT above the level at which growth due to autoactivation stops (*see* **Note 16**).
18. Extra confirmation of protein interactions using another selection marker gene helps to avoid false positives.
19. For mating, a single fresh yeast colony of each yeast strain can be resuspended in 50 μL of sterile MQ water.
20. It is not important if the first droplet dried or not, when spotting the second droplet on top of it.
21. Or droplets of yeast resuspended in sterile water.
22. It is advisable to make glycerol stocks of the collection of prey vectors (AD) for other screens at other times. The same is true for bait vectors (BD).
23. Make all possible combinations with the mating. For example, when having 10 bait vectors (BD) and 10 prey vectors (AD), this results in a matrix of 100 combinations. We advise to include also a tube with sterile water in the AD and BD collection to be screened. The water + yeast combinations should be negative after transferring yeast material

from the SD Complete plate to the SD –Leu/–Trp plate, and the water–water combination should give no growth from the SD Complete plate onward. When the number of combinations to be screened is large it is advisable to make use of electronic multi-channel pipettes or a pipetting robot to avoid errors in making the yeast combinations for the mating.

24. In general, it is advisable to perform a duplicate screen. Always test for interactions on both HIS and ADE selection. This results in theory in eight data points for each combination, four times with protein A as bait and B as prey (two scores from the ADE selection and two scores from the HIS selection) and four times reciprocally, with protein B as bait and A as prey. In case of autoactivation for one of the two proteins, it still results in four data points for the specific combination.
25. Which yeast strain is used for co-transformation is not important. For co-transformation a mixture of the three plasmids can be used. The volume of sterile MQ can be reduced to make up for the additional plasmids so that the total volume of the transformation reaction stays the same. Note that combining three plasmids in one transformation event sometimes gives unexpected recombinations, resulting in the presence of all selection markers, but lack of particular parts of the used vectors. Because of this, it is advisable to check obtained yeast colonies for the presence of the plasmids by yeast colony PCR. Preferably, primers should be used that span the coding region of the proteins of interest. Besides co-transformation, another option to obtain yeast with all three vectors is by mating of one yeast strain that contains one vector with another yeast strain containing two vectors.
26. Another option is to lift the colonies by a filter from the original plate.
27. It is important to trigger the yeast to express the bait and prey protein by a mild selection for protein–protein interaction. Preferably, selection pressure should be weak enough that a negative control still gives some background growth. This yeast can be used as negative control for the blue-white screen and should stay white during the course of the experiment. It is important to include such a negative control because upon long incubation times even wild-type yeast gives a bit of blue color in the LacZ assay. Alternatively, the same yeast strain can be used, but complemented for the *his3* mutation that is used for selecting protein–protein interaction events.

28. Chloroform is used to destroy the yeast cells and make them permeable for the X-GAL substrate.
29. Dissolve the RoBlue medium in the microwave and cool down to 42°C. Add X-GAL (0.5 mg/mL), mix, and pour a thin layer directly onto the yeast plate. Use about 15 mL per plate (diameter of 15 cm).
30. The stronger the protein interaction is, the quicker the color reaction will be. Strong interactors will result in blue coloring in 1 h; weaker interactors can take up to 24 h.
31. The PCR conditions may have to be adjusted depending on the Taq polymerase and buffer to be used. The correct $MgCl_2$ concentration in the PCR reaction is important and may have to be adjusted. The FW oligonucleotide is specific for the AD domain (for pDEST22 or previous restriction enzyme-based version of GAL4-AD plasmids) and the RV oligonucleotide is specific for the ADH terminator (for pDEST22 or previous version of GAL4-AD plasmids).
32. When glycerol stocks have to be prepared for a large number of yeast cultures, an option is to grow in 96-well plates in SD freezing medium. Use flat-bottom plates and use a maximum of 150 μL medium per well, close the plate with parafilm or foil for PCR plates, and grow overnight at 30°C with shaking at 250–300 rpm. Other yeast strains than PJ69-4α/A tend to grow slower in standard flat-bottom 96-well plates, and for these it is advisable to use plates with larger wells that allow to shake faster without getting cross-contamination.
33. Do not use liquid nitrogen to freeze yeast glycerol stocks. Put the tube or plate directly in the –80°C freezer. To use the stored frozen glycerol stock, put stock on ice and let it melt slowly; mix the culture and take a bit of yeast (5–10 μL) for inoculation and put the tube back in the –80°C freezer. Treat 96-well plates with yeast in SD freezing medium (*see* **Note 32**) the same.

Acknowledgments

The work in de Folter laboratory is financed by the Mexican Science Council (CONACyT 82826) and the EU-FP7 EVOCODE project. R.G.H. Immink is supported by the Centre for BioSystems Genomics (CBSG) program, which is part of the Netherlands Genomics Initiative (NGI).

References

1. Uetz, P., Giot, L., Cagney, G., Mansfield, T. A., Judson, R. S., Knight, J. R., Lockshon, D., Narayan, V., Srinivasan, M., Pochart, P., Qureshi-Emili, A., Li, Y., Godwin, B., Conover, D., Kalbfleisch, T., Vijayadamodar, G., Yang, M., Johnston, M., Fields, S., and Rothberg, J. M. (2000) A comprehensive analysis of protein-protein interactions in *Saccharomyces cerevisiae*. *Nature* ***403***, 623–627.
2. Ito, T., Chiba, T., Ozawa, R., Yoshida, M., Hattori, M., and Sakaki, Y. (2001) A comprehensive two-hybrid analysis to explore the yeast protein interactome. *Proc. Natl. Acad. Sci. USA* ***98***, 4569–4574.
3. Walhout, A. J. M., Sordella, R., Lu, X., Hartley, J. L., Temple, G. F., Brasch, M. A., Thierry-Mieg, N., and Vidal, M. (2000) Protein interaction mapping in *C. elegans* using proteins involved in vulval development. *Science* ***287***, 116–122.
4. Giot, L., Bader, J. S., Brouwer, C., Chaudhuri, A., Kuang, B., Li, Y., Hao, Y. L., Ooi, C. E., Godwin, B., Vitols, E., Vijayadamodar, G., Pochart, P., Machineni, H., Welsh, M., Kong, Y., Zerhusen, B., Malcolm, R., Varrone, Z., Collis, A., Minto, M., Burgess, S., McDaniel, L., Stimpson, E., Spriggs, F., Williams, J., Neurath, K., Ioime, N., Agee, M., Voss, E., Furtak, K., Renzulli, R., Aanensen, N., Carrolla, S., Bickelhaupt, E., Lazovatsky, Y., DaSilva, A., Zhong, J., Stanyon, C. A., Finley, R. L., Jr., White, K. P., Braverman, M., Jarvie, T., Gold, S., Leach, M., Knight, J., Shimkets, R. A., McKenna, M. P., Chant, J., and Rothberg, J. M. (2003) A protein interaction map of *Drosophila melanogaster*. *Science* ***302***, 1727–1736.
5. Stelzl, U., Worm, U., Lalowski, M., Haenig, C., Brembeck, F. H., Goehler, H., Stroedicke, M., Zenkner, M., Schoenherr, A., Koeppen, S., Timm, J., Mintzlaff, S., Abraham, C., Bock, N., Kietzmann, S., Goedde, A., Toksoz, E., Droege, A., Krobitsch, S., Korn, B., Birchmeier, W., Lehrach, H., and Wanker, E. E. (2005) A human protein-protein interaction network: a resource for annotating the proteome. *Cell* ***122***, 957–968.
6. Rual, J.-F., Venkatesan, K., Hao, T., Hirozane-Kishikawa, T., Dricot, A., Li, N., Berriz, G. F., Gibbons, F. D., Dreze, M., Ayivi-Guedehoussou, N., Klitgord, N., Simon, C., Boxem, M., Milstein, S., Rosenberg, J., Goldberg, D. S., Zhang, L. V., Wong, S. L., Franklin, G., Li, S., Albala, J. S., Lim, J., Fraughton, C., Llamosas, E., Cevik, S., Bex, C., Lamesch, P., Sikorski, R. S., Vandenhaute, J., Zoghbi, H. Y., Smolyar, A., Bosak, S., Sequerra, R., Doucette-Stamm, L., Cusick, M. E., Hill, D. E., Roth, F. P., and Vidal, M. (2005) Towards a proteome-scale map of the human protein-protein interaction network. *Nature* ***437***, 1173–1178.
7. Parrish, J. R., Yu, J., Liu, G., Hines, J. A., Chan, J. E., Mangiola, B. A., Zhang, H., Pacifico, S., Fotouhi, F., DiRita, V. J., Ideker, T., Andrews, P., and Finley, R. L., Jr. (2007) A proteome-wide protein interaction map for Campylobacter jejuni. *Genome Biol.* ***8***, R130.
8. Rain, J. C., Selig, L., De Reuse, H., Battaglia, V., Reverdy, C., Simon, S., Lenzen, G., Petel, F., Wojcik, J., Schachter, V., Chemama, Y., Labigne, A., and Legrain, P. (2001) The protein-protein interaction map of Helicobacter pylori. *Nature* ***409***, 211–215.
9. Fields, S., and Song, O. (1989) A novel genetic system to detect protein-protein interactions. *Nature* ***340***, 245–246.
10. Fields, S. (2009) Interactive learning: lessons from two hybrids over two decades. *Proteomics* ***9***, 5209–5213.
11. Shoemaker, B. A., and Panchenko, A. R. (2007) Deciphering protein-protein interactions. Part I. Experimental techniques and databases. *PLoS Comput. Biol.* ***3***, e42.
12. Causier, B., and Davies, B. (2002) Analysing protein-protein interactions with the yeast two-hybrid system. *Plant Mol. Biol.* ***50***, 855–870.
13. Bruckner, A., Polge, C., Lentze, N., Auerbach, D., and Schlattner, U. (2009) Yeast two-hybrid, a powerful tool for systems biology. *Int. J. Mol. Sci.* ***10***, 2763–2788.
14. Suter, B., Kittanakom, S., and Stagljar, I. (2008) Two-hybrid technologies in proteomics research. *Curr. Opin. Biotechnol.* ***19***, 316–323.
15. Schmid, M., Davison, T. S., Henz, S. R., Pape, U. J., Demar, M., Vingron, M., Scholkopf, B., Weigel, D., and Lohmann, J. U. (2005) A gene expression map of *Arabidopsis thaliana* development. *Nat. Genet.* ***37***, 501–506.
16. Laubinger, S., Zeller, G., Henz, S., Sachsenberg, T., Widmer, C., Naouar, N., Vuylsteke, M., Scholkopf, B., Ratsch, G., and Weigel, D. (2008) At-TAX: a whole genome tiling array resource for developmental expression analysis and transcript identification in *Arabidopsis thaliana*. *Genome Biol.* ***9***, R112.

17. Morsy, M., Gouthu, S., Orchard, S., Thorneycroft, D., Harper, J. F., Mittler, R., and Cushman, J. C. (2008) Charting plant interactomes: possibilities and challenges. *Trends Plant Sci.* **13**, 183–191.
18. Walter, M., Chaban, C., Schutze, K., Batistic, O., Weckermann, K., Nake, C., Blazevic, D., Grefen, C., Schumacher, K., Oecking, C., Harter, K., and Kudla, J. (2004) Visualization of protein interactions in living plant cells using bimolecular fluorescence complementation. *Plant J.* **40**, 428–438.
19. Immink, R. G. H., Gadella, T. W. J., Jr., Ferrario, S., Busscher, M., and Angenent, G. C. (2002) Analysis of MADS box protein-protein interactions in living plant cells. *Proc. Natl. Acad. Sci. USA* **99**, 2416–2421.
20. Shannon, P., Markiel, A., Ozier, O., Baliga, N. S., Wang, J. T., Ramage, D., Amin, N., Schwikowski, B., and Ideker, T. (2003) Cytoscape: a software environment for integrated models of biomolecular interaction networks. *Genome Res.* **13**, 2498–2504.
21. Iragne, F., Nikolski, M., Mathieu, B., Auber, D., and Sherman, D. (2005) ProViz: protein interaction visualization and exploration. *Bioinformatics* **21**, 272–274.
22. Breitkreutz, B. J., Stark, C., and Tyers, M. (2003) Osprey: a network visualization system. *Genome Biol.* **4**, R22.
23. Mrowka, R. (2001) A Java applet for visualizing protein-protein interaction. *Bioinformatics* **17**, 669–671.
24. Chang, A. N., McDermott, J., and Samudrala, R. (2005) An enhanced Java graph applet interface for visualizing interactomes. *Bioinformatics* **21**, 1741–1742.
25. de Folter, S., Immink, R. G. H., Kieffer, M., Parenicová, L., Henz, S. R., Weigel, D., Busscher, M., Kooiker, M., Colombo, L., Kater, M. M., Davies, B., and Angenent, G. C. (2005) Comprehensive interaction map of the Arabidopsis MADS box transcription factors. *Plant Cell.* **17**, 1424–1433.
26. de Folter, S., Shchennikova, A. V., Franken, J., Busscher, M., Baskar, R., Grossniklaus, U., Angenent, G. C., and Immink, R. G. H. (2006) A B_{sister} MADS-box gene involved in ovule and seed development in petunia and Arabidopsis. *Plant J.* **47**, 934–946.
27. Immink, R. G., Tonaco, I. A., de Folter, S., Shchennikova, A., van Dijk, A. D., Busscher-Lange, J., Borst, J. W., and Angenent, G. C. (2009) SEPALLATA3: the 'glue' for MADS box transcription factor complex formation. *Genome Biol.* **10**, R24.
28. Immink, R. G. H., Ferrario, S., Busscher Lange, J., Kooiker, M., Busscher, M., and Angenent, G. C. (2003) Analysis of the petunia MADS-box transcription factor family. *Mol. Gen. Genet.* **268**, 598–606.
29. Ferrario, S., Immink, R. G., Shchennikova, A., Busscher-Lange, J., and Angenent, G. C. (2003) The MADS box gene *FBP2* is required for SEPALLATA function in petunia. *Plant Cell* **15**, 914–925.
30. Ciannamea, S., Kaufmann, K., Frau, M., Tonaco, I. A., Petersen, K., Nielsen, K. K., Angenent, G. C., and Immink, R. G. (2006) Protein interactions of MADS box transcription factors involved in flowering in *Lolium perenne*. *J. Exp. Bot.* **57**, 3419–3431.
31. Koegl, M., and Uetz, P. (2007) Improving yeast two-hybrid screening systems. *Brief. Funct. Genomics Proteomics* **6**, 302–312.
32. Cusick, M. E., Klitgord, N., Vidal, M., and Hill, D. E. (2005) Interactome: gateway into systems biology. *Hum. Mol. Genet.* **14** *Spec. No. 2*, R171–R181.
33. Gietz, R. D. (2006) http://home.cc.umanitoba.ca/~gietz/
34. James, P., Halladay, J., and Craig, E. A. (1996) Genomic libraries and a host strain designed for highly efficient two-hybrid selection in yeast. *Genetics* **144**, 1425–1436.
35. Duttweiler, H. M. (1996) A highly sensitive and non-lethal beta-galactosidase plate assay for yeast. *Trends Genet.* **12**, 340–341.

Chapter 9

Mapping Functional Domains of Transcription Factors

Ling Zhu and Enamul Huq

Abstract

Transcription factors are modular in nature in all organisms. In general, they have a DNA binding domain, one or more transcription activation and/or repressor domain, and often a dimerization domain. In many cases, transcription factors also have other protein–protein interaction domain(s). Mapping these functional domains in transcription factors is critical in understanding their molecular function. In this chapter, protocols for mapping the DNA binding domain and the transcription activation domain of a bHLH class of transcription factor are described. In principle, these protocols can be applied to other classes of transcription factors for mapping their functional domains.

Key words: *Arabidopsis*, bHLH transcription factor, dimerization, domain mapping, DNA binding, transcription activation, repression domain.

1. Introduction

Regulation of transcription is one of the most critical steps in controlling gene expression. Transcription involves a large number of protein factors including basal factors, RNA polymerases, and a host of regulatory transcriptional activators (1). It turns out that the basic mechanisms of RNA synthesis among all three major classes of RNA polymerases as well as among diverse organisms are largely similar. However, different organisms have evolved with diverse temporal and spatial regulation of transcription. In general, the temporal and spatial regulations are mediated by different classes of DNA binding transcriptional activator proteins.

The transcriptional activator proteins are modular in nature (2). In general, they have a DNA binding domain (DBD) and one or more transcription activation and/or repressor domain(s). In

L. Yuan, S.E. Perry (eds.), *Plant Transcription Factors*, Methods in Molecular Biology 754,
DOI 10.1007/978-1-61779-154-3_9,

many cases, transcription factors also have a dimerization domain and additional protein–protein interaction domain(s). The DNA binding domain is responsible for recognizing a specific DNA sequence and specifies the target genes for each transcription factor. The transcription activation and/or repressor domain, when fused to a DNA binding domain, can regulate gene expression.

One of the central features of the DNA binding domains is that they can be classified into different families based on either their primary sequence or their structural similarities. Major classes of DNA binding domains include helix-turn-helix (HTH) proteins, homeodomains, zinc finger proteins, steroid receptors, leucine zipper (bZIP) proteins, and the basic helix-loop-helix (bHLH) proteins. In addition, there are other novel classes of DNA binding domains (3). Although, a simple basic region enriched in basic amino acids is responsible for DNA binding for bZIP and bHLH proteins, usually a secondary structural feature (α-helix in most cases) is responsible for DNA binding in other classes of DNA binding proteins.

Unlike DNA binding domains, the transcription activation domains have less primary amino acid sequence similarity. The transcription activation domains have been classified into acidic, glutamine-rich, proline-rich, serine/threonine-rich, and HOB1/HOB2 activation domains (2). The defining feature is the enrichment of the aforementioned amino acids in a specific region of a transcription factor (*see* **Note 1**). Whether the enrichment of these diverse amino acids in a region constitutes a structural feature still remains an open question.

In this chapter, we provide detailed protocols for mapping the DNA binding and transcription activation domains of PHYTOCHROME INTERACTING FACTOR 1 (PIF1), a bHLH class transcription factor from *Arabidopsis* (4, 5). DNA binding domains in bHLH factors have been characterized both in animal and in plant systems (6–8), and the protocol is based on these studies (*see* **Note 2**). In addition, PIF1 and other PIFs also interact with phytochrome A (phyA) and phytochrome B (phyB) with distinct *a*ctive *p*hy*A* (APA) and *a*ctive *p*hy*B* (APB) binding domains, respectively. Detailed methods for mapping these functional domains have been described elsewhere (7, 9, 10).

2. Materials

2.1. Particle Bombardment

1. Effector, reporter, and control constructs as illustrated in **Fig. 9.1**.
2. Appropriate plant tissue. In this study, we used *Arabidopsis* dark grown seedlings sown on MS medium with 0.8% sucrose with an overlying filter paper.

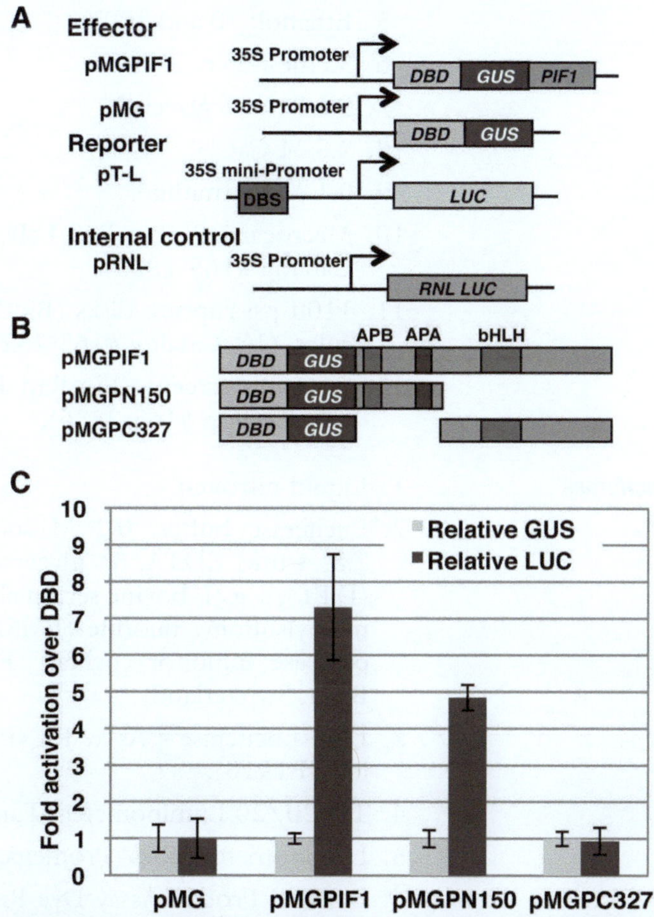

Fig. 9.1. Mapping of the transcriptionalactivation domain(s) of PIF1. (**a**) Constructs used for the experiment. The effector constructs were designed to express a GAL4 DNA binding domain (DBD) PIF1 fusion (pMGPIF1) or the GAL4 DNA binding domain alone (pMG). The reporter construct (pT-L) expresses a firefly luciferase (LUC) from the 35S minimal promoter fused to the GAL4 DNA binding site (DBS). The internal control (pRNL) expresses a Renilla luciferase (RNL LUC) from the 35S promoter. (**b**) PIF1 deletion constructs used to map the transcriptional activation domains. Each effector construct in (**a**) and (**b**) is fused to β-glucuronidase (GUS) to permit the determination of the expression level of the fusion proteins (*see* **Note 16**). (**c**) Three-day-old etiolated *Arabidopsis* seedlings were co-bombarded with the reporter, effector, and internal control constructs. Seedlings were treated for 15 min with far-red light and then incubated in darkness for 16 h. Means ± SE from four biological replicates are shown. Transcriptional activity was measured in seedling extracts by a dual-luciferase assay system (Promega, Madison, WI). Fold activation is expressed as transcriptional activation activity of DBD–GUS–PIF1 over transcriptional activity of DBD–GUS (*light gray bar*) and normalized with GUS activity for the amount of protein expressed by each construct (*dark gray bar*).

3. Biolistic PDS-1000/He and Hepta Systems (Bio-Rad Laboratories, Inc., Hercules, CA).
4. 1.0 μm gold microcarriers (Bio-Rad Laboratories, Inc., Hercules, CA. Catalog #165-2263).

5. Ethanol: 70 and 100%.
6. Sterile water.
7. Sterile 50% glycerol.
8. 2.5 M $CaCl_2$.
9. 0.1 M spermidine.
10. Macrocarriers (Bio-Rad Laboratories, Inc., Hercules, CA. Catalog #165-2335).
11. 1100 psi rupture disks (Bio-Rad Laboratories, Inc., Hercules, CA. Catalog #165-2329).
12. Stopping screens (Bio-Rad Laboratories, Inc., Hercules, CA. Catalog #165-2336).

2.2. Luciferase Assays

1. Liquid nitrogen.
2. Luciferase buffer: 0.2 M sodium phosphate buffer, pH 7.8, 4 mM EDTA, 5% glycerol (v/v), 2 mM dithiothreitol (DTT), 1 g/L bovine serum albumin (BSA), 2 mM phenylmethylsulfonyl fluoride (PMSF), and 10% (v/v) complete protease inhibitor cocktail (F. Hoffmann-La Roche Ltd, Basel, Switzerland).
3. Dual-Luciferase® Assay Kit (Promega, Madison, WI. Catalog #E1910).
4. TD-20/20 Luminometer (Turner Designs).
5. Luminometer tubes (Promega, Madison, WI).
6. Bio-Rad Protein Assay Dye Reagent Concentrate (Bio-Rad Laboratories, Inc., Hercules, CA. Catalog #500-0006).

2.3. GUS Assays

1. Reaction buffer: Dissolve 22 mg of 4-methylumbelliferyl-β-D-glucuronide (MUG) in 50 mL of luciferase extraction buffer shown in **Section 2.2,** step 2. Because both GUS and LUC assays are performed on the same extract, LUC extraction buffer is used here.
2. Stop buffer: 0.2 M Na_2CO_3.
3. 1 μM 4-methylumbelliferone (MU) standard solution: dissolve 0.0019 g 4-methylumbelliferone (MU) in stop buffer to make 10 mM 4-methylumbelliferone (MU). Dilute 10 mM MU to 1 μM 4-methylumbelliferone (MU) with stop buffer.
4. Polystyrene tubes (Sarstedt, Nümbrecht, Germany. Catalog #55.476.005).

2.4. Site-Directed Mutagenesis

1. QuikChange® Site-Directed mutagenesis kit (Stratagene, La Jolla, CA. Catalog# 200519) (*see* **Note 3**).
2. Custom oligonucleotides to introduce mutations.

3. *Dpn*I.
4. XL1-blue supercompetent cells.
5. LB agar plates with appropriate antibiotic.
6. QIAprep Spin Miniprep Kit (Qiagen, Inc., Valencia, CA. Catalog # 27106).
7. Standard reagents and equipment to sequence to verify constructs.

2.5. Expression of Protein Using the TNT System

1. TNT® Quick Coupled Transcription/Translation Systems (Promega, Madison, WI. Catalog # L1170).
2. Plasmid DNA or PCR-generated DNA template consisting of the transcription factor of interest and variations thereof cloned downstream of the T7 or SP6 RNA polymerase promoter. See the TNT manual for details.
3. 6X SDS gel loading buffer: 300 mM Tris–HCl, pH 6.8, 12% SDS, 0.6% bromophenol blue, 60% glycerol, 600 mM DTT.
4. 10% sodium dodecyl sulfate polyacrylamide gel electrophoresis (SDS-PAGE) gel:

	Lower gel	Upper gel
30% Acrylamide/bis solution	3.4 mL	0.83 mL
4X Buffer	2.5 mL	1.25 mL
Tetramethylethylenediamine (TEMED)	5 μL	5 μL
10% ammonium persulfate (APS)	50 μL	50 μL
H_2O	4.05 mL	2.87 mL
Total	10 mL	5 mL

5. 4X lower gel buffer: 1.5 M Tris–HCl, pH 8.8, 0.4% SDS.
6. 4X upper gel buffer: 0.5 M Tris–HCl, pH 6.8, 0.4% SDS.
7. Fixing solution 100 mL: Add 10 mL of glacial acetic acid to a mixture of 50 mL methanol and 40 mL H_2O.
8. 10% glycerol.
9. Whatman 3MM filter paper.
10. Standard equipment to run, dry, and image SDS-PAGE gels.

2.6. Electrophoretic Mobility Shift Assay (EMSA)

1. Custom oligonucleotides to generate probe.
2. 3 M sodium acetate (NaOAc), pH 5.2.
3. Ethanol.
4. 50 mM NaCl.

5. 10 mM Tris–HCl, pH 8.5.
6. Random primed DNA Labeling Kit (F. Hoffmann-La Roche Ltd, Basel, Switzerland. Catalog #11004760001).
7. 10X binding buffer: 200 mM HEPES, pH 7.9, 400 mM KCl, 10 mM EDTA, and 10 mM dithiothreitol (DTT).
8. 50% glycerol.
9. 10% NP-40.
10. Poly(dI-dC), 500 ng/μL.
11. 10X TBE buffer (1 L): 107.8 g Tris base, 55 g boric acid, and 7.44 g disodium EDTA·$2H_2O$.
12. 10X gel loading buffer: 250 mM Tris–HCl, pH 7.5, 0.2% bromophenol blue (w/v), and 40% glycerol (v/v).
13. 5% non-denaturing polyacrylamide gel:

30% acrylamide:bisacrylamide (29:1)	2.5 mL
10X TBE	0.75 mL
10% APS (w/v)	75 μL
TEMED	7.5 μL
H_2O	11.67 mL
Total volume	15 mL

14. Centri-Sep spin column (Princeton Separations, Adelphia, NJ. Catalog # CS-900).

3. Methods

3.1. Particle Bombardment

3.1.1. Seed Plating and Germination

1. Sterilize 0.01 g of *Arabidopsis* seeds for each sample and plate onto MS + 0.8% sucrose plate with filter paper as described in **Section 2.1**, step 2. Try to align the seeds one by one on the filter paper in a line and keep ~1 cm space between the lines.
2. Keep the plates at 4°C for 4 days for stratification.
3. Expose the seeds for 1 h to white light at room temperature to stimulate germination. Keep the plates open in the hood during that time to reduce water accumulation in the plates. Align all the plates and wrap with aluminum foil with paper towel at the bottom. Place plates vertically at 21°C for 60–72 h, so seedlings will grow vertically. The purpose of the paper towel is to absorb extra water from the medium when growing seedlings vertically.

4. The seedlings will grow flat on the filter paper. Approximately, 5–7 mm tall dark grown seedlings are suitable for particle bombardment experiment (*see* **Note 4**).

3.1.2. Preparation of Gold Microcarriers Coated with DNA

1. Gold microcarriers stock: weigh 30 mg gold microcarriers, rinse with 70% ethanol, and vortex for 3–5 min. Allow the microcarriers to soak in 70% ethanol for 15 min. Centrifuge at 2000×*g* in a microcentrifuge and rinse three times with sterile water. Centrifuge at 2000×*g* between each wash. Remove all the water, and add 500 μL of sterile 50% glycerol. Mix well and aliquot into 10 Eppendorf tubes. Each tube contains 3 mg of gold microcarriers. The stock can be stored at −20°C. For four shootings (four plates), 1 mg of gold microcarriers and 10 μg DNA (1 μg/μL) are needed.
2. When ready to shoot, aliquot the gold microcarriers from the stock to a new tube. Spin gold microcarriers briefly at 2000 rpm in a microfuge and remove the 50% glycerol.
3. Add 1 mL of H_2O to rinse the microcarriers without resuspension. Remove the H_2O carefully without disturbing the microcarriers.
4. For four shootings, resuspend 1 mg of microcarriers in 10 μL of DNA (1 μg/μL). If using more than one plasmid, the total amount of DNA remains 10 μg but an equimolar ratio is used. Add the following in the order specified: 67.5 μL of H_2O, 100 μL of 2.5 M $CaCl_2$, and 22.5 μL of 0.1 M spermidine.
5. Vortex immediately at 4°C for 30 min.
6. Spin at 500 rpm in a microfuge for 5 min at 4°C. Discard the supernatant, add 1 mL of 100% ethanol, and resuspend by pipetting up and down.
7. Centrifuge at 13,000 rpm in a microcentrifuge for 1 min at 4°C, and discard the supernatant.
8. Resuspend the microcarrier coated with DNA with an appropriate amount of 100% ethanol (12 μL/ shot).
9. The microcarriers coated with DNA can be kept on ice for 1–2 h.

3.1.3. Bombardment

1. Pipette 12 μL of the suspension as evenly as possible onto the center of a macrocarrier. Allow ethanol to evaporate.
2. Place 1100 psi rupture disk in the Biolistic unit. The distance between the rupture disk and the macrocarrier is 8–10 mm and between the macrocarrier and the stopping screen is 1 cm.
3. Place petri dishes containing flat grown seedlings in the PDS-1000/He and Hepta Systems approximately 6 cm below the stopping screen.

4. Evacuate chamber to 28–29 mmHg and bombard the target once.
5. After bombardment, keep the plates at 21°C for 16 h, collect the samples in 1.5 mL Eppendorf tubes, and freeze in liquid nitrogen for luciferase and GUS assays.

3.2. Luciferase Assays (See Note 5)

3.2.1. Preparation of Crude Extracts

1. Grind samples into powder in liquid nitrogen.
2. Add 300 μL of luciferase buffer per sample and vortex.
3. Centrifuge at 13,000 rpm in a microcentrifuge for 15 min.
4. Transfer 200 μL of the supernatant for each sample into new eppendorf tubes as cell lysate.
5. Take 5 μL of supernatant from each sample and determine the protein concentration using the Bio-Rad Laboratories, Inc., Hercules, CA Protein Assay (*see* **Section 3.2.3**).

3.2.2. Measurement of Luciferase Activity Using the Promega Dual-Luciferase® Assay Kit

1. Predispense 100 μL of LARII into luminometer tube.
2. Transfer 20 μL of sample, mixing by pipetting up and down.
3. Measure firefly luciferase activity.
4. Dispense 100 μL of Stop & Glo® reagent (included in the kit).
5. Measure Renilla luciferase activity.

3.2.3. Measurement of Protein Concentration

1. Make the standard curve of protein concentration:

Protein concentration (μg/mL)	0.5 mg/mL BSA (μL)	H_2O (μL)	Luciferase buffer (μL)	Reaction buffer (μL)
0	0	795	5	200
2.5	5	790	5	200
5	10	785	5	200
10	20	775	5	200
20	40	755	5	200

Mix the samples well and incubate for 5 min at room temperature before measuring protein concentration using a Beckman DU530 UV-VIS Life Science Spectrophotometer at 595 nm.

2. Reaction buffer: Bio-Rad Protein Assay Dye Reagent Concentrate. Add 795 μL of H_2O, 200 μL of reaction buffer, and 5 μL of protein crude extracts for each sample. Mix well by pipetting a few times and incubate for 5 min at room temperature before measuring the protein concentration using a

Beckman DU530 UV-VIS Life Science Spectrophotometer at 595 nm.

3.3. GUS Assays

3.3.1. Setting Up the Fluorometer (See **Note 6**)

1. Place blank tube (1000 μL of 0.2 M Na_2CO_3) and turn SPAN control fully clockwise. Use the ZERO control to set the display at 000.
2. Place 0.1 μM MU solution (100 μL of 1 μM MU + 900 μL of 0.2 M Na_2CO_3) and use SPAN control to adjust the display to show 1000.
3. Once finished with adjustments, keep the ZERO and SPAN settings.
4. To test the fluorometer, place

   0.01 μM MU (10 μL of 1 μM MU + 990 μL of 0.2 M Na_2CO_3) – display should show around 100 (±10).

   0.025 μM MU (25 μL of 1 μM MU + 975 μL of 0.2 M Na_2CO_3) – display should show 250 (±20).

   0.05 μM MU (50 μL of 1 μM MU + 950 μL of 0.2 M Na_2CO_3) – display should show 500 (±30).
5. The fluorometer is now ready to take the sample readings.

3.3.2. Measurement of GUS Activity

1. Put 900 μL of stop solution (0.2 M Na_2CO_3) in the polystyrene tubes (one set at a time).
2. Add 500 μL of reaction buffer in an Eppendorf tube and 50 μL of crude extracts for each sample. Mix together and immediately incubate at 37°C.
3. Start the timer. Take fluorescence readings at 5, 10, 15, 30, and 60 min (five readings; depending on the activity, the time course can be changed).
4. For every time point, remove 100 μL of reaction sample from 37°C, add to the 900 μL of stop buffer in polystyrene tubes, and place in the fluorometer. The display should read from 100 to 1000. If the numbers are out of the reading range, the GAIN can be set at 1X or 50X to decrease or increase the signal. Alternatively, the samples can be diluted to bring readings within range.
5. Calculate GUS activity from the slope of the time course.

3.3.3. Calculation of Transcriptional Activation Activity

1. Divide the luciferase data by the Renilla luciferase data for each sample to normalize the data for shooting efficiency.
2. Divide the luciferase: Renilla luciferase ratio by the protein concentration for each sample to normalize the data.
3. Divide the values obtained in step 2 for all the samples by the value for the DBD (GAL4 DNA binding domain; pMG

in **Fig. 9.1**) only. This will set the DBD control to 1 and show fold change activity for other constructs in relation to DBD alone.

4. Divide the GUS activity data by the protein concentration for each sample to normalize the data.
5. Divide the GUS data in step 4 for all the samples by the GUS value for the DBD only. This will set the DBD control to 1 and show fold changes in protein level for other constructs in relation to DBD only.
6. Divide the normalized transcriptional activation activity data in step 3 by the GUS data in step 5. This will show the fold change for each construct compared to DBD only with a normalized expression level for each construct (indicated by GUS expression levels). Repeat the process for at least three biological replicates and make an average with error bars. An example of the results produced is shown in **Fig. 9.1c**.

3.4. Site-Directed Mutagenesis

3.4.1. Primers for Site-Directed Mutagenesis Design

1. Use QuikChange Primer Design Program at http://www.Stratagene.com/qcrpimerdesign to design mutagenic primers. Primer lengths should be 25–45 bp. The desired mutation should be in the middle of the primers on both strands annealing with each other.
2. Primer melting temperature (T_m) should be equal to or higher than 78°C. The following formula can be used to calculate T_m:

   $T_m = 81.5 + 0.41(\%GC) - 675/N - \%$ mismatch (For introducing mutations, N is the primer length in bases)

   $T_m = 81.5 + 0.41(\%GC) - 675/N$ (For introducing insertions and deletions, N does not include the bases which are being inserted or deleted)

   Primers should contain at least 40% GC content and terminate in one or more G or C bases.

3.4.2. Site-Directed Mutagenesis Reaction (See **Note 7**)

1. Mutagenesis reaction setup:

dsDNA template (5–50 ng)	x μL
10X reaction buffer	5 μL
Primer 1 (125 ng)	x μL
Primer 2 (125 ng)	x μL
dNTP (2.5 mM)	5 μL
PfuTurbo DNA polymerase (2.5 U/μL)	1 μL
ddH_2O	x μL
Total	50 μL

2. Cycling parameters: 95°C 30 s→95°C 30 s, 55°C 1 min, 68°C 1 min/kb of plasmid length, repeat for 12–18 cycles (point mutations: 12 cycles; single amino acid changes: 16 cycles; multiple amino acid deletions or insertions: 18 cycles).
3. Digestion of the parental dsDNA.
   Add 1 μL of *Dpn*I (10 U/μL) into the reaction tube, mix well, and incubate the tube at 37°C for 1 h.

3.4.3. Transformation and Selection of Correct Plasmid

1. Transfer 1 μL of *Dpn*I-treated sample into XL1-Blue supercompetent cells by heat shock and plate on LB plates containing appropriate antibiotic (*see* **Note 8**).
2. Pick up single colonies from the plate, grow overnight at 37°C, miniprep plasmid DNA, and sequence to confirm the mutation.

3.5. Expression of Protein Using the TNT System

3.5.1. TNT® Quick Coupled Transcription/Translation Reactions

1. Take out the TNT® Quick Coupled Transcription/Translation reagents from –80°C. Rapidly thaw the TNT® Quick Master Mix and place on ice. Do not freeze and thaw the reagents more than twice. Aliquot the unused reagents into small aliquots and store at –80°C.
2. Add the components of reaction as listed below into 1.5 mL Eppendorf tubes. Mix gently by tapping with fingers, and centrifuge briefly if necessary.

	Standard reaction	**Standard reaction using [^{35}S] methionine**
TNT® Quick Master Mix	40 μL	40 μL
Methionine, 1 mM	1 μL	–
[^{35}S] methionine (1000 Ci/mmol at 10 mCi/mL)	–	2 μL
Plasmid DNA template(s) (0.5 μg/μL)	2 μL	2 μL
H_2O	7 μL	6 μL
Total	50 μL	50 μL

3. Incubate the reaction at 30°C for 60–90 min.

3.5.2. Analysis of the Translation Products Using [^{35}S] Methionine

1. Pipette a 10 μL aliquot from the reaction and add 2 μL of 6X SDS loading buffer. The remainder of the reaction can be stored at –80°C.
2. Mix and incubate at 95°C for 3 min.

3. Load samples onto a 10% SDS-PAGE gel. Using a constant current of 15 mA in the stacking gel and 30 mA in the separating gel, run gel until the bromophenol blue dye runs off the bottom of the gel.
4. Gel fixation: shake the gel in fixing solution for 30 min followed by shaking in 10% glycerol for 10 min (*see* **Note 9**).
5. Place the gel on a sheet of Whatman 3MM filter paper. Cover with plastic wrap and dry on a gel dryer.
6. Expose the gel in a PhosphorImaging screen overnight. Scan the image the next morning using a PhosphorImager. An example of the results produced is shown in **Fig. 9.2c**.

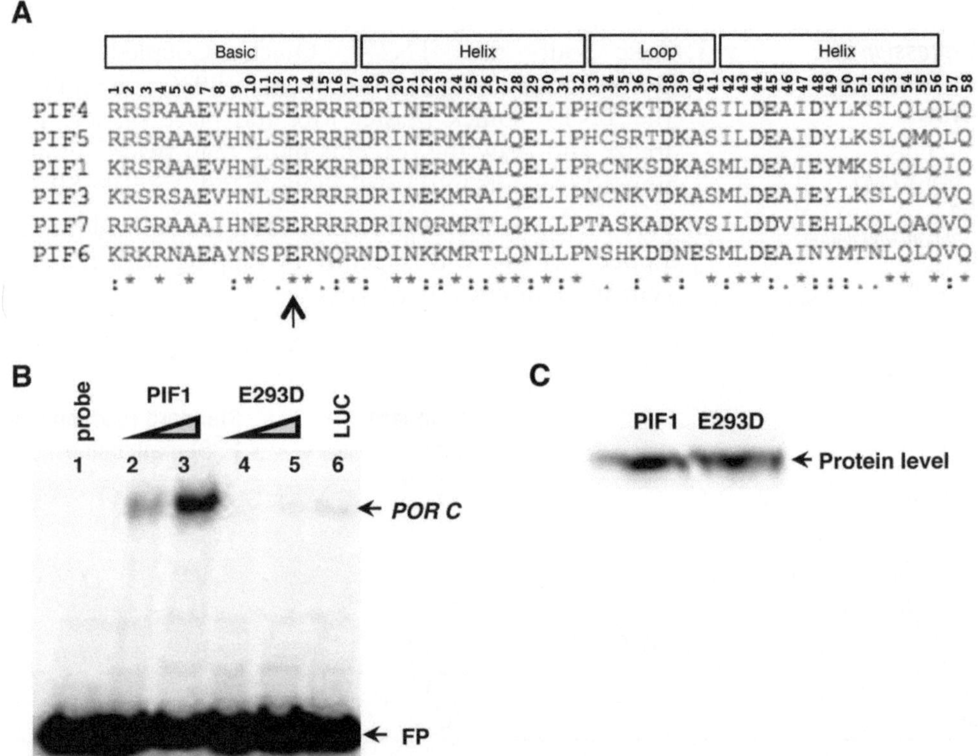

Fig. 9.2. Mapping of the DNA binding domain of PIF1, a bHLH transcription factor. (**a**) Alignment of the basic helix-loop-helix domain of the bHLH proteins PIF1, PIF3, PIF4, PIF5, PIF6, and PIF7. * indicates identical amino acids; : indicates conserved amino acids; and . indicates similar amino acids. *Arrow* indicates the conserved glutamic acid that is responsible for DNA binding. (**b**) The PIF1E293D mutant does not bind to a G-box DNA sequence element (*POR C*) (12, 13). In vitro translated PIF1 or PIF1E293D was incubated with a radiolabeled fragment of *POR C* in a DNA gel shift assay. *Lane 1*, free probe; *lanes 2 and 3*, increasing amount of wt PIF1, *lanes 4 and 5*, increasing amount of PIF1E293D mutant protein, and *lane 6*, unrelated luciferase (LUC) protein as a negative control. FP, free *POR C* probes. *POR C* with the *arrow* indicates probe that is "shifted" in the gel due to decreased mobility caused by association of the PIF1 protein with the probe. (**c**) Comparison of the levels of wild-type and mutant (E293D) PIF1 proteins produced by in vitro transcription and translation system.

3.6. Electrophoretic Mobility Shift Assays (EMSAs)

3.6.1. Preparation of Probes for EMSA

1. Mix 3.5 nmol of each single-stranded DNA oligo and heat it at 95°C for 5 min on a heat block.
2. Leave the block at room temperature afterward for 1–1.5 h to cool down to <35°C.
3. Add 1/10 volume 3 M NaOAc, pH 5.2 and 2.5X volume ethanol.
4. Store at –20°C for 20 min.
5. Centrifuge at 16,000×*g* in a microfuge at 4°C for 20 min.
6. Discard the supernatant and wash the pellet with 1 mL of 70% ethanol.
7. Dry the pellet and resuspend in 50 μL of 50 mM NaCl as high concentration double-stranded oligo.
8. Dilute the high concentration double-stranded oligo 1:100 in 10 mM Tris–HCl, pH 8.5.
9. Label diluted double-stranded oligo with Klenow (random primed DNA labeling) as follows:

Double-stranded oligo	10 μL
Klenow buffer	2.5 μL
2.5 mM dNTP (-dCTP)	2.5 μL
[α-^{32}P]CTP	2 μL
Klenow (2.5 U)	0.5 μL
H_2O	7.5 μL
Total volume	25 μL

10. Allow the reaction to proceed at RT for 0.5–1 h. Meanwhile, prepare the Centri-Sep column for cleaning labeled probe by adding 800 μL of water to the column, mix by vortexing, and allow it to settle for 30 min (*see* **Note 10**).
11. Drain excess water in the Centri-Sep spin column by gravity flow. Centrifuge at 3000 rpm in a microfuge for 2 min at room temperature and discard the flow-through and the collection tube.
12. Put the column in a fresh 1.5 mL Eppendorf tube and add labeled probe to the center of the column. Centrifuge at 3000 rpm in a microfuge for 2 min at room temperature. Keep the flow-through.
13. Prepare a dilution of 10,000 cpm/μL of radiolabeled probe for the binding reaction.

3.6.2. DNA Binding Reaction

1. Assemble the following reaction in a sterile Eppendorf tube:

10X binding buffer	1 μL
50% glycerol (v/v)	2 μL
10% NP-40 (v/v)	0.1 μL
500 ng/μl poly(dI-dC)	0.1 μL
Protein (0.5–5 μg)	2 μL
H_2O	2.8 μL
Total volume	8 μL

2. Incubate the reaction on ice for 10 min.
3. Add 2 μL of radiolabeled probe (10,000 cpm/μL) to the tube and keep on ice for 1 h (*see* **Note 11**).
4. Add 1 μL of room temperature gel loading 10X buffer to stop the reaction (*see* **Note 12**).

3.6.3. Preparation of Non-denaturing Polyacrylamide Gel

1. These instructions are for the use of a Hoefer SE-260 gel system (10 × 10 cm gel). It is critical that the glass and alumina plates for the gels are thoroughly cleaned with a rinsable detergent after each use (e.g., Alconox, New York, NY) and rinsed extensively with distilled water to remove the detergent (*see* **Note 13**).
2. Prepare a 1.5 mm thick, 5% gel by mixing 0.75 mL of 10X TBE buffer with 2.5 mL of 30% acrylamide/bis acrylamide solution (29:1), 11.67 mL of water, 75 μL of 10% ammonium persulfate solution, and 7.5 μL of TEMED. Pour the gel and insert the comb on the top. The gel should polymerize in about 30 min.
3. Prepare the running buffer by diluting 10 mL of 10X TBE buffer with 190 mL of water in a measuring cylinder. Cover with parafilm and invert a few times to mix.
4. Once the gel has set, carefully remove the comb and use a pipetman fitted with 200 μL tips to wash the wells with running buffer.
5. Add the running buffer to the upper (first) and lower (second) chambers of the gel unit. The gel is now ready (*see* **Note 14**).

3.6.4. Electrophoresis of Protein–DNA Complex

1. Pre-run the 5% non-denaturing polyacrylamide gel in 0.5X TBE buffer for 10 min at 10 V/cm before loading the samples (*see* **Note 15**).
2. After loading the samples, run the gel at room temperature in 0.5X TBE buffer at 10 V/cm until the bromophenol blue dye has run three-fourths of the gel.

3. Open the gel plates and place the gel on a sheet of Whatman 3MM filter paper. Cover with plastic wrap and dry on a gel dryer under vacuum.
4. Expose the gel in a PhosphorImaging screen overnight. Scan the image the next morning using a PhosphorImager. An example of the results produced is shown in **Fig. 9.2b**.

4. Notes

1. As mentioned in the **Section 1** preliminary information on a putative transcription activation domain of a new transcription factor can be obtained from the primary amino acid sequence depending on the abundance of certain amino acids in a region of a protein. These are good indications that a certain region containing high percentages of glutamine or proline or acidic amino acids may function as an activation domain. However, the presence or the absence of an activation domain needs to be experimentally verified because simply the presence or the absence of certain amino acids cannot predict an activation domain with certainty.
2. Extensive mutational analyses have been performed for bHLH proteins from animal systems to identify the amino acid residues contributing to DNA binding (6). These residues are highly conserved across species indicating that they play similar roles in bHLH proteins from all organisms. Therefore, based on sequence similarity, it is relatively easy to identify those residues and mutagenize them to create mutant proteins that do not bind to DNA (**Fig. 9.2a**). This method can be used for other transcription factor families to some extent depending on the availability of similar studies in those families.
3. The site-directed mutagenesis kit from Stratagene, La Jolla, CA, works well. However, a kit is not necessary for mutagenesis. Any proofreading enzyme with appropriate buffer and dNTP can be used to perform mutagenesis.
4. This protocol is designed to assess the transcriptional activation activity of PIF1, which is active at the seedling stage (4, 5), and is the reason why we used very young stage dark grown seedlings. However, this assay can be adapted for other stage plants and/or cell culture systems. Protoplasts from *Arabidopsis* or other plants can be transformed using electroporation or PEG methods using the plasmid combinations as shown in **Fig. 9.1a, b** and transcriptional activation activity can be examined.

5. The luminometer is often very tricky to work with. It is important that the setting is set correctly. Make sure to check the sensitivity used on the luminometer every time before use as it will affect the data. Performing four or more technical repeats helps to reduce the error bar. In addition, use of same weight samples helps to reduce large variations in readings.
6. The fluorometer is also very tricky to work with. Allow the fluorometer to warm up for at least 15 min. 1 mM MU stock solution can be stored at 4°C in the dark for at least 1 month. Old MUG solution will increase the fluorescence background. The reading may change slightly, so setting a consistent time for taking each reading such as 1 min after putting the tube into the fluorometer will help reduce variation. Similarly, performing four or more repeats will help minimize the error bar.
7. The reaction can be scaled down to 10 μL as only ~2 μL is needed for transformation. All the components should be scaled down for 10 μL including *Dpn*I enzyme in the subsequent step for digestion of parental plasmid DNA.
8. The plasmid DNA template for site-directed mutagenesis should be from a dam^+ *Escherichia coli* strain because *Dpn*I only cleaves methylated GATC sites.
9. Fixing the gel followed by incubation in 10% glycerol for 10 min will prevent the gel from cracking during drying. Gel cracking often separates the lanes on the gel image.
10. As an alternative to Centri-Sep columns, homemade columns can be prepared using glass wool, a Pasteur pipette and Sephadex G-25 beads. The Sephadex beads can be equilibrated in TE, pH 8.0, buffer. A small amount of glass wool can be inserted into a Pasteur pipette to clog the pipette, so that the beads will stay in the pipette. The homemade columns can be centrifuged in a table top centrifuge.
11. DNA binding reaction can be performed either on ice or at room temperature. However, incubation on ice may stabilize binding for some protein–DNA complexes.
12. For EMSA gel running, the loading buffer can be loaded in one corner lane only without adding to each sample or can be added to the probe only lane. The samples have 10% glycerol, so they will settle at the bottom of each lane after loading in a gel. This might avoid interference with the mobility of the protein–DNA complex.
13. It is *extremely* important to clean the EMSA gel apparatus including the glass and alumina plates extensively to

remove all the detergent. Because this is a native gel, presence of any detergent will destabilize the protein–DNA complex and may produce negative result.

14. Adding running buffer to the top chamber first allows one to detect any possible leak from the top chamber. The gel assembly can be fixed if a leak is detected before pouring the running buffer into the bottom chamber.
15. The protein–DNA complex runs according to the size of the protein. For larger proteins or for supershift assays using antibody or other interacting proteins, a 4% gel can be used for better separation as described (11).
16. It is important that all the constructs are made with an internal reporter such as GUS as shown in **Fig. 9.1a, b**. Some of the fragments may not show transcriptional activation activity possibly due to low and/or no expression or due to lack of transcriptional activation activity. The internal reporter allows one to distinguish between these two possibilities.

Acknowledgments

We thank Dr. Shen and members of the Huq laboratory for critical reading of this manuscript. This work was supported by grants from NSF (IOS-0822811 and IOS-0849287) to E.H.

References

1. Rajnish, J. A. and Hahn, S. (1996) Transcription: basal factors and activation. *Curr. Opin. Genet. Dev.* **6**, 151–158.
2. Johnson, P. F., Sterneck, E., and Williams, S. C. (1993) Activation domains of transcriptional regulatory proteins. *J. Nutr. Biochem.* **4**, 386–398.
3. Pabo, C. O. and Saur, R. T. (1992) Transcription factors: structural families and principles of DNA recognition. *Annu. Rev. Biochem.* **61**, 1053–1095.
4. Huq, E., Al-Sady, B., Hudson, M., Kim, C., Apel, K., and Quail, P. H. (2004) Phytochrome-interacting factor 1 is a critical bHLH regulator of chlorophyll biosynthesis. *Science* **305**(5692), 1937–1941.
5. Oh, E., Kim, J., Park, E., Kim, J. I., Kang, C., and Choi, G. (2004) PIL5, a phytochrome-interacting basic helix-loop-helix protein, is a key negative regulator of seed germination in Arabidopsis thaliana. *Plant Cell* **16**(11), 3045–3058.
6. Davis, R. L., Cheng, P. -F., Lassar, A. B., and Weintraub, H. (1990) The MyoD DNA binding domain contains a recognition code for muscle-specific gene activation. *Cell* **60**, 733–746.
7. Shen, H., Zhu, L., Castillon, A., Majee, M., Downie, B., and Huq, E. (2008) Light-induced phosphorylation and degradation of the negative regulator PHYTOCHROME INTERACTING FACTOR 1 depends upon its direct physical interactions with photoactivated phytochromes. *Plant Cell* **20**, 1586–1602.
8. Al-Sady, B., Kikis, E. A., Monte, E., and Quail, P. H. (2008) Mechanistic duality of transcription factor function in phytochrome signaling. *Proc. Natl. Acad. Sci. USA* **105**, 2232–2237.

9. Khanna, R., Huq, E., Kikis, E. A., Al-Sady, B., Lanzatella, C., and Quail, P. H. (2004) A novel molecular recognition motif necessary for targeting photoactivated phytochrome signaling to specific basic helix-loop-helix transcription factors. *Plant Cell* **16**(11), 3033–3044.
10. Al-Sady, B., Ni, W., Kircher, S., Schafer, E., and Quail, P. H. (2006) Photoactivated phytochrome induces rapid PIF3 phosphorylation prior to proteasome-mediated degradation. *Mol. Cell* **23**, 439–446.
11. Martinez-Garcia, J. F., Huq, E., and Quail, P. H. (2000) Direct targeting of light signals to a promoter element-bound transcription factor. *Science* **288**(5467), 859–863.
12. Su, Q., Frick, G., Armstrong, G., and Apel, K. (2001) POR C of Arabidopsis thaliana: a third light- and NADPH-dependent protochlorophyllide oxidoreductase that is differentially regulated by light. *Plant Mol. Biol.* **47**, 805–813.
13. Moon, J., Zhu, L., Shen, H., and Huq, E. (2008) PIF1 directly and indirectly regulates chlorophyll biosynthesis to optimize the greening process in Arabidopsis. *Proc. Natl. Acad. Sci. USA* **105**, 9433–9438.

Chapter 10

Bimolecular Fluorescence Complementation as a Tool to Study Interactions of Regulatory Proteins in Plant Protoplasts

Sitakanta Pattanaik, Joshua R. Werkman, and Ling Yuan

Abstract

Protein–protein interactions are an important aspect of the gene regulation process. The expression of a gene in response to certain stimuli, within a specific cell type or at a particular developmental stage, involves a complex network of interactions between different regulatory proteins and the *cis*-regulatory elements present in the promoter of the gene. A number of methods have been developed to study protein–protein interactions in vitro and in vivo in plant cells, one of which is bimolecular fluorescence complementation (BiFC). BiFC is a relatively simple technique based upon the reconstitution of a fluorescent protein. The interacting protein complex can be visualized directly in a living plant cell when two non-fluorescent fragments, of an otherwise fluorescent protein, are fused to proteins found within that complex. Interaction of tagged proteins brings the two non-fluorescent fragments into close proximity and reconstitutes the fluorescent protein. In addition, the subcellular location of an interacting protein complex in the cell can be simultaneously determined. Using this approach, we have successfully demonstrated a protein–protein interaction between a R2R3 MYB and a basic helix-loop-helix MYC transcription factor related to flavonoid biosynthetic pathway in tobacco protoplasts.

Key words: Bimolecular fluorescent complementation, split YFP, transcription factors, protein–protein interaction, protoplast expression.

1. Introduction

Regulation of gene expression is a complex process controlled by a network of interactions between different regulatory proteins and cofactors. An in-depth analysis of protein–protein interactions is crucial for generating valuable information about the protein function and the regulatory network involved in the expression of gene(s) in a specific cell type, in response to certain stimuli or

L. Yuan, S.E. Perry (eds.), *Plant Transcription Factors*, Methods in Molecular Biology 754,
DOI 10.1007/978-1-61779-154-3_10, © Springer Science+Business Media, LLC 2011

at a particular developmental stage in a pathway in plants. The information generated from such studies can help in manipulating important biochemical pathways leading to the production of crops with higher yield potential, improved nutritional value or useful compounds for human health.

Over the years, several methods have been developed to study protein–protein interactions in vivo, as well as in vitro, in plants. The commonly used methods are yeast two-hybrid assay (1), co-immunoprecipitation (2), surface plasmon resonance spectroscopy (3), bioluminescence resonance energy transfer (BRET; 4) and fluorescence resonance energy transfer (FRET; 5). In the past few years, a new technique called bimolecular fluorescence complementation (BiFC) has been widely used to study protein–protein interactions in living cells. The BiFC assay was initially developed to study protein–protein interactions in living mammalian cells (6, 7). Subsequently, this technique has been adapted for plant science research and is being widely used to study protein–protein interactions in plant cells (reviewed in 5, 8 and 9).

A set of BiFC vectors has been designed and validated by demonstrating known interactions between different classes of proteins in plant protoplasts, as well as in transgenic plants (10–13). The BiFC assay has gained popularity over the years, because it is a relatively simple technique based upon the reconstitution of a fluorescent complex. When two non-fluorescent fragments of an otherwise fluorescent protein, such as the yellow fluorescent protein (YFP), are brought into close proximity by the interaction of two proteins fused to the fragments, the functionality of the fluorescent protein is restored (**Fig. 10.1a, b**). The other advantage of BiFC is that it simultaneously allows the determination of the subcellular location of an interacting protein complex in a living cell (8, 9).

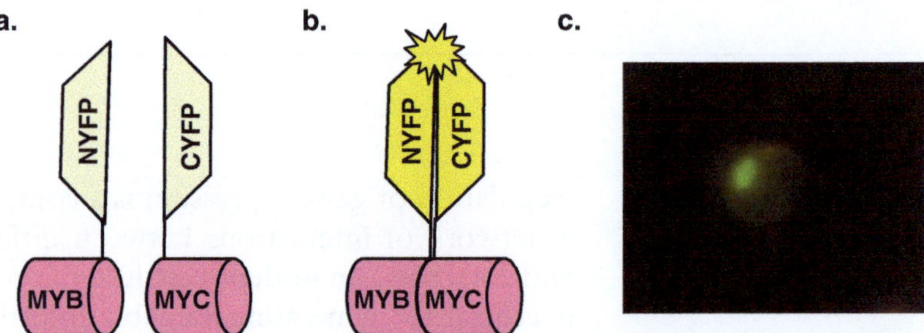

Fig. 10.1. A schematic representation of bimolecular fluorescence complementation (BiFC) to study protein–protein interactions between two transcription factors. (**a**) The MYB and the MYC (bHLH) transcription factors are fused to the N (NYFP) and C (CYFP) terminal regions of yellow fluorescent protein, respectively. (**b**) Interaction between MYB and MYC (bHLH) results in reconstitution of a functional fluorophore and visualization of fluorescence. (**c**) BiFC assay showing interaction of two transcription factors: tobacco R2R3MYB, An2, and maize bHLH, Lc, in a tobacco protoplast (YFP fluoresces green).

Additionally, a series of multi-color BiFC (mcBiFC) vectors have been developed using spectral variants of the fluorescent proteins that allow one to study the interaction of multiple proteins simultaneously in a living plant cell (14, 15). The association of different spectral variants of the fluorescent proteins, through the interacting protein partners fused to them, can lead to different emission spectra which can be visualized in a single cell. The mcBiFC technique has been successfully used to demonstrate interaction between *Agrobacterium tumefaciens* VirE2 protein and *Arabidopsis* nuclear import apparatus proteins (AtImpa-1 and AtImpa-4) (14), and concurrent interaction of different calcineurin B-like proteins (CBLs) with their interacting protein kinases (CIPK) *in planta* (15).

We are interested in studying the transcriptional regulation of the flavonoid biosynthetic pathway in plants. In all plants studied to date, three families of transcription factors, namely the R2R3 MYB, basic helix-loop-helix (bHLH; MYC) and the WD40 proteins, regulate the structural genes of this pathway. We describe here a method to visualize the in vivo interaction of two classes of regulatory proteins from the flavonoid pathway in plant protoplasts using the BiFC assay.

2. Materials

2.1. Plasmids and Primers

1. Transcription factor template plasmids: Plasmids containing the full-length cDNAs of the transcription factors of interest are used as templates in PCR amplification (16). In this study, the tobacco R2R3 MYB transcription factor, *An2*, and the maize bHLH factor, *Lc*, are used.
2. pNYFP and pCYFP BiFC expression plasmids: Both the pNYFP and the pCYFP vectors contain the *CaMV35S* promoter followed by a multi-cloning site (MCS), the N/C-terminal part of yellow fluorescent protein (YFP) and nopaline synthase (*nos*) terminator. The pNYFP and pCYFP vectors contain amino acids 1–155 and amino acids 156–239 of YFP, respectively. The protein of interest is cloned into the MCS in correct orientation and reading frame to generate a translational fusion with NYFP or CYFP (*see* **Note 1**).
3. Custom, synthetic oligonucleotides.

2.2. Generation of YFP Fusion Constructs

1. *Taq* DNA polymerase: Easy-A high fidelity PCR cloning enzyme (Agilent Technologies, USA). Store at –20°C.
2. dNTPs: 100 mM dNTPs (USB Affymetrix, USA). Prepare aliquots of a 10 mM stock solution of dNTPs in sterile milli-Q water for regular use. Store at –20°C.

3. Appropriate restriction enzymes for generation of clones (New England Biolabs). Store at –20°C.
4. T4 DNA ligase (New England Biolabs). Store at –20°C.
5. UltraPure, low-melting agarose (Invitrogen, USA).
6. Buffer used in gel electrophoresis: 50X Tris–Acetate–EDTA (TAE) Buffer (242 g of Tris base; 100 mL of 0.5 M EDTA, pH 8.0; and 57.1 mL of glacial acetic acid per 1 L solution). Autoclave and store at room temperature. Prepare a 1X TAE in milli-Q water for regular use.
7. 1% (w/v) Ethidium bromide (EtBr) in water. Wear gloves when handling. Store at 4°C.
8. Owl separation system (Thermo Scientific, USA) for gel electrophoresis.
9. Gel purification of PCR products: Wizard SV Gel and PCR cleanup system (Promega, USA).
10. *Escherichia coli* strain for transformation: K12 TB1 cells (F^- *ara* Δ(*lac-proAB*) [Φ*80dlac* Δ(*lacZ*)*M15*] *rpsL*(Str^R) *thi hsdR*) (New England BioLabs, USA) or similar cells.
11. Luria-Bertani (LB) medium: 10 g/L sodium chloride (NaCl), 10 g/L Bacto tryptone and 5 g/L Bacto yeast extract. Add 15 g/L of Bacto agar to LB medium to make LB agar plates. Dissolve in water and autoclave.
12. Plasmid isolation from *E. coli* cells: Wizard Plus SV minipreps (Promega, USA).
13. Reagents and instrument for DNA sequencing: GenomeLab DTCS Quick Start kit, GenomeLab separation gel, GenomeLab separation buffer (Beckman Coulter, USA) and Beckman Coulter CEQ-8000 sequencer. Store the DTCS Quick Start Kit at –20°C and the separation buffer and gel are stored at 4°C.

2.3. Media for Tobacco Cell Suspension Culture, Protoplast Isolation, Electroporation and Microscopy

1. Cell suspension culture medium: Murashige and Skoog's medium with minimal organics (Sigma, USA, Cat. No. M6899) *plus* 204 mg/L potassium phosphate monobasic (KH_2PO_4), 0.5 mg/L pyridoxine HCl, 0.5 mg/L nicotinic acid, 0.5 mg/L thiamine HCl, 0.2 mg/L 2,4-dichlorophenoxyacetic acid (2,4-D) and 0.1 mg/L kinetin (*see* **Note 2**). Adjust medium pH to 5.8 before autoclaving.
2. MMC solution: 91.1 g/L mannitol, 1.95 g/L 2-morpholinoethane sulphonic acid (MES) and 1.47 g/L $CaCl_2 \cdot 2H_2O$. Dissolve in milli-Q water, adjust to pH 5.6 and autoclave. Store at room temperature.
3. Enzyme solution: 0.75% (w/v) Cellulase Onozuka RS (Yakult Honsha, Japan) and 0.075% (w/v) pectinase

(Sigma). Dissolve the enzymes in MMC solution on a stirring plate and filter sterilize using Millipore syringe filter (0.45 μm size; Millipore, USA) (*see* **Note 3**). Hold at room temperature.

4. Sucrose solution: 250 g/L sucrose, 1.95 g/L MES and 1.47 g/L $CaCl_2 \cdot 2H_2O$. Dissolve in milli-Q water, adjust to pH 5.6 and autoclave. Store at room temperature.
5. Electroporation buffer: 91.1 g/L mannitol, 5.21 g/L KCl and 975 mg/L MES. Dissolve in milli-Q water, adjust to pH 5.6 and autoclave. Store at room temperature.
6. Protoplast culture medium: 91.1 g/L mannitol, 1.95 g/L MES, 1.47 g/L $CaCl_2 \cdot 2H_2O$, 120 mg/L $MgSO_4 \cdot 7H_2O$, 27 mg/L KH_2PO_4, 101 mg/L KNO_3, 2 mg/L KI and 30 g/L sucrose. Dissolve in milli-Q water, adjust to pH 5.6 and autoclave. Store at room temperature.
7. Electroporator: Gene Pulser II apparatus with the Capacitance Extender II (Model 165-2107) (Bio-Rad, USA).
8. Electroporation cuvettes: 0.4 cm gap.
9. Sterile 50 mL conical tubes.
10. Microscopy: Inverted microscope (Eclipse TE200, Nikon, Japan) setup for fluorescence microscopy.

3. Methods

3.1. Generation of YFP Fusion Constructs

1. The full-length coding regions of the transcription factors of interest (*An2* and the MYB-interacting region of *Lc* (aa 1–212) for this study) are amplified by PCR using gene-specific forward and reverse primers from the respective template plasmids containing the full-length cDNAs. PCR amplifications are carried out in a total volume of 25 μL containing: 10 ng of plasmid template, 0.4 μM primer pair, 200 μM dNTPs and 0.5 U of Easy-A High Fidelity PCR cloning enzyme. PCR is performed on a Peltier Thermal Cycler (Bio-Rad, USA) using the following program: preheat 94°C for 2 min; 30 cycles of 94°C for 30 s, 55°C for 30 s and 72°C for 1 min; followed by an incubation at 72°C for 5 min.
2. Prepare a 1% agarose gel in 1X TAE buffer with EtBr, final concentration of approximately 0.00005% (v/v), and pour the gel into a gel tray with comb. Once the gel has set, carefully remove the comb, add running buffer (1X TAE) to the unit, load the PCR product into a well or two, load a 1 kb DNA ladder in another well, connect it to the power supply and run at 95 V until bands are well separated. The PCR

products are purified using Wizard SV Gel and PCR cleanup system and quantified.

3. Digest the products with appropriate restriction enzymes (*Xho*I and *BamHI* in this study) for 2 h at 37°C followed by heat inactivation at 70°C for 15 min and keep on ice. About 1 μg of the pNYFP and pCYFP vectors are also digested with same restriction enzymes for 2 h at 37°C and then gel purified using Wizard SV Gel and PCR cleanup system.
4. The digested *An2* and *Lc* PCR products are ligated to the pNYFP and pCYFP vectors. Ligation reactions are set up in a total volume of 20 μL containing 1 μL of T4 ligase in buffer with ATP at 20°C for 2 h. The ligation mix is transformed into chemically competent TB1 cells, plated on LB plates containing 100 μg/mL ampicillin and incubated at 37°C overnight.

3.2. Plasmid Isolation, Digestion to Check Clones and Sequencing

1. The transformants are grown in 2 mL of LB medium containing 100 μg/mL ampicillin at 37°C overnight. Plasmids are isolated from the overnight cultures using Wizard Plus SV minipreps and eluted in water (about 30 μL). The plasmids are then digested with appropriate restriction enzymes (*Xho*I and *BamHI*) at 37°C for 1 h and checked on an agarose gel for the presence of the desired insert. Plasmids containing the correct sized fragments are sequenced.
2. Sequencing PCR is performed in a total volume of 20 μL containing 300 ng of plasmid, 1 μL of sequencing primer (10 μM stock solution) and 8 μL of reaction mix (GenomeLab DTCS Quick Start kit). The sequencing PCR products are purified and dissolved in sample loading solution as per the manufacturer's instructions. Sequencing to verify the integrity of the clones is performed in a Beckman Coulter sequencer, CEQ-8000.

3.3. Isolation of Tobacco Cell Suspension Protoplasts, Electroporation of Plasmids and Visualization

Tobacco cell suspension cultures (cultivar Xanthi) are maintained by subculturing to fresh medium at 4 day intervals (5 mL of culture is transferred to 50 mL of medium). Cells for protoplast isolation are typically split from mother cultures during subculturing and harvested for protoplasts on the third day. Three-day-old cell cultures are used for protoplast isolation (*see* **Note 4**).

1. Cells are harvested in a 50 mL tube by centrifugation at 260*g* for 4 min in a Sorvall Legend RT Plus with a swinging-bucket rotor at 25°C.
2. The culture medium is replaced with 30 mL of enzyme solution (*see* **Note 3**), and the cell suspension is transferred to a 250 mL flask, incubated at 26°C in dark with slow shaking (50 rpm).

3. Pre-chill labelled electroporation cuvettes on ice for later use.
4. After 2 h, the cell suspension is transferred to a 50 mL tube, centrifuged at 260*g* for 3 min (*see* **Note 5**). Then the enzyme solution is carefully removed.
5. The protoplasts are washed once by very gently resuspending in 20 mL of MMC solution and centrifuging again at 260*g* for 3 min. Supernatant is carefully removed and then the protoplasts are gently resuspended in 10 mL of MMC solution.
6. In a new 50 mL tube, 20 mL of 25% sucrose solution is added. Protoplasts are subsequently layered carefully onto the sucrose solution and centrifuged for 4 min at 260*g*. Upon settling, protoplasts form a layer at the interface of the sucrose and MMC solutions, which is carefully recovered. They are transferred to a fresh tube and resuspended in 10 mL of electroporation buffer.
7. An aliquot of 750 μL, containing approximately 2×10^6 protoplasts, is transferred to a pre-chilled electroporation cuvette (0.4 cm gap) along with 5 μg of both the pNYFP and pCYFP plasmids containing the appropriately paired transcription factors (*see* **Note 6**). The cuvette is then electroporated at 200 V and 950 μF and placed back onto ice.
8. Protoplasts are immediately transferred to a new tube, centrifuged at 260*g* for 3 min and electroporation buffer is carefully removed. The protoplasts are resuspended in 1 mL of culture medium and plated onto a 25 mm plate coated with 1 mL of agarose. Incubate protoplasts at 26°C in the dark for 18–20 h (*see* **Note 7**).
9. After incubation overnight, a small amount of protoplasts are transferred to a clean microscope slide, covered with a coverslip and moved to the stage of an inverted microscope setup for fluorescence microscopy (*see* **Note 8**).
10. Use the bright field (white light) to bring the protoplasts into focus (200X–400X) and then switch to filtered light. Scan the slides for fluorescence. Negative controls should give minimal to no fluorescence, while interacting partners should glow brightly. In the example given in this protocol, these proteins are localized and interact in the nucleus. Thus, the nuclei of transformed cells will fluoresce green (**Fig. 10.1C**), indicating protein–protein interaction.

4. Notes

1. Neither a translational enhancer nor a 5′ UTR is present in the expression plasmids and only the coding sequence of the desired gene is used – starting with the ATG. These split YFP vectors do not contain a linker or spacer sequence. For some proteins, successful reconstitution of YFP can be highly particular and may require swapping of the proteins to be fused, including a linker for added flexibility or switching from a N-terminal fusion to a C-terminal fusion or vice versa.
2. Filter sterilized stock solutions for phytohormones are as follows: 10 mg/mL 2,4-D in 0.1 M NaOH and 10 mg/mL kinetin in 0.1 M NaOH. Store at –20°C.
3. Use freshly prepared enzyme solution for the protoplast isolation.
4. Three-day-old tobacco cell suspensions are ideal for obtaining good and viable protoplasts.
5. Protoplasts are delicate and must be handled very carefully. Sterile disposable transfer pipets are ideal for manipulating protoplasts. Centrifugation at high speeds will damage the protoplasts. Sterile technique is important, but not critical as the assay only spans 2 days.
6. Empty pNYFP and pCYFP plasmids or one empty and the other one fused to a protein fragment must be used as a negative control in a BiFC experiment in order to detect false positives.
7. We use a small cardboard box to incubate the electroporated protoplasts overnight.
8. The protoplasts can handle short excursions in light. Fluorescent viewing must be done in the dark. YFP/eYFP has an excitation maximum at 512 nm and an emission maximum at 529 nm. Best results will be obtained using filter sets specifically designed for YFP (excitation 490–500 nm, dichromatic mirror 510 nm, emission 520–560 nm). Filter sets for FITC are also suitable for BiFC using YFP.

Acknowledgements

We thank Dr. E. Grotewold of the Ohio State University for the split YFP vectors. We also express our appreciation to Dr. S. Wessler of the University of Georgia for the *Lc* cDNA.

This work is supported by a grant from the Kentucky Tobacco Research and Development Center to L.Y.

References

1. Causier, B. and Davies, B. (2002) Analyzing protein-protein interactions with the yeast two-hybrid system. *Plant Mol. Biol.* **50**, 855–870.
2. Masters, S. C. (2004) Co-immunoprecipitation from transfected cells. *Methods Mol. Biol.* **261**, 337–348.
3. Cooper, M. A. (2003) Label-free screening of bio-molecular interactions. *Anal. Bioanal. Chem.* **377**, 834–842.
4. Pfleger, K. D. G. and Eidne, K. A. (2006) Illuminating insights into protein-protein interactions using bioluminescence resonance energy transfer (BRET). *Nat. Methods* **3**, 165–174.
5. Bhat, R. A., Lahaye, T., and Panstruga, R. (2006) The visible touch: *in planta* visualization of protein-protein interactions by fluorophore-based methods. *Plant Methods* **2**, 12.
6. Hu, C. D., Chinenov, Y., and Kerppola, T. K. (2002) Visualization of interactions among bZIP and Rel family proteins in living cells using bimolecular fluorescence complementation. *Mol. Cell* **9**, 789–798.
7. Hu, C. D. and Kerppola, T. K. (2003) Simultaneous visualization of multiple protein interactions in living cells using multicolor fluorescence complementation analysis. *Nat. Biotechnol.* **21**, 539–545.
8. Ohad, N., Shichrur, K., and Yalovsky, S. (2007) The analysis of protein-protein interactions in plants by bimolecular fluorescence complementation. *Plant Physiol.* **145**, 1090–1099.
9. Weinthal, D. and Tzfira, T. (2009) Imaging protein-protein interactions in plant cells by bimolecular fluorescence complementation assay. *Trends Plant Sci.* **14**, 59–63.
10. Bracha-Drori, K., Shichrur, K., Katz, A., Oliva, M., Angelovici, R., Yalovsky, S., and Ohad, N. (2004) Detection of protein-protein interactions in plants using bimolecular fluorescence complementation. *Plant J.* **40**, 419–427.
11. Walter, M., Chaban, C., Schutze, K., Batistic, O., Weckermann, K., Nake, C., Blazevic, D., Grefen, C., Schumacher, K., Oecking, C., Klaus Harter, K., and Kudla, J. (2004) Visualization of protein interactions in living plant cells using bimolecular fluorescence complementation. *Plant J.* **40**, 428–438.
12. Martin, K., Kopperud, K., Chakrabarty, R., Banerjee, R., Brooks, R., and Goodin, M. M. (2009) Transient expression in *Nicotiana benthamiana* fluorescent marker lines provides enhanced definition of protein localization, movement and interactions *in planta*. *Plant J.* **59**, 150–162.
13. Schütze, K., Harter, K., and Chaban, C. (2009) Bimolecular fluorescence complementation (BiFC) to study protein-protein interactions in living plant cells. *Methods Mol. Biol.* **479**, 189–202.
14. Lee, L.-Y., Fang, M.-J., Kuang, L.-Y., and Gelvin, S. B. (2008) Vectors for multi-color bimolecular fluorescence complementation to investigate protein-protein interactions in living plant cells. *Plant Methods* **4**, 24.
15. Waadt, R., Schmidt, L. K., Lohse, M., Hashimoto, K., Bock, R., and Kudla, J. (2008) Multicolor bimolecular fluorescence complementation (mcBiFC) reveals simultaneous formation of alternative CBL/CIPK complexes *in planta*. *Plant J.* **56**, 505–516.
16. Pattanaik, S., Kong, Q., Zaitlin, D., Werkman, J. R., Claire, H. X., Patra, B., and Yuan, L. (2010) Isolation and functional characterization of a floral-tissue specific R2R3 MYB regulator from tobacco. *Planta* **231**, 1061–1076.

Chapter 11

Isolation of Transcription Factor Complexes from *Arabidopsis* Cell Suspension Cultures by Tandem Affinity Purification

Jelle Van Leene, Dominique Eeckhout, Geert Persiau, Eveline Van De Slijke, Jan Geerinck, Gert Van Isterdael, Erwin Witters, and Geert De Jaeger

Abstract

Defining protein complexes is critical to virtually all aspects of cell biology because most cellular processes are regulated by stable or more dynamic protein interactions. Elucidation of the protein–protein interaction network around transcription factors is essential to fully understand their function and regulation. In the last decade, new technologies have emerged to study protein–protein interactions under near-physiological conditions. We have developed a high-throughput tandem affinity purification (TAP)/mass spectrometry (MS) platform for cell suspension cultures to analyze protein complexes in *Arabidopsis thaliana*. This streamlined platform follows an integrated approach comprising generic Gateway-based vectors with high cloning flexibility, the fast generation of transgenic suspension cultures, TAP adapted for plant cells, and tandem matrix-assisted laser desorption ionization MS for the identification of purified proteins. Recently, we evaluated the GS tag, originally developed to study mammalian protein complexes, that combines two IgG-binding domains of protein G with a streptavidin-binding peptide, separated by two tobacco etch virus cleavage sites. We found that this GS tag outperforms the traditional TAP tag in plant cells, regarding both specificity and complex yield. Here, we provide detailed protocols of the GS-based TAP platform that allowed us to characterize transcription factor complexes involved in signaling in response to the plant phytohormone jasmonate.

Key words: *Arabidopsis thaliana*, transcription factor, cell suspension culture, protein complex, protein interaction, tandem affinity purification.

L. Yuan, S.E. Perry (eds.), *Plant Transcription Factors*, Methods in Molecular Biology 754,
DOI 10.1007/978-1-61779-154-3_11, © Springer Science+Business Media, LLC 2011

1. Introduction

Jasmonates (JA) are plant hormones that regulate many aspects of plant defense and development. Perception of the JA signal activates a signal transduction cascade that leads to an extensive reprogramming of gene expression. A number of players in this signaling pathway have already been identified. MYC2, a bHLH transcription factor, has a central role in this system and acts both as an activator and as a repressor of downstream JA responses (1). In the absence of jasmonoyl-isoleucine (JA-Ile), the most important if not the only bioactive JA (2), the JASMONATE ZIM-DOMAIN (JAZ) proteins bind MYC2 and inhibit its activity (3). Upon perception of JA-Ile, the F-box protein CORONATINE INSENSITIVE 1 (COI1), which is part of the SCF^{COI1} E3 ubiquitin ligase complex, specifically recognizes the JAZ proteins and targets them to the 26S proteasome for degradation (4). Still, many aspects of this signaling pathway remain unresolved.

Because tandem affinity purification (TAP) (5) combined with mass spectrometry (MS)-based protein identification had proven to be a powerful approach that led to the first genome-wide screens for protein complexes in yeast (*Saccharomyces cerevisiae*) (6–8), we developed a streamlined TAP platform for *Arabidopsis thaliana* cell suspension cultures (9). The GS tag that combines two IgG-binding domains of protein G with a streptavidin-binding peptide to study mammalian protein complexes (10) has recently been evaluated in plant cells. This GS tag has been found to outperform the traditional TAP tag in specificity as well as yield (11).

For the simultaneous visualization of the dynamics of JAZ complex assembly, TAP was carried out with cell cultures elicited or not elicited with JA. Because JAZ degradation on treatment with JA has been reported to be very rapid (12), we first investigated JAZ degradation kinetics in our model system. A cycloheximide chase experiment revealed an estimated half-life of 1.4 min for a JAZ1–firefly luciferase (fLUC) fusion protein on treatment with JA (13). Next, JAZ1 was carboxy-terminally fused to the GS tag (*see* **Sections 2.1** and **3.1** and produced under the control of a cauliflower mosaic virus (CaMV) 35S promoter in *Arabidopsis* cells (*see* **Sections 2.2** and **3.2**). Cultures without and after addition of JA were purified elaborately (*see* **Sections 2.3** and **3.3**). To avoid full degradation of the tagged bait, cell cultures producing JAZ1–GS were harvested 1 min after treatment with JA. Purified proteins were concentrated, separated by polyacrylamide gel electrophoresis (PAGE), and stained with Coomassie blue (*see* **Sections 2.4** and **3.4** and **Fig. 11.2**). Finally, full lanes were cut in slices from top to bottom. The proteins were eluted and prepared for sequencing by tandem matrix-assisted laser desorption ionization (MALDI) MS (*see* **Sections 2.5** and **3.5**). Several proteins already shown to interact with JAZ1 were identified (**Fig. 11.2** and **Table 11.1**), validating the approach. First, JAZ1 formed a complex with JAZ12, matching the reported dimerization of JAZ proteins (14). Second, interaction of JAZ1 with MYC3, a close relative to MYC2, was observed. Last, COI1 occurred in a complex with JAZ1, but only in the presence of JA, showing that TAP is applicable for the identification of ligand-mediated protein–protein interactions. We focused on an uncharacterized protein isolated in the TAP analysis (**Fig. 11.2** and **Table 11.1**), designated Novel Interactor of JAZ (NINJA), and encoded by the locus At4g28910. To investigate NINJA function in JA signaling, we set up a new TAP experiment with NINJA as bait. NINJA was present in a complex with the Groucho/Tup1-type co-repressor TOPLESS (TPL) and its homologues TPR2 and TPR3, independently of JA elicitation (**Table 11.1**). Furthermore, MYC3 and the group-II TIFY proteins JAZ12 and photoperiod 2 (PPD2) were also identified, confirming the previous TAP data obtained with JAZ1. Based on these results, NINJA was analyzed functionally, demonstrating that it acts as a transcriptional repressor, the activity of which is mediated by a functional TPL-binding EAR repression motif. Accordingly, both NINJA and TPL proteins function as negative regulators of jasmonate responses (13).

2. Materials

2.1. Cloning of the Bait Tag Fusion Constructs in GS Tag Expression Vectors

1. *E. coli* DH5α genotype $F^{-}\varphi80dlacZ\Delta M15$ Δ(lacZYA-argF)U169 deoR, recA1 endA1 hsdR17(r_k^- m_k^+) phoA supE44 λ^- thi-1 gyrA96 relA1.
2. Platinum® *Taq* DNA Polymerase High Fidelity (Invitrogen, Carlsbad, CA).
3. Gateway® BP clonase II enzyme mix (Invitrogen).
4. Gateway® LR clonase II plus enzyme mix (Invitrogen).
5. High Pure PCR product purification kit (Roche Diagnostics, Brussels, Belgium).
6. All necessary entry (pEN-L4-2-R1, pEN-R2-GS tag-L3, and pEN-L4-2-L3) and destination vectors (pKCTAP and pKNGSTAP) are available at http://www.psb.ugent.be/gateway/index.php.

Table 11.1
MALDI-TOF/TOF MS identification of proteins interacting with JAZ1 (AT1G19180) and NINJA (AT4G28910). Three experimental repeats were performed per bait. Protein complexes were purified from untreated (mock) cell cultures expressing JAZ1 or NINJA GS tag fusions or from cultures treated for 1 min with jasmonate (+JA). (Reproduced from (13), with permission from © Nature Publishing Group)

Bait	Specifications	Prey locus	Prey name	Found/3 exp	Protein score	Expect	Best ion score	Expect
JAZ1	Mock	AT4G28910	NINJA	3	158	5.20E-12	51	1.83E-04
JAZ1	Mock	AT5G46760	MYC3	1	216	8.20E-18	43	1.33E-03
JAZ1	Mock	AT5G20900	JAZ12	1	173	1.60E-13	51	1.67E-04
JAZ1	+JA (1 min)	AT5G46760	MYC3	2	133	1.60E-09	44	1.15E-03
JAZ1	+JA (1 min)	AT4G28910	NINJA	2	108	5.20E-07	/	/
JAZ1	+JA (1 min)	AT5G20900	JAZ12	2	97	7.30E-06	42	1.62E-03
JAZ1	+JA (1 min)	AT2G39940	COI1	2	65	1.10E-02	28	2.70E-02
NINJA	Mock	AT5G27030	TPR3	3	429	4.10E-39	139	7.70E-14
NINJA	Mock	AT1G15750	TOPLESS	2	362	2.10E-32	134	3.70E-13
NINJA	Mock	AT4G14720	PPD2	2	99	4.60E-06	41	2.40E-03
NINJA	Mock	AT3G16830	TPR2	1	298	5.20E-26	93	1.10E-08
NINJA	Mock	AT5G46760	MYC3	1	91	2.70E-05	73	8.90E-07
NINJA	Mock	AT5G20900	JAZ12	1	65	9.70E-03	58	2.00E-05

2.2. Transformation and Upscaling of the Cell Suspension Culture

1. *Agrobacterium tumefaciens* C58C1Rif^R pMP90.
2. Yeast extract broth (YEB) medium: 5 g/L of beef extract, 1 g/L of yeast extract, 5 g/L of peptone, 5 g/L of sucrose, and 2 mM $MgSO_4$.
3. Cell suspension culture PSB-D of *A. thaliana* (L.) Heynh. ecotype Landsberg *erecta*, derived from MM2d (15).
4. 1 mg/mL α-naphthaleneacetic acid (NAA) stock solution: dissolve 10 mg NAA in 10 mL of 0.1 M NaOH. Filter sterilize with 0.22-μm filter (Millipore, Bedford, MA).
5. 1 mg/mL of kinetin stock solution: dissolve 10 mg kinetin in 10 mL of 0.1 M NaOH. Filter sterilize with 0.22-μm filter (Millipore).
6. MSMO medium: 4.43 g/L of Murashige & Skoog Basal salts with minimal organics (Sigma-Aldrich, St. Louis, MO), 30 g/L of sucrose, 0.5 mg/L of NAA, 0.05 mg/L of kinetin, pH 5.7, adjusted with 1 M KOH. NAA and kinetin are added after sterilization.
7. MSMO agar (0.8% w/v): dissolve 8 g of plant tissue culture agar (Lab M Ltd., Bury, UK) per 1 L of MSMO medium and autoclave. NAA, kinetin, and antibiotics should be added after the medium has cooled down to approximately 60°C.
8. Antibiotics: kanamycin, spectinomycin, rifampicin, gentamicin, carbenicillin, and vancomycin.
9. Acetosyringone.
10. 3M™ Micropore™ medical tape.
11. Erlenmeyer flasks (25, 100, 250, 500, and 2500 mL).
12. Sterile Multiwell™ 6-well plate (BD Biosciences, Franklin Lakes, NJ).
13. Filter discs grade 3 hw (Sartorius, Göttingen, Germany) and sintered glass pore filter.
14. 1.5-mL pellet mixer (Treff, Degersheim, Switzerland).
15. 10% sodium dodecyl sulfate (SDS).
16. SDS separating buffer: 1 M Tris–HCl, pH 8.7.
17. SDS stacking buffer: 1 M Tris–HCl, pH 6.8.
18. 30% acrylamide/bisacrylamide mix (ratio 29:1). Handle with care, because it is a neurotoxin in its monomeric form.
19. 10% ammonium persulfate (APS).
20. *N,N,N,N′*-tetramethyl-ethylenediamine (TEMED).
21. 10X TGS Running buffer: 250 mM Tris, pH 8.3, 1.92 M glycine, 1% SDS.

22. Dual color Prestained Precision Plus Protein Standard (Bio-Rad, Hercules, CA).
23. Transfer buffer: 25 mM (3.03 g/L) Tris, 192 mM glycine (14.4 g/L), and 15% MeOH.
24. Polyvinylidene difluoride (PVDF) membrane (Immobilon-P; Millipore).
25. Whatman paper.
26. 10X TBS: 12.11 g/L Tris–HCl, pH 8.0, 87.66 g/L NaCl.
27. 1X TBS-T: 100 mL of 10X TBS added to 900 mL of H_2O and 1 mL of Triton X-100.
28. Gel apparatus such as a mini-Protean III and protein gel blot apparatus such as a mini Trans-Blot cell (Bio-Rad).
29. Blocking buffer: 3% Skim Milk (BD Biosciences, Difco^{TM}), Difco catalog # 232100 in 1X TBS-T.
30. Peroxidase anti-peroxidase soluble complex antibody produced in rabbit (Sigma-Aldrich).
31. Western Lightning^{TM} *Plus*-ECL (PerkinElmer, Waltham, MA).
32. High-performance chemiluminescence films (GE Healthcare, Little Chalfont, UK).

2.3. Complex Purification

1. Extraction buffer: 25 mM Tris–HCl, pH 7.6, 15 mM $MgCl_2$, 5 mM ethylene glycol-bis(β-aminoethyl ether)*N*,*N*,*N*′,*N*′-tetraacetic acid (EGTA), 150 mM NaCl, 15 mM *p*-nitrophenylphosphate, 60 mM β-glycerophosphate, 0.1% (v/v) Nonidet P-40 (NP-40), 0.1 mM sodium vanadate, 1 mM NaF, 1 mM dithiothreitol (DTT), 1 mM phenylmethylsulfonylfluoride (PMSF), 10 μg/mL leupeptin, 10 μg/mL aprotinin, 5 μg/mL antipain, 5 μg/mL chymostatin, 5 μg/mL pepstatin, 10 μg/mL soybean trypsin inhibitor, 0.1 mM benzamidine, 1 μM *trans*-epoxysuccinyl-L-leucylamido-(4-guanidino)butane (E64), 5% (v/v) ethylene glycol.
2. IgG Sepharose wash buffer: 10 mM Tris–HCl, pH 8.0, 150 mM NaCl, 0.1% (v/v) NP-40, and 5% (v/v) ethylene glycol.
3. Rotating wheel.
4. Tobacco etch virus (TEV) buffer: 10 mM Tris–HCl, pH 8, 150 mM NaCl, 0.1% (v/v) NP-40, 0.5 mM EDTA, 1 μM E64, 1 mM PMSF, and 5% ethylene glycol.
5. Streptavidin elution buffer: TEV buffer with 20 mM desthiobiotin.
6. IgG Sepharose 6 Fast Flow (GE Healthcare).

7. Streptavidin Sepharose High Performance (GE Healthcare Cat. No. 17-5113-01).
8. Bradford Protein Assay Dye Reagent Concentrate (Bio-Rad).
9. AcTEV^{TM} Protease (Invitrogen).
10. Ultra Turrax T25 homogenizer mixer (IKA Works, Wilmington, NC).
11. Ultra-speed Centrifuge L7-55 and a SW28 rotor (Beckman Coulter, Brea, CA).
12. Ultra-Clear centrifuge tubes (25 × 89 mm) (Beckman Coulter).
13. 0.45-μm Minisart RC 25 syringe filter (Sartorius).
14. Polyprep columns (Bio-Rad).
15. Mobicol columns (MoBiTec, Göttingen, Germany).
16. Falcon tubes (50 mL).
17. Eppendorf tubes (1.5 mL).

2.4. Sample Preparation

1. Trichloroacetic acid solution, 6.1 N (Sigma-Aldrich).
2. 50 mM HCl/acetone solution, ice cold.
3. NuPAGE® LDS sample buffer (Invitrogen).
4. NuPAGE® Sample Reducing Agent (10X) (Invitrogen).
5. NuPAGE® MES SDS running buffer (for Bis–Tris gels only) (20X) (Invitrogen).
6. NuPAGE® Antioxidant (Invitrogen).
7. NuPAGE® Novex 4–12% Bis–Tris gel 1.0 mm, 10 well (Invitrogen).
8. SeeBlue® Plus2 prestained protein standard marker (Invitrogen).
9. Gel fix solution: 50% ethanol, 2% H_3PO_4.
10. Stain solution concentrate (SSC): 34% MeOH (v/v), 17% $(NH_4)_2SO_4$ (w/v), and 3% H_3PO_4 (v/v).
11. Coomassie Brilliant Blue G-250 (Sigma-Aldrich).
12. Salt solution for storing the gel: 20% (w/v) NaCl.

2.5. Identification of Co-purified Proteins by Mass Spectrometry

1. 50 mM NH_4HCO_3, freshly made.
2. Dithiothreitol, DTT.
3. Iodoacetamide.
4. Trypsin Gold, mass spectrometry grade (Promega, Madison, WI).
5. ULC/MS-grade water (Biosolve, Valkenswaard, The Netherlands).

6. ULC/MS-grade acetonitrile (Biosolve).
7. Microtiter plate with conically shaped wells (Biozyme).
8. Clear, adhesive seals for microtiter plates.
9. Sonicator.
10. Microcolumn solid phase tips: Cleanup C18 pipette tips, 200-nL bed volume (Agilent Technologies, Santa Clara, CA).
11. MALDI target plate Opti-TOFTM384 Well Insert (Applied Biosystems, Foster City, CA).
12. Digest buffer: 12.5 μg trypsin/mL in 50 mM NH_4HCO_3:10% CH_3CN (v/v).
13. Internal peptide standard: 20 fmol/μL Glu1 Fibrinopeptide B (Sigma-Aldrich), 20 fmol/μL des-Pro2-Bradykinin (Sigma-Aldrich), and 20 fmol/μL Adrenocorticotropic Hormone Fragment 18-39 human (Sigma-Aldrich).
14. MALDI matrix. Recrystallize commercial α-cyano-4-hydroxycinnamic acid (Sigma-Aldrich) by sequentially washing and decanting the crystals in ULC/MS-grade water and 50% ULC/MS-grade acetonitrile. Dissolve the crystals until saturation in a 95% acetonitrile solution, add water to a 1:5 ratio, collect and dry the crystals. From these α-cyano-4-hydroxycinnamic acid crystals, make a saturated solution in 50% CH_3CN (v/v):0.1% CF_3COOH (v/v) spiked with an internal peptide standard.
15. MALDI tandem MS instrument: 4800 Proteomics Analyzer (Applied Biosystems).
16. GPS Explorer 3.6 (Applied Biosystems).
17. Database search engine Mascot 2.2 (Matrix Science, Boston, MA).

3. Methods

3.1. Cloning of the Bait Tag Fusion Constructs in GS Tag Expression Vectors

The TAP vectors are based on the Gateway technology and can be ordered at http://www.psb.ugent.be/gateway/index.php. Two different vector systems are available, one for fusion of the tag to the C-terminus (**Fig. 11.1a**) and one for fusion of the tag to the N-terminus of the bait protein (**Fig. 11.1b**).

3.1.1. Primer Design

1. For C-terminal TAP tag fusions, the promoter (*see* **Note 1**) is cloned in pDonrP4P1R. The promoter sequence is amplified by polymerase chain reaction (PCR) with the *att*B4 forward (fwd) primer, 5′-GGGG-ACA-ACT-TTG-TAT-

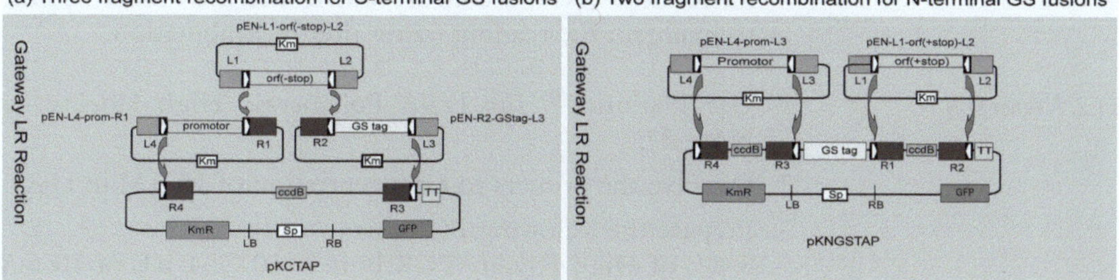

Fig. 11.1. Overview of the Gateway cloning strategy for TAP tag fusion constructs. TT, cauliflower mosaic virus 35S transcription terminator; CcdB, toxic killer gene for negative selection; KmR, neomycin phosphotransferase II selection marker; GFP, green fluorescent protein visible marker; LB and RB, *left* and *right* T-DNA borders, respectively (Reprinted from (11), with permission from Elsevier Ltd).

AGA-AAA-GTT-G-(template-specific sequence)-3′, and the *att*B1R reverse (rev) primer, 5′-GGGG-AC-TGC-TTT-TTT-GTA-CAA-ACT-TG-(template-specific sequence)-3′. Approximately 20 nucleotides are taken as template-specific sequence, preferably ending with a guanine (G) or a cytosine (C) nucleotide.

2. For N-terminal TAP tag fusions, the promoter is cloned in pDonrP4P3 (*see* **Note 1**). The promoter sequence is amplified by PCR with the *att*B4 fwd primer, 5′-GGGG-ACA-ACT-TTG-TAT-AGA-AAA-GTT-G-(template-specific sequence)-3′, and the *att*B3R rev primer, 5′-GGGG-AC-AAC-TTT-GTA-TAA-TAA-AGT-TG-(template-specific sequence)-3′. Approximately 20 nucleotides are taken as template-specific sequence, preferably ending with a guanine (G) or a cytosine (C) nucleotide.

3. The open reading frame (ORF) of the bait protein (*see* **Note 2**) is cloned in pDonr221 (multisite Gateway works also with pDonr201 or pDonr207, but less efficiently in our hands). The ORF sequence is amplified by PCR with the *att*B1 fwd primer, 5′-GGGG-ACA-AGT-TTG-TAC-AAA-AAA-GCA-GGC-GNN-(template-specific sequence)-3′, and the *att*B2 rev primer, 5′-GGGG-AC-CAC-TTT-GTA-CAA-GAA-AGC-TGG-GTN-(template-specific sequence)-3′. Approximately 20 nucleotides are taken as template-specific sequence, preferably ending with a guanine (G) or a cytosine (C) nucleotide. For C-terminal TAP tag fusions, one extra nucleotide (N) is incorporated into the reverse primer to maintain the reading frame after recombination, and the stop codon is not included. In the forward primer, the KOZAK sequence CCACC is added before the start codon of the template-specific sequence to ensure optimal translation initiation. For N-terminal fusions, the stop codon is included in the reverse primer and two additional

nucleotides (NN) are incorporated into the forward primer to maintain the reading frame after recombination.

3.1.2. Adaptor PCR

1. Use Platinum® *Taq* DNA Polymerase High Fidelity (*see* **Note 3**).
2. Dissolve the primers to a concentration of 10 μM in H_2O.
3. Prepare the following PCR mix:
   5 μL of High Fidelity PCR buffer (10X), 1 μL of 10 mM dNTP mixture, 2 μL of 50 mM $MgSO_4$, 1 μL of 10 μM fwd primer, 1 μL of 10 μM rev primer, *x* μL of template DNA (with '*x*' volume as required), 0.2 μL of Platinum® *Taq* High Fidelity, and *y* μL of nanopure H_2O (with '*y*' volume such that a final volume of 50 μL is obtained).
4. Mix gently, centrifuge briefly, and transfer the mixture to a PCR tube.
5. Place the tube in a thermal cycler and run the following program:
   Step 1: 95°C for 2 min (min)
   Step 2: 95°C for 30 s
   Step 3: 55°C for 30 s
   Step 4: 68°C for at least 2 min (use 1 min per kb)
   Step 5: repeat steps 2–4 for 35 cycles
   Step 6: 68°C for 5 min
   Step 7: 4°C until end
6. Check the PCR product by analyzing 10 μL of the mixture via agarose gel electrophoresis.
7. Clean the PCR product with the 'High Pure PCR product purification kit.'

3.1.3. Creation of Entry Clones by Gateway BP Reaction

1. Prepare the following Gateway BP reaction mixture (*see* **Fig. 11.1** and **Note 4**):
   2 μL BP clonase II (5X)
   50 fmol donor vector (for instance, pDonr221)
   50 fmol PCR product (*att*B1-B2 flanked for pDonr221)
   x μL H_2O to a final volume of 10 μL
2. Mix gently, centrifuge briefly, and incubate the mixture for at least 2 h or overnight at 25°C.
3. Add 1 μL of proteinase K and incubate for 10 min at 37°C.
4. Transform by heat shock transformation:
   Thaw 50 μL of heat shock-competent *E. coli* DH5α on ice, add 5 μL of prechilled BP reaction, and tap gently to mix. Do not pipette up and down. Incubate the mixture for 20 min on ice. Put in a warm water bath or a thermomixer

at 42°C for 42 s. Immediately transfer back on ice for 2 min. Add 950 μL of SOC medium and incubate for 1 h at 37°C. Spin cells down in a microfuge (4000 rpm for 5 min) and resuspend pellet in 100 μL of SOC.

5. Plate all 100 μL on LB^+ with 50 μg/mL kanamycin (Km) (replace Km with 12.5 μg/mL gentamicin (Gm) if pDonr207 is used) and incubate overnight at 37°C.
6. Screen for colonies with the correct construct by colony PCR with GW M13 fwd primer, GTAAAAC GACGGCCAGTCTTA, and GW M13 rev primer, CCAGGAAACAGCTATGACCAT, and analyze by agarose gel electrophoresis.
7. Prepare plasmid DNA according to a standard DNA preparation protocol and check the sequence of the ORF with the same primers.

3.1.4. Creation of GS Tag Expression Vectors by Multisite Gateway LR Reaction

1. Prepare the following Gateway LR reaction mixture (*see* **Fig. 11.1** and **Note 4**):

   For C-terminal GS tag fusions

   2 μL LR clonase II Plus (Plus is necessary for multisite LR reactions)

   10 fmol pKCTAP (78 ng) (*see* **Note 5**)

   10 fmol pEN-L4-2-R1 (25 ng) (entry vector with promoter)

   10 fmol pEN-L1-ORF-L2 (entry vector with ORF without stop codon)

   10 fmol pEN-R2-GStag-L3 (entry vector with GS TAP tag) (*see* **Note 6**)

   x μL H_2O to a final volume of 10 μL

   For N-terminal GS tag fusions

   2 μL LR clonase II Plus

   10 fmol pKNGSTAP (93 ng) (GS TAP tag is in the backbone of this destination vector)

   10 fmol pEN-L4-2-L3 (25 ng) (entry vector with promoter)

   10 fmol pEN-L1-ORF-L2 (entry vector with ORF with stop codon)

   x μL H_2O to a final volume of 10 μL

2. Mix gently, centrifuge briefly, and incubate overnight at 25°C.
3. Add 1 μL of proteinase K and incubate 10 min at 37°C.
4. Transform by heat shock as described in step 4 of **Section 3.1.3**.

5. Plate all 100 μL on LB$^+$ with 75 μg/mL spectinomycin (Sp) and incubate overnight at 37°C.
6. Screen for colonies with the correct construct by colony PCR with the following primers: for C-terminal TAP tag fusions, 35S promoter fwd (CCACTATCCTTCGCAA GACCC) and 35S terminator rev1 (GGTGATTTTTGCG GACTCTAGCAT) and for N-terminal TAP tag fusions, NGSTAP fwd (TGAGCTGGAGCAGCTACGGGC) and 35S terminator rev2 (TGCGGACTCTAGCATGGCCG).
7. Prepare plasmid DNA according to a standard DNA preparation protocol.

3.2. Transformation and Upscaling of the Cell Suspension Culture

3.2.1. Transformation of Expression Vectors in A. tumefaciens

1. Inoculate 2 mL of yeast extract broth (YEB) medium containing 100 μg/mL of rifampicin (Rif) and 40 μg/mL of Gm with *A. tumefaciens* C58C1Rif^R pMP90. Grow at 28°C for 24 h.
2. Centrifuge the bacteria down at 14,000 rpm for 2 min in a microfuge.
3. Wash the cell pellet four times with decreasing amounts of ice-cold nanopure water: 2 mL, 1 mL, 500 μL, and 200 μL. Centrifuge for 2 min at 14,000 rpm between washing steps in a microfuge.
4. Keep cells on ice for a maximum of 1 h.
5. Transform by electroporation as follows:

   Add approximately 100 ng of plasmid DNA to 40 μL of bacteria; put the mixture on ice for 1 min.

   Transfer DNA/bacteria mix to a prechilled electroporation cuvette, taking care to avoid air bubbles in the cuvette, and electroporate at 2.5 kV in a 0.2-cm cuvette.

   After electroporation add 960 μL of YEB medium and transfer to a 12-mL tube.

   Shake for 2 h at 28°C at 150 rpm.

   Centrifuge the cells down at 4000 rpm for 5 min in a microfuge and resuspend the pellet in 100 μL of YEB.
6. Plate 10 and 100 μL on YEB plates containing 100 μg/mL Rif, 40 μg/mL Gm, and 100 μg/mL Sp.

3.2.2. Transformation of A. thaliana Cell Suspension Culture PSB-D (See **Note 7**)

1. Day 1: inoculate 2 mL of YEB medium containing 100 μg/mL Rif, 40 μg/mL Gm, and 100 μg/mL Sp with transformed Agrobacteria in a 12-mL tube and shake overnight at 28°C and 150 rpm. Dilute a 7-day-old PSB-D cell suspension culture 1:5 in 100 mL of fresh MSMO (*see* **Note 8**).
2. Day 2: inoculate 20 mL of YEB medium containing 100 μg/mL Rif, 40 μg/mL Gm, and 100 μg/mL Sp with

the 2 mL of Agrobacteria preculture and shake overnight at 28°C and 150 rpm.

3. Day 3: centrifuge the bacterial culture at 3220×*g* for 15 min. Discard the supernatant. Resuspend the cell pellet with 40 mL of MSMO by vortexing and spin the cells down. Discard the supernatant. Resuspend the cell pellet in 40 mL of MSMO by vortexing, measure the OD_{600}, and spin the cells down. Resuspend the cell pellet with MSMO to an OD_{600} of 1.0. Prepare the following transformation mix in triplicate in a 6-well culture plate: 3 mL of a 2-day-old PSB-D cell suspension culture, 200 μL of washed Agrobacteria (OD_{600} 1.0), 6 μL of acetosyringone (100 mM stock). Tape the 6-well plate with 3 M micropore medical tape and co-cultivate the mixture for 48 h at 25°C with gentle agitation at 130 rpm on an orbital shaker in the dark.
4. Day 6: transfer 3 mL of the transformation mixture into a 25-mL Erlenmeyer flask containing 8 mL of MSMO containing 50 μg/mL Km, for selection of transformed cells, 500 μg/mL carbenicillin (Cb), and 500 μg/mL vancomycin (Vm) (MSMO/KVC) to remove the Agrobacteria (*see* **Note 9**). Incubate at 25°C with gentle agitation at 130 rpm for 9 days in the dark.
5. Day 15: transfer 9-day-old culture completely into a 100-mL Erlenmeyer flask containing 25 mL of MSMO/KVC. Incubate at 25°C with gentle agitation at 130 rpm for 7 days in the dark.
6. Day 22: allow cells to settle down to the bottom of the Erlenmeyer flask by applying no agitation for 5 minutes, pipette as much cells as possible from the bottom (approximately 7 mL), transfer culture into 35 mL (*see* **Note 10**) of MSMO/KVC and grow for 7 days at 25°C with gentle agitation at 130 rpm in the dark.
7. Day 29: allow cells to settle down and transfer 7 mL of the 7-day-old settled cells into 35 mL of MSMO containing 25 μg/mL Km (MSMO/K) in a 100-mL Erlenmeyer flask. After 7 days, subculture weekly by transferring 7 mL into 45 mL of fresh MSMO/K in a 100-mL Erlenmeyer flask.

3.2.3. Expression Analysis

1. Harvest plant cells from a 25-mL culture 2 days after subculturing on a sintered glass pore filter with a filter disc paper.
2. Half-fill a 1.5-mL Eppendorf tube with filtered cell material, freeze in liquid nitrogen, and store at –80°C.
3. Add 200 μL of extraction buffer and homogenize with a 1.5-mL pellet mixer.

4. Freeze in liquid nitrogen and thaw on ice to an icy slurry. Thawing can be accelerated by vortexing, but do not thaw mixture to liquid.
5. Centrifuge 15 min at 4°C at 14,000 rpm in a microfuge.
6. Transfer the supernatant to a new prechilled Eppendorf tube.
7. Repeat steps 5 and 6.
8. Determine the protein concentration by the Bradford assay (*see* step 1 of **Section 3.3.2**).
9. Resolve 60 μg of total protein extract by SDS-PAGE on 0.75 mm 12% SDS minigels (Mini-Protean III; Bio-Rad).

   SDS-PAGE:

   Clean glass plates for the gels consecutively with detergent, distilled H_2O, and 70% ethanol and air-dry the plates. Mount a big glass plate with fixed 0.75-mm spacer with a small glass plate on top in the gel-casting stand. Prepare the 12% separating gel by mixing 3.75 mL of separating buffer, pH 8.7, with 4 mL of acrylamide/bis solution, 1.04 mL of H_2O, 100 μL of 10% SDS, 100 μL of 10% APS, and 10 μL of TEMED. Pipette about 3.75 mL for one gel between the glass plates and gently overlay with H_2O. Let the gel polymerize for approximately 30 min.

   Pour off the water and rinse the top of the gel by drying with Whatman paper.

   Prepare the 5% stacking gel by mixing 1.25 mL of stacking buffer, pH 6.8, with 1.7 mL of acrylamide/bis solution, 6.84 mL of water, 100 μL of 10% SDS, 100 μL of 10% APS, and 10 μL of TEMED. Pipette approximately 1 mL of this mixture on top of the separating gel and immediately insert the comb. The stacking gel should also be allowed to polymerize for about 30 min. Prepare running buffer by diluting 100 mL of the 10X TGS running buffer with 900 mL of H_2O in a measuring cylinder and mix. Carefully remove the comb after polymerization of the stacking gel and wash the wells thoroughly with H_2O. Mount the gel in the gel cassette and place it into the buffer tank. Add 1X running buffer to the inner and outer chambers and load 60 μg of each sample in a well. Load 5 μL of the dual color prestained Precision Plus Protein standard. Connect the electrodes to a power supply and run the gel at 180 V for approximately 1 h until the front reaches the bottom of the gel. The gel is now ready for protein gel blotting.
10. Transfer resolved proteins to an Immobilon-P membrane by protein gel blotting and detect baits fused to the GS tag by immunodetection as follows:

Protein gel blotting:

After the proteins have been separated by SDS-PAGE they are transferred to PVDF membranes with the Mini Trans-Blot cell (Bio-Rad). Remove the stacking gel from the separating gel. Prepare the blot sandwich in a tray filled with prechilled transfer buffer. Mount starting from the negative side from the cassette (black) as follows: a sponge, two Whatman papers, the gel, the membrane, two Whatman papers, and a sponge all submerged in transfer buffer and close the cassette. Pretreat the Immobilon-P membrane by equilibration first for 5 min in 100% MeOH, followed by 5 min in transfer buffer. Make sure that no air bubbles are trapped in between the different layers of the sandwich. Place the cassette in the transfer tank so that the gel is positioned at the negative electrode and the membrane at the positive one.

Cool the system with a frozen water container and ice-cold transfer buffer. Put the lid on the blotting tank, attach to the power supply, and allow transfer at 270 mA for at least 1 h. Once the transfer is completed, take the cassette out of the tank and disassemble. Mark the visually blotted colored molecular weight markers with a pencil on the membrane because they fade during the immunodetection process.

Immunodetection:

Incubate the Immobilon-P membrane in blocking buffer overnight at 4°C or 1 h at room temperature on an orbital shaker. Make sure that the membrane is completely submerged. Discard the blocking buffer. Incubate the membrane for 1 h at room temperature with peroxidase–anti-peroxidase 1:2500 dilution in 10 mL of blocking buffer on an orbital shaker. Remove the antibody solution and wash the membrane once for 15 min and four times for 5 min with 1X TBS-T solution. Make sure that the membrane is completely submerged. Mix equal amounts of the two chemiluminescent reagents from the ECL kit (0.125 mL mixture/cm^2). After the final wash, carefully dip the membrane dry to remove the remaining TBS-T buffer and incubate the membrane for 1 min in the chemiluminescent mixture with vigorous shaking. Remove the blot from the mixture and dip dry again with a paper towel. Place the membrane between a plastic foil that is attached in a film cassette. Expose the membrane to ECL film in a dark room for 1 and 10 min. Develop the autoradiogram.

3.2.4. Upscaling of Transgenic PSB-D Culture

1. Add 40 mL of a 7-day-old PSB-D culture to 180 mL of fresh MSMO in a 500-mL Erlenmeyer flask.

2. Divide the culture 7 days after subculturing into two 2.5-L Erlenmeyer flasks (110 mL each), each containing 900 mL of fresh MSMO.
3. After 7 days, subculture 10 times 200 mL of the culture into 800 mL of fresh MSMO in ten 2.5-L Erlenmeyer flasks.
4. Pool 10 L of culture in a 10-L flask after the appropriate treatment and at the appropriate day after subculturing (*see* **Note 11**) and divide into ten 1-L flasks.
5. Wait at least 15 min to allow the plant cells to settle down and decant approximately 800 mL of the medium.
6. Harvest plant cells from the remaining 200 mL on a sintered glass filter, wrap in aluminum foil, weigh, and freeze in liquid nitrogen (*see* **Note 12**).
7. Store at –80°C.

3.3. Complex Purification (See Note 13)

3.3.1. Total Protein Extract Preparation

1. Grind 15 g of cells to homogeneity in liquid nitrogen with mortar and pestle (*see* **Note 14**).
2. Divide the ground cell material into two prechilled Falcon tubes and add 5 mL of extraction buffer to each Falcon tube. Prepare the crude protein extract by mixing three times for 30 s with an Ultra Turrax T25 homogenizer mixer (*see* **Note 15**). Pour the mixture into a prechilled centrifugation tube and centrifuge at 36,900×*g* for 20 min. Transfer the supernatant to prechilled Ultra-Clear centrifuge tubes and centrifuge at 178,000×*g* for 45 min at 4°C in an Ultra-speed centrifuge.
3. Collect the supernatant and pass the extract through a 0.45-μm syringe filter and collect in a prechilled 50-mL Falcon tube.

3.3.2. Tandem Affinity Purification (See Note 16)

1. Measure protein concentration via the Bradford assay as follows:

   Make three different dilutions in nanopure H_2O (1:5, 1:10, and 1:20).

   Add 2.5 (1:5), 5 (1:10), and 10 μL (1:20) of the diluted extract to three different Eppendorf tubes (0.5 μL extraction buffer for control), corresponding to 0.5 μL of the extract.

   Add H_2O to 800 μL.

   Add 200 μL of the Bradford reagent.

   Mix and incubate for 15 min in a cuvette.

   Measure the OD_{595}.

   Calculate the average of the three values and multiply by 2.

   Read the total protein concentration in mg/mL on a standard curve determined with different bovine serum albumin (BSA) concentrations. Typical values are in the range of 10–15 mg/mL.

2. Transfer 100 μL of human IgG Sepharose 6 Fast Flow beads (100 μL effective beads = 133 μL 75% slurry) to a 15-mL Falcon tube and equilibrate three times 10 column volumes of extraction buffer as follows:
   Add 1 mL of extraction buffer and wash the beads by gently tapping the tube.
   Centrifuge the beads at 450×*g* for 1 min.
   Carefully remove the supernatant (*see* **Note 17**).
   Repeat twice.
3. Add 200 mg of total protein extract to the 100 μL of IgG Sepharose 6 Fast Flow resin and incubate for 1 h on a rotating wheel at 4°C.
4. Transfer the binding mixture to a polyprep column and wash with 10 mL of IgG Sepharose wash buffer and 5 mL of TEV buffer (*see* **Note 18**).
5. Close the lower cap of the polyprep column, add 400 μL of TEV buffer, and transfer the beads to a 1.5-mL Eppendorf tube.
6. Add 100 units (10 μL) of AcTEV protease. Incubate for 1 h at 16°C on a rotating wheel. After 30 min, add 100 units of AcTEV protease.
7. Transfer the mixture to a Mobicol column and collect the eluate in a 1.5-mL Eppendorf tube.
8. Pass an additional 400 μL of TEV buffer over the beads and collect together with the previous eluate.
9. Equilibrate 100 μL of Streptavidin Sepharose resin (100 μL effective beads = 400 μL 25% slurry) by washing three times with 10 column volumes of TEV buffer.
10. Transfer the Streptavidin beads to a 1.5-mL Eppendorf tube and add the TEV eluted fraction.
11. Incubate the mixture on a rotating wheel for 1 h at 4°C.
12. Transfer the binding mixture to a polyprep column, let the unbound fraction flow through, and wash the beads with 10 mL of TEV buffer.
13. Elute the column very slowly by passing 1 mL of Streptavidin elution buffer over the beads and collect in a 1.5-mL Eppendorf tube.

3.4. Sample Preparation

3.4.1. Concentration of Purified Protein Complexes via Trichloroacetic Acid (TCA) Precipitation

1. Add 333 μL of 100% trichloroacetic acid (TCA) to the 1 mL of eluate to obtain a final concentration of 25%.
2. Mix gently by tapping the tube and keep on ice overnight (*see* **Note 19**).
3. Centrifuge the precipitated proteins down at 14,000 rpm at 4°C in a microfuge.

4. Carefully remove the supernatant and wash the pellet twice with 500 μL of ice-cold 50 mM HCl/acetone solution.
5. Air-dry the pellet for 1 h and store at –80°C.

3.4.2. Resolution of Proteins by NuPAGE Electrophoresis

1. Resuspend the proteins in sample buffer:
   Add 7.5 μL of NuPAGE LDS sample buffer (4X).
   Add 3 μL of NuPAGE Reducing Agent (10X).
   Add 19.5 μL of nanopure H_2O.
2. Pipette up and down and heat the samples at 70°C for 10 min.
3. Prepare a 1X SDS running buffer by adding 40 mL of 20X NuPAGE MES SDS running buffer to 760 mL of nanopure H_2O.
4. Take a precast 4–12% gradient NuPAGE Bis-Tris gel, remove the white strip, and place the gel in the NuPAGE® SDS running system.
5. Fill the upper buffer chamber with 200 mL of 1X NuPAGE SDS running buffer and add 500 μL of NuPAGE antioxidant solution.
6. Fill the lower buffer chamber with 600 mL of 1X NuPAGE SDS running buffer.
7. Load 27 μL of the sample and store the remaining 3 μL at –20°C for additional analysis, for instance, protein gel blot analysis.
8. Load 5 μL of SeeBlue Plus2 prestained protein standard marker.
9. Run the gel at 200 V for 47 min.

3.4.3. Visualization of Proteins by Coomassie Brilliant Blue G-250 Staining

1. Fix the gel in 50% EtOH/2% H_3PO_4 for 2 h to overnight (*see* **Fig. 11.2**).
2. Remove the fixing solution and wash the gel three times for 20 min in nanopure H_2O.
3. Replace the water with SSC and equilibrate the gel for 30 min to 1 h.
4. Add Coomassie Brilliant Blue G-250 powder (50 mg/50 mL SSC solution).
5. Shake gently for 48 h on an orbital shaker.
6. Transfer the gel to nanopure H_2O for destaining, take a gel image for documentation, and shake until further processing or, alternatively, store in 20% NaCl at 4°C.

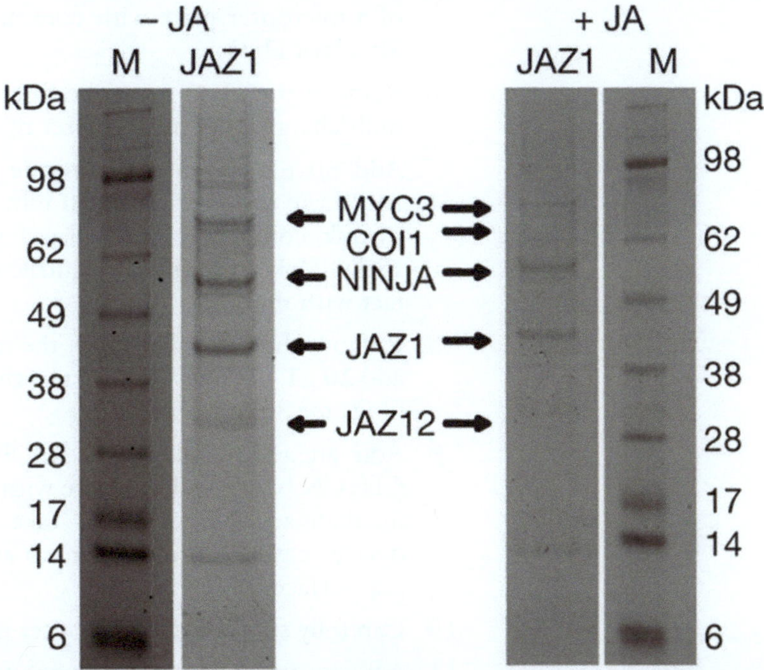

Fig. 11.2. NuPAGE gel separation and Coomassie G visualization of proteins co-purified with JAZ1-GS from *Arabidopsis* cells, after mock treatment (–JA) or elicitation (+JA). *Arrows* mark the positions of proteins identified by MS. Lane M shows the molecular mass markers (Reprinted from (13), with permission from © Nature Publishing Group).

3.5. Identification of Co-purified Proteins by Mass Spectrometry

3.5.1. Protein S-S Reduction, Alkylation, Proteolysis, Peptide Isolation, and Peptide Spotting (See **Note 20**)

1. If the gel slab has been stored in 20% NaCl, remove the salt solution and wash the gel three times for 10 min in nanopure H_2O on an orbital shaker. Repeat the washing step until the Coomassie Brilliant Blue dye is removed. Remove the liquid and equilibrate the gel in 50 mM NH_4HCO_3 for 10 min (*see* **Table 11.1**).
2. Remove the liquid and replace it with a solution containing 6.66 mM DTT and 50 mM NH_4HCO_3. Incubate for 45 min at 57°C. Make sure the gel is submerged at all times. This step reduces the disulfide bonds.
3. Remove the liquid and replace it with a solution containing 50 mM iodoacetamide and 50 mM NH_4HCO_3. Incubate for 40 min at 30°C in the dark. Make sure the gel is submerged at all times. This step alkylates the cysteine residues and prevents the formation of disulfide bonds. This alkylation step adds 57.021 Da to the mass of every cysteine residue.
4. Wash the gel slabs three times in HPLC-grade H_2O.
5. Cut complete lanes from the gels into 48 equal slices. Cut each slice into 1 mm^3 parts and collect each slice in a well

of a microtiter plate with conical-shaped wells containing 50 μL of H_2O.

6. Remove the H_2O from the microtiter plate wells using a multichannel pipette and filter tips.
7. Add 50 μL of 95% acetonitrile per well, seal the plate, and shake carefully for 5–10 min (until the gel plugs have shrunk completely and turned white and have become sticky. Make sure the acetonitrile does not come into contact with the seal).
8. Remove the acetonitrile, put the microtiter plate on ice and add 20 μL of Digest buffer, seal the plate, and rehydrate gel plugs for 30 min at 4°C.
9. Add an additional 10 μL of 50 mM NH_4HCO_3:10% CH_3CN (v/v), seal the plate with a clear adhesive seal, and incubate at 37°C for 3.5 h. Take care to use a top-heating device (e.g., a thermocycler) to avoid condensation at the seal surface.
10. Carefully sonicate the microtiter plate for 5 min.
11. Activate and wash the microcolumn cleanup tips with 95% acetonitrile (three times 5 μL). Concentrate the resulting peptides from each well on a microcolumn solid phase C18 tip by pipetting up and down (10 μL) a few times in each well; remove the loaded tips from the pipette, bring 10 μL of H_2O on top of its C18 bed, and desalt the peptides by pushing the H_2O through the C18 bed.
12. Elute the peptides directly onto a MALDI target plate using 1.5 μL of 50% CH_3CN:0.1% CF_3COOH solution saturated with α-cyano-4-hydroxycinnamic acid, spiked with internal peptide standards.

3.5.2. Acquisition of Mass Spectra

A MALDI tandem MS instrument is used to acquire peptide mass fingerprints and subsequent 1-kV collision-induced dissociation fragmentation spectra of selected peptides. Each MALDI plate is calibrated according to the manufacturer's specifications. All peptide mass fingerprinting (PMF) spectra are internally calibrated with three internal standards at m/z 963.516 (des-Pro2-Bradykinin), m/z 1570.677 (Glu1-Fibrinopeptide B), and m/z 2465.198 (Adrenocorticotropic Hormone Fragment 18-39), resulting in an average mass accuracy of 5 ± 10 ppm for each peptide mixture spot on the analyzed MALDI targets. With the individual PMF spectra, up to 16 peptides, exceeding a signal-to-noise ratio of 20, that pass through a mass exclusion filter (9) are submitted to fragmentation analysis (MS/MS).

3.5.3. MS-Based Protein Homology Identification

PMF spectra and the peptide sequence spectra of each sample are processed using the accompanying software suite (GPS Explorer

3.6, Applied Biosystems) with parameter settings as described (9). Data search files are generated according to the settings presented previously (9) and submitted for protein homology identification against The Arabidopsis Information Resource (TAIR) 9.0 by using a local database search engine (Mascot 2.2). Protein homology identifications with a relative score exceeding 95% probability are retained. Preferentially, identifications resulting from combined searches, PMF and MSMS, are retained. Additionally, PMF-only identifications are retained, but an additional restriction is implemented here. To reduce the number of false-positive identifications, hits for which 50% or more of the matched peptides have a tryptic misscleavage are discarded. The false-positive discovery rate is estimated by running each query against a TAIR 9 decoy database.

4. Notes

1. We use the constitutive 35S cauliflower mosaic virus promoter. The 35S promoter cloned into pDonrP4P1R is available from http://www.psb.ugent.be/gateway/index.php as pEN-L4-2-R1 or in pDonrP4P3 as pEN-L4-2-L3.
2. The ORF can be amplified from a cDNA library or from existing full-length cDNA clones. Most available cDNA clones can be found at the TAIR web site. Alternatively, the genomic sequence from start codon to stop codon, including introns, can be amplified from genomic DNA.
3. Use a *Taq* DNA polymerase with 3′–5′ exonuclease (proofreading) activity. For colony PCR, no proofreading activity is required.
4. In all Gateway reactions it is important to mix equimolar amounts of DNA. Use the following formula to convert femtomoles (fmol) of DNA to nanograms (ng) of DNA: ng = (x fmol)(N)(660 fg/fmol)(1 ng/10^6 fg), where x is the number of femtomoles and N is the size of the DNA in bp.
5. pKCTAP contains a kanamycin resistance gene and a green fluorescent protein (GFP) expression cassette allowing selection of transformed cells with kanamycin. Transformation efficiency can be screened based on the GFP signal observed with a fluorescence microscope. Alternatively, pHCTAP can be used as destination vector that allows selection of transformed plant cells with 20 μg/mL hygromycin. Instead of the GFP expression cassette, this vector contains a red fluorescent protein (RFP) expression cassette.

6. The traditional TAP tag has been replaced by the GS tag, because the latter gives rise to higher protein tag fusion accumulation levels and a lower background than purification with the traditional TAP tag.
7. If transcription factors are used that are light dependent, the PSB-L culture, derived from MM1 (15), can be used. The PSB-L culture grows under a day (16 h)/night (8 h) regime at 22°C on a shaker at 150 rpm.
8. The cell suspension culture PSB-D is maintained by weekly subculturing 7 mL of a 7-day-old culture into 45 mL of fresh MSMO in a 100-mL Erlenmeyer flask. Growth conditions are always shaking on an orbital shaker at 130 rpm at 25°C in the dark. Healthy PSB-D cells are characterized by a yellow color, while unhealthy cells often appear white. To screen for bacterial or fungal contamination, allow plant cells to settle down and check the clarity of the medium.
9. Never close an Erlenmeyer flask completely with the cap, but rotate the cap loosely on the flask and prevent contamination by sealing the cap on the flask with 3 M micropore medical tape.
10. If inefficient growth is observed (low cell density), decrease volume to 20–25 mL of fresh MSMO/KVC.
11. The optimal time point of harvesting may vary for different transcription factors and can be determined first in a time course experiment in which the transcription factor production is evaluated. For transcription factors that are active in proliferating cells, cultures are harvested 2 days after subculturing. For the TAP experiments on the JAZ1 cultures, cultures were not (mock) and treated for 1 min before harvesting with 100 μM JA (Duchefa, Haarlem, The Netherlands) 6 days after subcultivation.
12. Approximately 10–15 g of cells is typically harvested from 1 L of a 2-day-old culture.
13. From this stage on, extreme care should be taken to avoid ambient protein contamination (e.g., keratin contamination originating from skin, hair, woolen clothing, and dust) and contamination of detergents, plasticizers, or polymers, such as polyethylene glycol. Wear powder-free and non-latex gloves at all times during protein sample preparation. Wash the gloves with water before using them to handle the samples and equipment. Thoroughly wash anything that will come into contact with the sample (i.e., gel apparatus, staining trays, gel excising implements, gel storage equipment, and recipients) to remove keratins and other contaminants. Vials should preferably be washed with ULC/MS-grade acetonitrile:ULC/MS-grade water (95:5) (v/v) (do not use 100% acetonitrile to avoid leeching of plasticizers)

and then ULC/MS-grade water. Sample vials should be free of plasticizers, antistatic agents, and contaminants that can compromise the high-sensitivity MS analysis. Anything that touches the gel or sample is a possible source of contamination. Avoid storing gel in Saran wrap or similar material and instead use new, cleaned plastic or glass gel trays. Avoid using molecular weight cut-off filters (e.g., Centricon filters); if they must be used, thoroughly wash them. These are sources for synthetic polymer contamination in samples. Do not use vials with rubber o-rings or gaskets. Avoid using parafilm to seal vials.

14. Typically, the purification is done in duplicate; therefore, make 400 mg of total protein extract from 25 to 30 g of starting material and divide the extract for two parallel purifications.
15. After every mixing step, cool down the Falcon tubes on ice for 30 s to prevent heating of the sample and protein degradation. From this point onward, all steps – except the TEV elution – should be carried out at 4°C. Prechill all recipients on ice.
16. If an antibody is available against the protein of interest, isolate a small fraction from all purification steps and analyze the purification yield by protein gel blotting.
17. To prevent bead loss, a syringe with needle (0.45 mm) can be used.
18. Always avoid air bubbles over the beads and do not let the beads become dry.
19. A white cloud appears when proteins precipitate.
20. From here on, use MS-grade water and solvents. NH_4HCO_3 should be of high quality and the buffer should be made fresh prior to use, since it easily decomposes into its volatile reaction products, carbon dioxide and ammonia.

References

1. Lorenzo, O., Chico, J. M., Sánchez-Serrano, J. J., and Solano, R. (2004) *JASMONATE INSENSITIVE1* encodes a MYC transcription factor essential to discriminate between different jasmonate-regulated defense responses in Arabidopsis. *Plant Cell* **16**, 1938–1950.
2. Katsir, L., Schilmiller, A. L., Staswick, P. E., He, S. Y., and Howe, G. A. (2008) COI1 is a critical component of a receptor for jasmonate and the bacterial virulence factor coronatine. *Proc. Natl. Acad. Sci. USA* **105**, 7100–7105.
3. Chini, A., Fonseca, S., Fernández, G., Adie, B., Chico, J. M., Lorenzo, O., García-Casado, G., López-Vidriero, I., Lozano, F. M., Ponce, M. R., Micol, J. L., and Solano, R. (2007) The JAZ family of repressors is the missing link in jasmonate signalling. *Nature* **448**, 666–671.
4. Thines, B., Katsir, L., Melotto, M., Niu, Y., Mandaokar, A., Liu, G., Nomura, K., He, S. Y., Howe, G. A., and Browse, J. (2007) JAZ repressor proteins are targets of the SCF^{COI1}

complex during jasmonate signalling. *Nature* **448**, 661–665.

5. Rigaut, G., Shevchenko, A., Rutz, B., Wilm, M., Mann, M., and Séraphin, B. (1999) A generic protein purification method for protein complex characterization and proteome exploration. *Nat. Biotechnol.* **17**, 1030–1032.
6. Gavin, A.-C., Aloy, P., Grandi, P., Krause, R., Boesche, M., Marzioch, M., Rau, C., Jensen, L. J., Bastuck, S., Dümpelfeld, B., Edelmann, A., Heurtier, M.-A., Hoffman, V., Hoefert, C., Klein, K., Hudak, M., Michon, A.-M., Schelder, M., Schirle, M., Remor, M., Rudi, T., Hooper, S., Bauer, A., Bouwmeester, T., Casari, G., Drewes, G., Neubauer, G., Rick, J. M., Kuster, B., Bork, P., Russell, R. B., and Superti-Furga, G. (2006) Proteome survey reveals modularity of the yeast cell machinery. *Nature* **440**, 631–636.
7. Gavin, A.-C., Bösche, M., Krause, R., Grandi, P., Marzioch, M., Bauer, A., Schultz, J., Rick, J. M., Michon, A.-M., Cruciat, C.-M., Remor, M., Höfert, C., Schelder, M., Brajenovic, M., Ruffner, H., Merino, A., Klein, K., Hudak, M., Dickson, D., Rudi, T., Gnau, V., Bauch, A., Bastuck, S., Huhse, B., Leutwein, C., Heurtier, M.-A., Copley, R. R., Edelmann, A., Querfurth, E., Rybin, V., Drewes, G., Raida, M., Bouwmeester, T., Bork, P., Seraphin, B., Kuster, B., Neubauer, G., and Superti-Furga, G. (2002) Functional organization of the yeast proteome by systematic analysis of protein complexes. *Nature* **415**, 141–147.
8. Krogan, N. J., Cagney, G., Yu, H., Zhong, G., Guo, X., Ignatchenko, A., Li, J., Pu, S., Datta, N., Tikuisis, A. P., Punna, T., Peregrín-Alvarez, J. M., Shales, M., Zhang, X., Davey, M., Robinson, M. D., Paccanaro, A., Bray, J. E., Sheung, A., Beattie, B., Richards, D. P., Canadien, V., Lalev, A., Mena, F., Wong, P., Starostine, A., Canete, M. M., Vlasblom, J., Wu, S., Orsi, C., Collins, S. R., Chandran, S., Haw, R., Rilstone, J. J., Gandi, K., Thompson, N. J., Musso, G., St Onge, P., Ghanny, S., Lam, M. H. Y., Butland, G., Altaf-Ul, A. M., Kanaya, S., Shilatifard, A., O'Shea, E., Weissman, J. S., Ingles, C. J., Hughes, T. R., Parkinson, J., Gerstein, M., Wodak, S. J., Emili, A., and Greenblatt, J. F. (2006) Global landscape of protein complexes in the yeast *Saccharomyces cerevisiae*. *Nature* **440**, 637–643.
9. Van Leene, J., Stals, H., Eeckhout, D., Persiau, G., Van De Slijke, E., Van Isterdael, G., De Clercq, A., Bonnet, E., Laukens, K., Remmerie, N., Henderickx, K., De Vijlder, T., Abdelkrim, A., Pharazyn, A., Van Onckelen, H., Inzé, D., Witters, E., and De Jaeger, G. (2007) A tandem affinity purification-based technology platform to study the cell cycle interactome in *Arabidopsis thaliana*. *Mol. Cell. Proteomics* **6**, 1226–1238.
10. Bürckstümmer, T., Bennett, K. L., Preradovic, A., Schütze, G., Hantschel, O., Superti-Furga, G., and Bauch, A. (2006) An efficient tandem affinity purification procedure for interaction proteomics in mammalian cells. *Nat. Methods* **3**, 1013–1019.
11. Van Leene, J., Witters, E., Inzé, D., and De Jaeger, G. (2008) Boosting tandem affinity purification of plant protein complexes. *Trends Plant Sci.* **13**, 517–520.
12. Grunewald, W., Vanholme, B., Pauwels, L., Plovie, E., Inzé, D., Gheysen, G., and Goossens, A. (2009) Expression of the *Arabidopsis* jasmonate signalling repressor *JAZ1/TIFY10A* is stimulated by auxin. *EMBO Rep.* **10**, 923–928.
13. Pauwels, L., Fernández Barbero, G., Geerinck, J., Tilleman, S., Grunewald, W., Cuéllar Pérez, A., Chico, J. M., Vanden Bossche, R., Sewell, J., Gil, E., García-Casado, G., Witters, E., Inzé, D., Long, J. A., De Jaeger, G., Solano, R., and Goossens, A. (2010) NINJA connects the co-repressor TOPLESS to jasmonate signalling. *Nature* **464**, 788–791.
14. Chini, A., Fonseca, S., Chico, J. M., Fernández-Calvo, P., and Solano, R. (2009) The ZIM domain mediates homo- and heteromeric interactions between Arabidopsis JAZ proteins. *Plant J.* **59**, 77–87.
15. Menges, M., and Murray, J. A. H. (2002) Synchronous Arabidopsis suspension cultures for analysis of cell-cycle gene activity. *Plant J.* **30**, 203–212.

Chapter 12

Assaying Transcription Factor Stability

Jasmina Kurepa and Jan A. Smalle

Abstract

Similar to the activities of transcription factors (TFs) in other eukaryotes, activities of many plant TFs are determined via regulated proteolysis by the ubiquitin/26S proteasome system. Thus, to fully understand the function of a TF, it is important to determine the fate of the active TF protein and unravel the environmental and intrinsic signals that control its total cellular level. Here we describe how to determine whether a TF of interest is targeted to the 26S proteasome for degradation. The given method combines analyses of the effects of translational inhibition and the inhibition of proteasome activity. An important requirement for these experiments is to monitor in parallel the effects of translational and proteasomal inhibition on the abundance of the TF and (1) on ubiquitin, which becomes rapidly depleted upon translational inhibition (2), on polyubiquitinated proteins, which accumulate upon successful inhibition of the 26S proteasome, and (3) on glutamine synthase, a very stable protein that is used as a general metabolic control. The method described here can be used to test TF stability under a variety of conditions and in different genetic backgrounds.

Key words: Immunoblot analyses, cycloheximide, proteasome inhibitor, transcription factor stability, glutamine synthase, ubiquitin.

1. Introduction

Studies with various eukaryotic species have shown that correlations between mRNA and protein abundances are generally weak, suggesting that post-translational controls such as proteolysis exert a significant influence on the final outcome of gene expression (1–4). Post-translational controls of transcription factor (TF) activities are of particular interest. Many TFs have been shown to be unstable proteins, with their activities tightly linked to their proteolysis rates. In plants, for example, many unstable TFs with essential regulatory roles in various signal transduction

L. Yuan, S.E. Perry (eds.), *Plant Transcription Factors*, Methods in Molecular Biology 754,
DOI 10.1007/978-1-61779-154-3_12, © Springer Science+Business Media, LLC 2011

pathways have been identified, and the stability of these TFs has been shown to be signal dependent (5–15).

In all eukaryotes, the 26S proteasome is the main protease responsible for the degradation of cytosolic and nuclear proteins (16–19). The 26S proteasome is a central component of the ubiquitin (Ub)—proteasome system, and it predominantly degrades proteins that have been modified with a poly-Ub chain (20, 21). The modification is generally initiated by the perception and transduction of a cellular or environmental signal that induces a specific interaction of the putative target protein with a Ub ligase (i.e., E3). The poly-Ub chain is covalently linked to the protein through the concerted action of a Ub activase (El) and a Ub-conjugase (E2). Because the 26S proteasome specifically interacts with the poly-Ub chain, polyubiquitination effectively converts a protein into a proteasome substrate (22–24).

Degradation of a TF by the 26S proteasome can be assayed by combining translational inhibition with suppression of proteasome activity. Translational inhibition over a time course combined with immunoblotting analysis using TF antisera is used to determine whether a TF is being degraded under the condi-

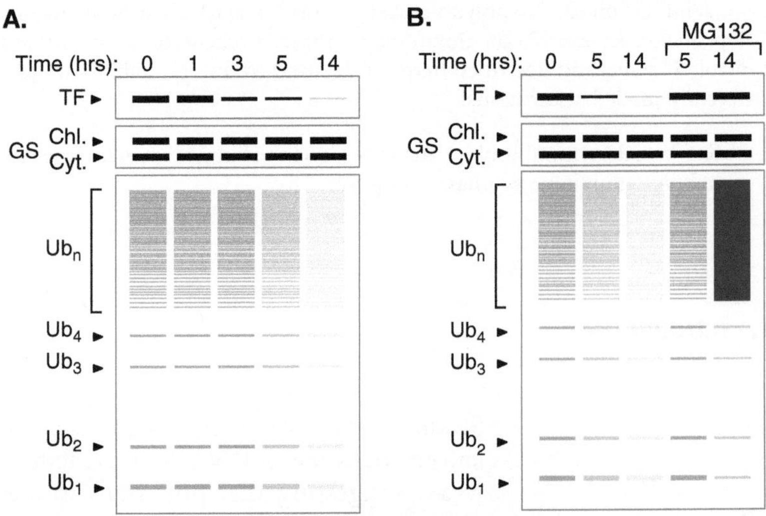

Fig. 12.1. Schematic representation of TF stability analysis. (**a**) Schematic immunoblots illustrating the CHX chase analysis. Using antisera against the TF and against glutamine synthase (GS) and Ub as controls, a conclusion can be drawn on whether the TF is a proteolysis target under the incubation conditions used. (**b**) If the TF abundance is decreased relative to a stable protein (in this example, cytosolic (Cyt.) and chloroplastic (Chl.) isoforms of GS) after CHX treatment, then the role of the 26S proteasome is tested by including MG132 in the CHX chase at the time points that lead to at least a 50% decrease in protein abundance. Suppression of the CHX depletion effect by MG132 constitutes as proof for the involvement of the proteasome in TF proteolysis. Ub_1, Ub monomer; Ub_2, Ub dimer; Ub_3 and Ub_4, Ub trimer and tetramer; and Ub_n, polyubiquitinated proteins.

tions used. If protein instability is detected, then the role of the 26S proteasome can be tested by combining translational inhibition with inhibition of 26S proteasome activity. The involvement of the proteasome in this depletion is confirmed by suppression of the depletion effect by the proteasome inhibitor MG132 (**Fig. 12.1**). Essential components of this experiment are the use of control antisera that allow the unequivocal monitoring of the effects of the treatments as well as the inclusion of a prolonged time point that allows testing the effectiveness of the inhibitors and conditions used. The combination of antisera against glutamine synthase and Ub can be used as universal controls for protein stability assays (25).

2. Materials

2.1. Growth and Treatment of Arabidopsis Seedlings

1. Murashige and Skoog (MS) basal salt mixture (PhytoTechnology Laboratories®, Shawnee Mission, KS).
2. Agar, phyto (Research Product International Corp., Mt. Prospect, IL).
3. Sucrose, crystalline/certified ACS (Thermo Fisher Scientific Inc.).
4. *Arabidopsis* seed.
5. 70% ethanol.
6. Bleach.
7. Sterile water.
8. Cycloheximide (MP Biomedicals, Solon, OH) is dissolved at 50 mg/mL in dimethyl sulfoxide (DMSO). The stock solution is stable for months when stored at –20°C. Cycloheximide (CHX) is highly toxic and degraded by alkali (pH > 7). Gloves should be worn at all times, and the working solutions should be collected and treated with an alkaline solution prior to disposal. All work surfaces, containers, and gloves should also be decontaminated by washing with an alkali solution such as soap.
9. MG132 (Z-Leu-Leu-Leu-CHO; Enzo Life Sciences International, Inc., Plymouth Meeting, PA) is dissolved at 100 mM in DMSO and kept at –80°C protected from light. MG132 is considered harmful, and concentrated solutions should be diluted prior to disposal (*see* **Note 1**).
10. 1.7 mL tubes.
11. Liquid nitrogen.

2.2. Sodium Dodecyl Sulfate-Polyacrylamide Gel Electrophoresis (SDS-PAGE)

1. 40% acrylamide solution containing 38.96% (w/v) acrylamide and 1.04% (w/v) bis-acrylamide for a monomer to cross-linker ratio of 37.5:1 (Research Product International Corp.). Acrylamide is toxic, and gloves should be worn at all times.
2. 4X separating gel buffer: 1.5 M Tris–HCl, pH 8.8, 0.4% SDS (*see* **Note 2**). The solution is stable for months in the refrigerator. Prior to casting the gel, heat the solution to room temperature.
3. 4X stacking gel buffer: 0.5 M Tris–HCl, pH 6.8, 0.4% SDS. Store in a refrigerator and heat to room temperature before use (*see* **Note 3**).
4. 7% ammonium persulfate (APS) in water. Aqueous solutions of APS are unstable and are best prepared fresh. Alternatively, the solutions can be distributed into single-use aliquots and stored at –20°C (*see* **Note 4**).
5. *N*,*N*,*N*′,*N*′-tetramethylethylenediamine (TEMED; Fisher BioReagents) (*see* **Note 4**).
6. Isopropanol or isobutanol.
7. 2X Laemmli sample buffer which contains 0.125 M Tris–HCl, pH 6.8, 4% SDS, 20% glycerol, 0.004% bromophenol blue, and 10% 2-mercaptoethanol. Sample buffer is made without 2-mercaptoethanol and stored at room temperature. 2-mercaptoethanol is added to an aliquot needed for the experiment just before use.
8. Sand, 50–70 mesh particle size (Sigma).
9. Microcentrifuge tube pestles.
10. Prestained markers such as EZ-Run prestained protein marker (Fisher BioReagents).
11. SDS-PAGE equipment (e.g., Thermo Scientific Owl dual-minigel vertical electrophoresis system).
12. Tris–glycine (TG) buffer (25 mM Tris, 192 mM glycine) can be made up as a 10X stock solution and is stable at room temperature for at least 6 months.
13. SDS-PAGE running buffer: dilute 10X TG buffer to 1X with water and add SDS to 0.1%.

2.3. Immunoblotting (Western Blotting)

1. TG-methanol transfer buffer (25 mM Tris, 192 mM glycine, and 20% methanol): dilute 10X TG buffer with water and methanol (ACS grade) just before use.
2. Whatman 3MM paper.
3. Supported nitrocellulose membrane Hybond-C Extra (GE, Piscataway, NJ).

4. Protein transfer equipment (e.g., Thermo Scientific Owl VEP-2 mini tank electroblotting system).
5. Ponceau S (Sigma, St. Louis, MO) stain: 0.2% Ponceau S (w/v) in 3% glacial acetic acid ACS grade (*see* **Note 5**). The stain is stable at room temperature almost indefinitely and can be reused.
6. Phosphate buffer saline (PBS) with Tween 20 (PBST: 3.2 mM Na_2HPO_4, 0.5 mM KH_2PO_4, 1.3 mM KCl, 135 mM NaCl, pH 7.4, and 0.1% Tween 20). PBS is prepared as a 20X stock (at this concentration, it has a longer shelf life than 10X or 1X solution). 20X stock is diluted and supplemented with Tween 20 before use (*see* **Note 6**).
7. Membrane blocking buffer: 10% nonfat dry milk in PBST or 3% albumin from bovine serum (BSA) in PBST (*see* **Note 7**).
8. Primary and secondary antibody dilution buffer: PBST with 2% nonfat dry milk or 1% BSA.
9. Anti-Ub antibody from Santa Cruz Biotechnology (*see* **Note 8**).
10. Anti-glutamine synthase antibody (Agrisera AB, Vännäs, Sweden).
11. Antisera against the TF or against a tag which is fused to the TF of interest.
12. Goat anti-rabbit secondary antibody conjugated to alkaline phosphatase (Santa Cruz Biotechnology, Santa Cruz, CA).
13. Alkaline phosphatase- or horseradish peroxidase-conjugated secondary antibody for detection of anti-TF or anti-tag antibody.
14. One-step nitro blue tetrazolium/5-bromo-4-chloro-3-indolyl phosphate (NBT/BCIP) substrate for alkaline phosphatase (Thermo Fisher Scientific) (*see* **Note 9**).
15. Thermo Scientific Pierce ECL Western Blotting Substrate for horseradish peroxidase.
16. X-ray film and X-ray film developer and fixer (Kodak) or a CCD camera-equipped ChemiDoc^{TM} XRS Gel Documentation system (Bio-Rad, Hercules, CA).

3. Methods

To determine if a specific TF is a proteasome target, one needs to test (1) how stable the protein is and (2) how the chemical or genetic inhibition of proteasome activity affects the protein stability. Here we outline a protein stability test combined with chemical inhibition of proteasome activity, because this route can be easily adapted to any plant species of interest and to any compound that is a candidate signal for the alteration of TF stability.

The first step of the analyses is a CHX chase in which the abundance of the TF is determined in the presence of a translation inhibitor. After CHX chase immunoblots are analyzed, one can determine the time point at which the abundance of the TF of interest is reduced by ~50%. In the hypothetical example shown in **Fig. 12.1a**, a 5 h long CHX treatment leads to a reduction in TF abundance by ~50%. The second step is a three-point CHX chase analysis combined with 26S proteasome inhibition. The abundance of the TF between the CHX-treated and CHX+MG132-treated samples is compared at three time points: at the beginning of the treatment, at the "half-life" time point (e.g., 5 h), and after a prolonged (e.g., overnight) treatment (**Fig. 12.1b**). Both treatments can be combined with a compound that is a likely regulator of the stability of the TF of interest.

An essential component of these analyses is to insure that the CHX and MG132 treatments indeed effected protein translation and proteasome-dependent degradation. To test this, in parallel to the analyses of TF abundance, one ought to follow the abundance of (1) total cellular Ub, the best control protein for the CHX treatment because its depletion does not depend on the activity of one particular E3 ligase, (2) polyubiquitinated proteins, i.e., Ub conjugates that accumulate in the presence of MG132, and (3) glutamine synthase, a stable enzyme which is not degraded by the proteasome (18, 25, 26). The below procedure describes TF stability analysis in light-grown intact *Arabidopsis* seedlings without additional treatments with signaling compounds (e.g., hormones) or stresses. The growth and treatment conditions can be modified to serve the project needs.

3.1. Growth and Treatment of Arabidopsis Seedlings

1. Make standard *Arabidopsis* MS/2 medium (half-strength Murashige and Skoog basal salt mixture with 1% sucrose) and adjust pH to 5.7 with 2 M KOH (*see* **Note 10**). Set aside a 100 mL aliquot of liquid MS/2 and add Phyto agar (0.8%) to the remaining MS/2 medium. Sterilize the medium by autoclaving and pour the agar-containing MS/2 into Petri plates. The solidified MS/2 plates should be stored at 4°C to minimize the possibility of contamination.
2. To sterilize *Arabidopsis* seeds, pour seeds into a 1.7 mL tube, add 1 mL of 70% ethanol, let stand for 5 min, and pipet the ethanol out. Add 1 mL of 50% commercial bleach (Clorox), shake, and let stand for 20 min. Pipet out the bleach solution and rinse the seeds in 1 mL of sterile water three times. The sterilized seeds can be stored at 4°C in the dark for

2–4 days to break dormancy. Alternatively, sterilized seeds can be sown immediately onto MS/2 plates (e.g., by pipetting the seeds in sterile water onto the agar surface with a 1 ml pipet), and the plates can be stored at 4°C in the dark (*see* **Note 11**).

3. Transfer the plates with seeds to a controlled environment growth chamber and grow for 7–10 days. The growth chamber temperature can be set between 22 and 25°C, and the light regimes include continuous light or long days (16 h light/8 h dark).
4. Prepare the treatment solutions in liquid MS/2. Wear gloves and prepare a container for CHX waste.

3.1.1. CHX Chase Analyses

1. Take 10 mL of liquid MS/2 medium and add CHX from a DMSO stock solution to a final concentration of 200 μM.
2. Pipet out 1 mL aliquots of the CHX-MS/2 solutions into the 1.7 mL tubes and prepare two tubes with 1 mL of MS/2 medium with DMSO as "Start" and "End" controls (*see* **Note 12**). Label the CHX tubes with the intended harvest time (e.g., "1 h" and "3 h").
3. Gently take the seedlings out of the agar medium (not to break the roots), measure fresh weight, and submerge the seedlings into the treatment medium. Note the fresh weight of each sample and ensure that all tubes contain a similar mass of seedlings (~50 or ~100 mg).
4. Immediately freeze the "Start" control in liquid nitrogen (LN_2) and transfer the rest of the tubes with seedlings to the environmental grown chamber. Harvest the samples after 1, 3, 5, and 14 h (*see* **Note 13**) by taking out the seedlings from the tube with tweezers, gently blotting them onto a paper napkin, and freezing in LN_2. Harvest the "End" control after a 14 h incubation.
5. Store the samples at –80°C until the whole series of treatments is finished. Proceed with protein sample preparation and immunoblotting (*see* **Sections 3.2** and **3.3**).

3.1.2. CHX Chase Analyses Combined with 26S Proteasome Inhibition

1. Make a 200 μM CHX master mix in liquid MS/2 and distribute 1 mL aliquots into four 1.7 mL tubes. Add MG132 to a final concentration of 100 μM to two of the tubes and vortex for a couple of seconds (*see* **Note 14**). Prepare the "Start" and "End" controls by adding the combined volume of DMSO added by CHX and MG132 stocks.
2. Add a similar mass of seedlings to each tube and transfer the tubes to the growth chamber.
3. Freeze the "Start" control immediately in LN_2 and collect the "half-life" and "overnight" treatment samples by taking

out the seedlings, gently blotting them onto a paper napkin, and freezing in LN_2. Harvest the "End" control after overnight incubation.

4. Store the samples at –80°C until protein extraction.

3.2. Sodium Dodecyl Sulfate-Polyacrylamide Gel Electrophoresis (SDS-PAGE)

3.2.1. Protein Sample Preparation

1. Prepare the 2X Laemmli sample buffer by adding 2-mercaptoethanol to the sample buffer stock and prepare sand and a disposable or Teflon pestle (e.g., Kontes pellet pestle or Scienceware microcentrifuge tube sample pestle, Fisher Scientific). Preheat a dry bath to 95°C.
2. Transfer all the samples from the –80°C freezer to LN_2.
3. Take out a sample from LN_2, open the tube immediately, add 2 volumes of 2X Laemmli sample buffer (e.g., if the fresh weight of the seedlings in the sample is 50 mg, add 100 μL of sample buffer), add a bit of sand, and grind until no pieces of tissue are visible. Secure the lids by using lid-lock microcentrifuge tubes or by putting lid-lock clips, and immediately place the tubes into the preheated dry bath for 5 min. Cool the samples to room temperature, spin them at maximum speed in the microcentrifuge for 5 min, and load onto an SDS-PAGE gel for separation (*see* **Note 15**).

3.2.2. SDS-PAGE

1. Wear gloves during the whole procedure to avoid exposure to acrylamide and to avoid contaminating the electrophoresis equipment or gels with dust and proteins (e.g., keratin).
2. Prepare three 0.75 mm thick gels: one with 12% acrylamide separating gel for the Ub blot, one with 10% acrylamide for the glutamine synthase blot, and the third gel with the percentage of acrylamide required for the analyses of the TF of interest. Prepare the separating gel by mixing the appropriate volume of 4X separating buffer, 40% acrylamide stock, and water. Add APS (67 μL of the 7% stock for 5 mL of the separating gel) and TEMED (4 μL for 5 mL of gel) only when you are ready to pour the gel. Pour the separating gel mix (either tilt the glass plates and pour directly from the mixing tube or transfer the gel with a Pasteur pipet) and leave space for the stacking gel. Overlay with ~100 μL of isopropanol or isobutanol (*see* **Note 16**). The gel will polymerize within 30 min.
3. Pour off the isopropanol (isobutanol) and rinse the top of the gel with distilled water. Remove the remaining drops of water with a Kimwipe or similar absorbent paper towels.
4. Clean the comb with ethanol. While the comb is drying, prepare 4% acrylamide stacking gel by mixing the appropriate volumes of 4X stacking buffer, 40% acrylamide, and water.

Add APS (45 μL for 3 mL) and TEMED (4 μL for 3 mL) when ready to pour. The gel will polymerize within 15 min.

5. When stacking gel is polymerized, remove the comb carefully making sure not to tear the well ears. Place the gel in the electrophoresis chamber and add SDS-PAGE running buffer to the inner and outer reservoirs. Make sure that there are no bubbles in the wells and that the well ears are straight. If necessary, release the bubbles and straighten the ears using a Hamilton syringe.
6. Load the prestained markers (e.g., 2 μL of EZ-Run prestained protein marker per lane), load 15 μL of the samples (*see* **Note 17**), and run the gel at 200 V (constant voltage) until the bromophenol blue front migrates to the bottom of the gel.
7. Turn off the power supply and only then disconnect the electrodes.

3.3. Immunoblotting (Western Blotting)

3.3.1. Protein Transfer Using Wet (Tank) Blotting

1. Place the transfer cassette in a shallow tray so that the black (cathode) side is lying at the bottom of the tray and the red (anode) side is toward you (*see* **Note 18**). Open the cassette, lay the foam sheet on the black cathode side of the cassette, and pour enough TG-methanol buffer to cover the foam pad. Cut out two pieces of Whatman 3 MM paper large enough to cover the foam and put one piece of Whatman paper on the foam pad. Make sure the paper is evenly soaked in the transfer buffer.
2. Disassemble the glass plates and cut off the stacking gel. Carefully transfer the separating gel onto the Whatman paper and make sure that no bubbles are trapped between the gel and the paper.
3. Cut out a piece of supported nitrocellulose membrane which is a bit larger than the gel. Pre-wet the membrane by immersing it into the transfer buffer and lay it on top of the gel. To minimize trapping bubbles, begin laying the membrane on one side so that the membrane itself pushes the bubbles toward the other edge.
4. Pre-wet the second piece of Whatman 3 MM paper and place it on the membrane. To ensure that no bubbles are trapped, role a pipet or a test tube over the sandwich.
5. Place the second piece of the foam pad onto the Whatman paper and close the transfer cassette.
6. Fill the transfer tank with transfer buffer and immerse the cassette.
7. Transfer at constant current for 400 mAh (e.g., 2 h at 200 mA).

3.3.2. Quality Control of the Transfer (Ponceau S Staining)

1. Take out the Whatman paper sandwich containing the gel and the membrane.
2. Place the membrane in a clean glass container and add enough Ponceau S stain to cover the membrane. Let stand at room temperature for 1 min, remove the stain, and rinse in distilled water.
3. A properly stained membrane has a white background and red-stained proteins. Take a photograph of the membrane as proof of loading.

3.3.3. Detection of the Epitope of Interest

1. The membrane that will be probed with anti-Ub antibodies needs to be autoclaved prior to incubation with the primary antibodies (*see* **Note 19**). Place the membrane in a Pyrex tray between two sheets of Whatman 3MM paper and add TG-methanol (transfer) buffer enough to cover the membrane. Place a 1 L bottle with water on top of the paper–membrane–paper sandwich (*see* **Note 20**) and autoclave for 10 min on liquid cycle. Proceed with the blocking step.
2. Place the autoclaved membrane and all the Ponceau S stained membranes into a container and add blocking solution (*see* **Note 21**). Block the membranes a minimum of 15 min on a shaker at room temperature (*see* **Note 22**).
3. Pour off the blocking solution and immediately add an appropriate dilution of primary antibody. Anti-Ub antibodies are used at a 1:1000 dilution and anti-glutamine synthase antibodies are used at 1:5000 dilution. The dilution of the anti-TF antibody needs to be determined for each case.
4. Incubate and shake the membranes with primary antibodies overnight at 4°C (*see* **Note 23**).
5. Pour off the primary antibody solutions and wash the membranes with PBST three times, 15 min each.
6. Prepare the secondary antibody solutions (1:2500 dilution of the goat anti-rabbit alkaline phosphatase-conjugated antibody for Ub and glutamine synthase and the same dilution of the appropriate secondary antibody for the TF). Incubate with the membranes for minimum 1 h at room temperature.
7. Pour off the secondary antibody solutions and wash the membranes three times for 15 min with PBST.
8. Add the appropriate enzyme substrate:
   (a) For the alkaline phosphatase-conjugated secondary antibodies, add the One-step NBT/BCIP substrate, and photograph the membrane at different development times (e.g., as soon as the purple precipitate appears and a couple of minutes after that).
   (b) For the peroxidase-conjugated secondary antibodies

(1) Pre-warm the ECL solutions A and B to room temperature, mix equal volumes of A and B, and pour over the membrane. It is essential to cover the membrane evenly with the developing solution (e.g., by tilting the membrane or by gently spreading the developer with a Pasteur pipet).

(2a) X-ray film developing: Tilt the membrane to drain the developing solution, wrap it in Saran Wrap, and expose in the darkroom. The customary first exposure time is 30 s. Develop and fix the exposed film and estimate whether the exposure time needs to be shortened or extended. When film is dry, scan it on a flat-bed scanner or take a photograph on a white-light box.

(2b) Signal detection using ChemiDoc documentation system: Place the membrane onto the documentation system platform, cover with the developing solution, focus the camera, open the iris maximally, and ensure that there are no filters between the CCD camera and the membrane. Capture image using time-lapse exposure.

3.4. Data Analyses

3.4.1. Densitometric Analyses Using ImageJ (See Note 24)

1. Open the image of the immunoblot and go to the "Analyze" menu. The steps for 1D gel analysis are listed under "Analyze>Gels."
2. Highlight the rectangular selection tool and outline the first lane. Go to "Analyze>Gels>Select First Lane" or use the keyboard shortcut Command key – #1. The first lane will be outlined in the image file and labeled as 1 (**Fig. 12.2a** – step 1).
3. Drag the rectangle to the next lane and select "Analyze>Gels>Select Next Lane" or use the keyboard shortcut Command key – #2. The selected lane is outlined and labeled (**Fig. 12.2a** – step 2). Drag the rectangle to the next lane and repeat.
4. When all lanes are selected, go to "Select Analyze>Gels>Plot Lanes" or use the keyboard shortcut Command key – #3. That command will generate a densitometric plot (**Fig. 12.2a** – step 3).
5. Highlight the straight line tool and define the peak to be measured (e.g., draw the base lane and add the drop lines; **Fig. 12.2a** – step 4).
6. When all peaks are defined, highlight the wand tool and click inside of the peak. A new "Result" window will open listing the area of each peak (**Fig. 12.2a** – step 5).
7. Use the peak area measurements to calculate the relative abundance of the protein in each sample (*see* **Note 25**).

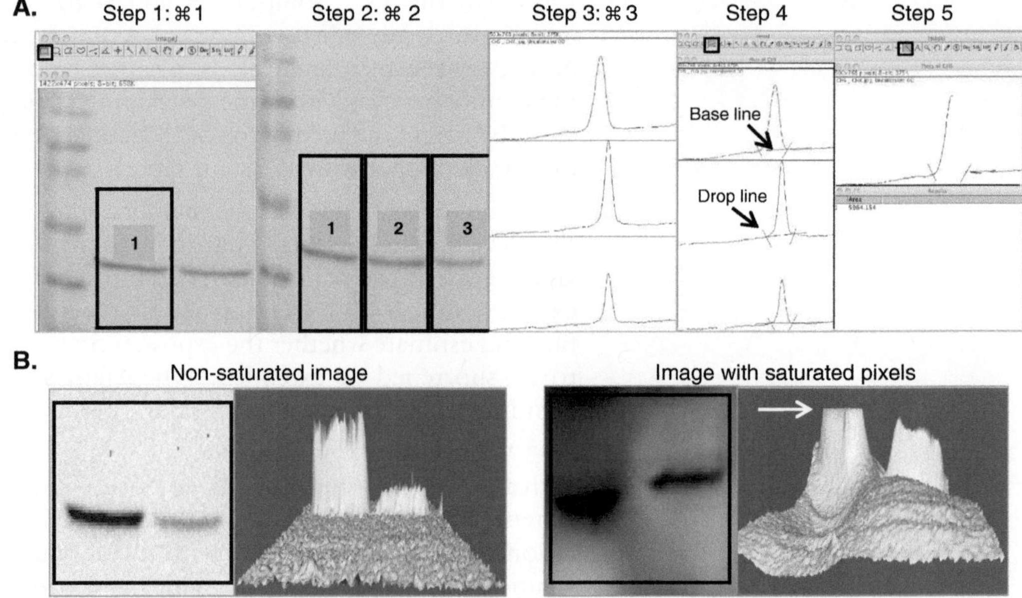

Fig. 12.2. Immunoblotting data analyses. (**a**) Steps of the densitometric determination of signal strengths using ImageJ. (**b**) Saturation tests using Quantity One software (Bio-Rad) ought to precede the quantification of signal strengths. Only images without saturated pixels should be quantified.

3.4.2. Analyses Using Quantity One Software (Bio-Rad)

1. Open the image and inspect for saturated pixels. Go to "View>3D Viewer," select the image area of interest, and click within the selection. A new window will open showing the 3D image of the signal (**Fig. 12.2b**). All signals with peaks that appear cut off are saturated and should not be used for quantification.
2. Mark lanes ("Lane>Frame Lane").
3. Mark bands of interest ("Band>Create Bands").
4. Go to Reports and select "All Lines Report" to view a table with relative band intensities (*see* **Note 25**).

4. Notes

1. MG132 is the most frequently used 26S proteasome inhibitor, and it reversibly inhibits the chymotrypsin-like activity of the proteasome. Many other reversible and irreversible, synthetic and naturally occurring proteasome inhibitors are available through a number of life science product providers, and any of them may be used in addition to MG132 for the proteasome inhibition experiments.

2. As customary for protein analyses, water used in all experiments should be ultrapure with resistivity of 18.2 MΩ/cm at room temperature.
3. The 4X stacking gel solution can be supplemented with 0.015% bromophenol blue to help visualize the gel wells. Bromophenol blue, which is present in the sample buffer, will not interfere with the electrophoresis.
4. Quality of the APS and TEMED solutions is critical for the proper polymerization of the gel. Both compounds should be purchased in small quantities and kept in a desiccator.
5. There are a number of protocols used to make the Ponceau S staining solution (0.5% Ponceau S (w/v) in 1% acetic acid; 0.1% Ponceau S (w/v) in 5% acetic acid; 0.3% Ponceau S (w/v) in 3% trichloroacetic acid). All staining solutions work well.
6. Tris-buffered saline (TBS) with Tween 20 (TBST: 10 mM Tris–Cl, pH 7.5, 150 mM NaCl, 0.15% Tween 20) can be used instead of the PBST buffer. Although the choice between these two buffers seems to be primarily based on the preference of laboratory members, these buffers can actually influence the quality of the analyses. For each new antibody, both buffers should be tested prior to the protein stability experiments.
7. Similarly to the basic immunoblotting buffers (PBST vs. TBST), the optimal blocking solution for each antibody should be determined empirically.
8. Almost all companies that produce or supply life science research products offer anti-Ub antibodies. Any polyclonal or monoclonal anti-Ub antibody can be used in these experiments (*see* also **Note 19**).
9. Buffers and substrates for the alkaline phosphatase detection are easy to make, but gloves should be worn at all times because NBT powder and solutions are toxic. The developing buffer (100 mM Tris–HCl, pH 9.5, 100 mM NaCl, 5 mM $MgCl_2$) is made as a 10X stock, and when kept at 4°C, it is stable for at least 1 year. The NBT stock solution contains 0.1 g NBT/mL 70% dimethylformamide (DMF), and the BCIP solution contains 0.05 g BCIP/mL DMF. Both solutions are kept at 4°C, and the NBT solution should be protected from light. To get the working solution, add 66 μL of NBT and 66 μL of BCIP stock solutions to 15 mL of 1X developing buffer.
10. Additional protocols and guidelines for growth of *Arabidopsis* plants can be found on the Arabidopsis Biological Resource Center at http://www.biosci.ohio-state.edu/pcmb/Facilities/abrc/handling.htm.

11. The chase experiments can also be done using plants grown in liquid cultures. For liquid cultures, sterilized seeds are pipeted into a 250 mL Erlenmeyer flask containing 50 mL of liquid MS/2 medium. The "liquid" cultures need to be shaken slowly (20–40 rpm) in an environmental growth chamber.
12. The "Start" and "End" controls are essential to test the possible circadian regulation of protein abundance or the effects of seedling submergence.
13. The given times are just guidelines; the optimal time course should be determined for every putative proteasome target.
14. The MG132 stock solutions when added to MS/2 medium form a white precipitate which will disperse by vigorous vortexing. Do not add plants until after vortexing.
15. Protein samples can also be kept at –20°C in the freezer for at least 1 month. However, it needs to be empirically determined for each protein of interest if it is stable under these conditions.
16. Some protocols suggest overlaying the separating gel with water. Water will indeed straighten the gel, but it will also make a shallow concentration gradient of acrylamide on top of the gel. In many cases, this gradient will not influence the final result, but – if possible – isopropanol or isobutanol should be used.
17. Fifteen microliters of a sample prepared as described is sufficient for the analyses of Ub, polyubiquitinated proteins, and glutamine synthase. This protein amount will allow the detection of many TFs as well. However, for low abundant proteins, one should titrate the correct protein amount.
18. Most of the Western blotting transfer apparatuses have clearly colored or labeled sides of the transfer cassette. If the cathode and anode side of the transfer cassette are not color coded or labeled, it is worth permanently and clearly labeling them to prevent retro-blotting.
19. Ub is a small, globular protein, and its heat inactivation is thought to expose latent antigenic sites. If the antiserum used to detect Ub is made against a surface epitope, the enhancement of the signal by autoclaving might not be large. However, if the anti-Ub antibody was made against a peptide or denatured Ub, the heat denaturation of the protein on nitrocellulose membrane prior to immunoblotting is essential. The increase in sensitivity of detection for Ub blots was described by Swerdlow et al. (27).
20. The membrane can also be autoclaved in water instead of transfer buffer. If the membrane is autoclaved without

weights (e.g., 1 L bottle with water on top of the sandwich), it will crease which may lead to uneven staining during the NBT/BCIP detection.

21. The remaining Ponceau S will wash off in the blocking solution and will not interfere with antibody binding.
22. Blocking times can be adjusted from 15 min to overnight. Overnight incubations are best done at 4°C to prevent coagulation of milk proteins.
23. Considering the time needed for protein sample preparation, gel casting and running, blotting, Ponceau S staining, and blocking, it is often convenient to incubate the membrane with the primary antibody overnight. Since many antisera can be reused (e.g., anti-glutamine synthase antisera can be frozen at –20°C after the first probing and can be reused up to three times; anti-Ub antisera cannot be reused), it is most prudent to do the overnight incubation at 4°C to prevent denaturation of the IgGs and coagulation of milk proteins.
24. ImageJ is a public domain program written in Java (so, it runs on any platform). It can be downloaded from http://rsbweb.nih.gov/ij/features.html.
25. Quantification of signal intensities from a single immunoblot using any of the methods will give only guidelines for future experiments. Relative signal intensities from two or preferably more immunoblots representing two or more independent treatments need to obtained, averaged, and analyzed for the final result.

Acknowledgments

This work was supported by the KTRD Center in Lexington, KY, and by grants from NSF (# 0919991) and from the National Research Initiative of the USDA Cooperative State Research, Education and Extension Service (#2005-35304-16043).

References

1. Greenbaum, D., Colangelo, C., Williams, K., and Gerstein, M. (2003) Comparing protein abundance and mRNA expression levels on a genomic scale. *Genome Biol.* **4**, 117.
2. Kislinger, T., Cox, B., Kannan, A., Chung, C., Hu, P., Ignatchenko, A., Scott, M. S., Gramolini, A. O., Morris, Q., Hallett, M. T., Rossant, J., Hughes, T. R., Frey, B., and Emili, A. (2006) Global survey of organ and organelle protein expression in mouse: combined proteomic and transcriptomic profiling. *Cell* **125**, 173–186.
3. Fu, N., Drinnenberg, I., Kelso, J., Wu, J.-R., Pääbo, S., Zeng, R., and Khaitovich, P.

(2007) Comparison of protein and mRNA expression evolution in humans and chimpanzees. *PLoS One* **2**, e216.
4. Nie, L., Wu, G., Culley, D. E., Scholten, J. C., and Zhang, W. (2007) Integrative analysis of transcriptomic and proteomic data: challenges, solutions and applications. *Crit. Rev. Biotechnol.* **27**, 63–75.
5. Gagne, J. M., Smalle, J., Gingerich, D. J., Walker, J. M., Yoo, S. D., Yanagisawa, S., and Vierstra, R. D. (2004) Arabidopsis EIN3-binding F-box 1 and 2 form ubiquitin-protein ligases that repress ethylene action and promote growth by directing EIN3 degradation. *Proc. Natl. Acad. Sci. USA* **101**, 6803–6808.
6. Guo, H., and Ecker, J. R. (2003) Plant responses to ethylene gas are mediated by $SCF^{EBF1/EBF2}$-dependent proteolysis of EIN3 transcription factor. *Cell* **115**, 667–677.
7. Potuschak, T., Lechner, E., Parmentier, Y., Yanagisawa, S., Grava, S., Koncz, C., and Genschik, P. (2003) EIN3-dependent regulation of plant ethylene hormone signaling by two *Arabidopsis* F box proteins: EBF1 and EBF2. *Cell* **115**, 679–689.
8. Schwager, K. M., Calderon-Villalobos, L. I., Dohmann, E. M., Willige, B. C., Knierer, S., Nill, C., and Schwechheimer, C. (2007) Characterization of the VIER F-BOX PROTEINE genes from Arabidopsis reveals their importance for plant growth and development. *Plant Cell* **19**, 1163–1178.
9. Osterlund, M. T., Hardtke, C. S., Wei, N., and Deng, X. W. (2000) Targeted destabilization of HY5 during light-regulated development of Arabidopsis. *Nature* **405**, 462–466.
10. Gray, W. M., Kepinski, S., Rouse, D., Leyser, O., and Estelle, M. (2001) Auxin regulates SCF^{TIR1}-dependent degradation of AUX/IAA proteins. *Nature* **414**, 271–276.
11. Kepinski, S., and Leyser, O. (2004) Auxin-induced SCF^{TIR1}-Aux/IAA interaction involves stable modification of the SCF^{TIR1} complex. *Proc. Natl. Acad. Sci. USA* **101**, 12381–12386.
12. Yang, X., Lee, S., So, J. H., Dharmasiri, S., Dharmasiri, N., Ge, L., Jensen, C., Hangarter, R., Hobbie, L., and Estelle, M. (2004) The IAA1 protein is encoded by AXR5 and is a substrate of SCF^{TIR1}. *Plant J.* **40**, 772–782.
13. Zenser, N., Ellsmore, A., Leasure, C., and Callis, J. (2001) Auxin modulates the degradation rate of Aux/IAA proteins. *Proc. Natl. Acad. Sci. USA* **98**, 11795–11800.
14. Chini, A., Fonseca, S., Fernández, G., Adie, B., Chico, J. M., Lorenzo, O., García-Casado, G., López-Vidriero, I., Lozano, F. M., Ponce, M. R., Micol, J. L., and Solano, R. (2007) The JAZ family of repressors is the missing link in jasmonate signalling. *Nature* **448**, 666–671.
15. He, J. X., Gendron, J. M., Yang, Y., Li, J., and Wang, Z. Y. (2002) The GSK3-like kinase BIN2 phosphorylates and destabilizes BZR1, a positive regulator of the brassinosteroid signaling pathway in Arabidopsis. *Proc. Natl. Acad. Sci. USA* **99**, 10185–10190.
16. Borissenko, L., and Groll, M. (2007) Diversity of proteasomal missions: fine tuning of the immune response. *Biol. Chem.* **388**, 947–955.
17. DeMartino, G. N., and Gillette, T. G. (2007) Proteasomes: machines for all reasons. *Cell* **129**, 659–662.
18. Kurepa, J., Toh-e, A., and Smalle, J. A. (2008) 26S proteasome regulatory particle mutants have increased oxidative stress tolerance. *Plant J.* **53**, 102–114.
19. Hanna, J., and Finley, D. (2007) A proteasome for all occasions. *FEBS Lett.* **581**, 2854–2861.
20. Smalle, J. A., and Vierstra, R. D. (2004) The ubiquitin 26S proteasome proteolytic pathway. *Annu. Rev. Plant Biol.* **55**, 555–590.
21. Dreher, K., and Callis, J. (2007) Ubiquitin, hormones and biotic stress in plants. *Ann. Bot. (Lond.)* **99**, 787–822.
22. Crews, C. M. (2003) Feeding the machine: mechanisms of proteasome-catalyzed degradation of ubiquitinated proteins. *Curr. Opin. Chem. Biol.* **7**, 534–539.
23. Pickart, C. M. (2000) Ubiquitin in chains. *Trends Biochem. Sci.* **25**, 544–548.
24. Thrower, J. S., Hoffman, L., Rechsteiner, M., and Pickart, C. M. (2000) Recognition of the polyubiquitin proteolytic signal. *EMBO J.* **19**, 94–102.
25. Kurepa, J., Karangwa, C., Duke, L. S., and Smalle, J. A. (2010) Arabidopsis sensitivity to protein synthesis inhibitors depends on 26S proteasome activity. *Plant Cell Rep.* **29**, 249–259.
26. Kurepa, J., and Smalle, J. A. (2008) Structure, function and regulation of plant proteasomes. *Biochimie* **90**, 324–335.
27. Swerdlow, P. S., Finley, D., and Varshavsky, A. (1986) Enhancement of immunoblot sensitivity by heating of hydrated filters. *Anal. Biochem.* **156**, 147–153.

Chapter 13

How to Assess the Intercellular Trafficking of Transcription Factors

Munawar Ahmad, Won Kyong Cho, Yeonggil Rim, Lijun Huang, and Jae-Yean Kim

Abstract

Non-cell-autonomous (NCA) control of plant development is an emerging field. Transcription factors (TFs) are the most important plant proteins involved in development and cell fate determination. In plants specialized intercellular symplastic channels, called plasmodesmata (PD), facilitate and regulate the NCA action of TFs. NCA-TFs move from cell to cell either selectively or non-selectively depending upon the specific interactions with PD or the pathway proteins. Here we describe different approaches to establish the role of TFs in NCA control of its function and the characteristic movement behavior.

Key words: Transcription factors, non-cell-autonomous, plasmodesmata, selective trafficking.

1. Introduction

1.1. Non-cell-autonomous Proteins

Intercellular trafficking of macromolecules, such as transcription factors (TFs), is emerging as an interesting field for the understanding of systems biology. In higher plants the symplastic cell-to-cell channels – plasmodesmata (PD) – provide the pathway for trafficking of macromolecules. Previous studies reported that the maize homeobox transcription factor, KNOTTED1 (KN1), could move from cell to cell in the maize leaf (1). Since then, several TFs which play important roles in plant development have been reported to traffic from cell to cell (2–4). It was shown that mobility of SHORT ROOT (SHR) is necessary for determination of endodermal cell fate (2). The movement of TFs could be through either a selective pathway or simple diffusion. Selectively trafficking TFs have the capability to dilate the PD size

L. Yuan, S.E. Perry (eds.), *Plant Transcription Factors*, Methods in Molecular Biology 754,
DOI 10.1007/978-1-61779-154-3_13,

exclusion limit (SEL) (5). Selective trafficking of proteins relies upon trafficking signals within the proteins that involve protein–protein interaction (1, 6, 7). In contrast non-selectively trafficking proteins simply diffuse through PD without specific protein–protein interaction (8). In *Arabidopsis*, more than 1500 TFs have been predicted (9). However, only a few TFs have been studied for their functions as non-cell-autonomous (NCA) trafficking proteins. Thus, the development of a high-throughput screening system to isolate potential non-cell-autonomously trafficking TFs would be valuable. We have established two efficient high-throughput systems to screen NCA-TFs and selectively trafficking TFs.

The first approach utilizes GAL4/UAS trans-activation system (10, 11) to express TF–mCherry fusions in-trans under the control of GAL4 expressed specifically in root endodermis and cortex as marked by cell-autonomous GFPer. This system is very useful to screen a large number of non-cell-autonomous movement proteins efficiently, but it cannot determine the mode of trafficking. Therefore, to establish whether the putative trafficking TFs follow a selective or non-selective pathway the trichome rescue system is employed subsequently (6). These systems will be described here separately (*see* **Note 1**).

1.2. Intercellular Movement Assay Using the GAL4/UAS System

The basic rationale is explained in **Fig. 13.1**. The enhancer trap line J0571 expresses GAL4 in endodermis and cortex in root which causes the tissue-specific expression of GFPer (green fluorescence protein with endoplasmic reticulum retention signal) from the same construct shown in **Fig. 13.1b** (*see* **Note 2**). The target TF is cloned under the control of upstream activation sequence (UAS) in the Gateway destination vector pCCL293

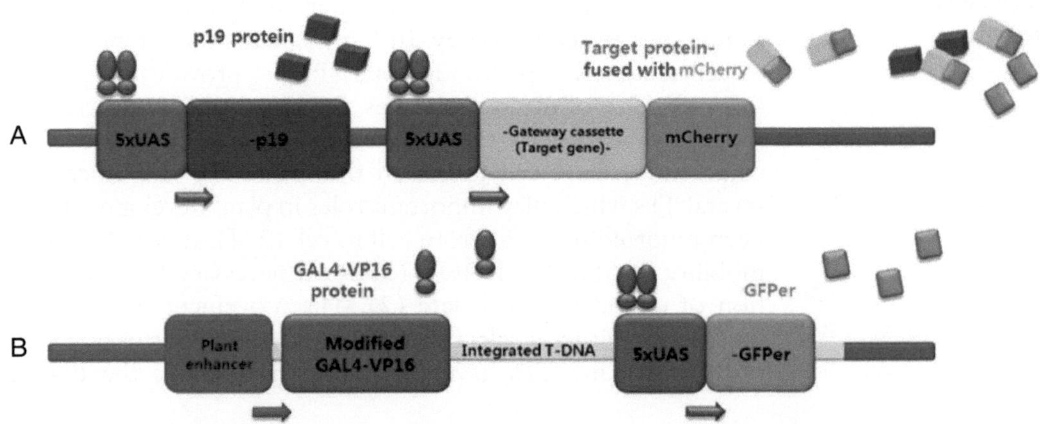

Fig. 13.1. Scheme for GAL4/UAS screening system. (**a**) Expression vector for TF–mCherry fusion. (**b**) Enhancer trap construct. GAL4 expressed under the control of plant enhancer controls the expression of downstream GFPer and also in trans-expression of p19 and TF–mCherry fusion from Gateway destination vector pCCL293.

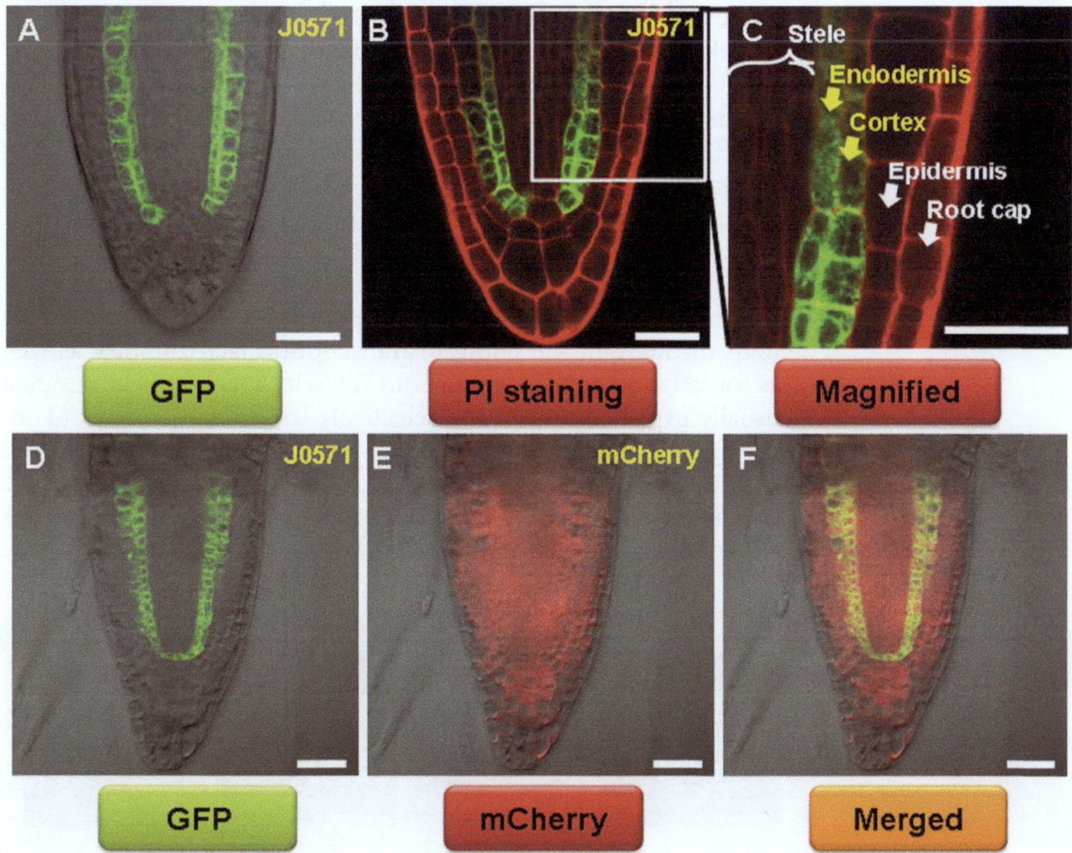

Fig. 13.2. Confocal laser scanning microscope images of young root tips. (**a–c**) Enhancer trap line J0571. (**d–f**) Expression of UAS::mCherry vector in line J0571. GFPer (*green color*: endodermis and cortex) and PI staining (*red color*: cell boundary staining) in **b**, c and mCherry (*red color*: whole root tip) in **e**, **f**.

(**Fig. 13.1a**). In order to reduce the silencing of construct expression, a viral silencing suppressor p19 (GI:9663887) was added to the modified Gateway destination vector (**Fig. 13.1a**; vector available upon request). The expression vector introduced in line J0571 expresses TF–mCherry fusion protein in the cortex and endodermal cells. Fluorescence GFP and mCherry can be readily seen with confocal laser scanning microscope as shown in **Fig. 13.2**. The TF–mCherry fusion will move out of the GFPer marked cells if the TF is a NCA-TF.

1.3. Intercellular Selective Protein Trafficking Assay Using Trichome Rescue System

The trichome rescue assay is based on the gain of cell-to-cell trafficking function by GL1, a member of MYB transcription factor family when fused to a selectively trafficking protein or polypeptide. GL1 is a cell-autonomous transcription factor (6) which regulates the downstream expression of GL2 (12) and is required for trichome formation (13). Endogenously both GL1 and GL2 are expressed in epidermal layers (12, 14). *gl1* mutants are glabrous, i.e., the trichomes on the leaf surface are absent.

The basic rationale of this system is illustrated in **Fig. 13.3**. When GL1 is expressed specifically in sub-epidermal layer using the RbcS2b promoter it fails to rescue trichome development in the *gl1* mutant background (**Fig. 13.3c**), but when fused to a selectively trafficking protein or peptide it was able to functionally complement the *gl1* phenotype by rescuing trichome development (6). This technique is useful for screening of selectively trafficking TFs. We have recently modified the vector system for high-throughput screening of transcription factor library as well as for deletion analysis to sort out the selectively trafficking signal usually expected within the selectively trafficking proteins (1, 6, 7). Transgenic plants are selected on MS medium supplemented with hygromycin B and observed for trichome rescue either at the seedling stage with a dissection microscope or at later stages with a magnifying lens (*see* **Note 3**).

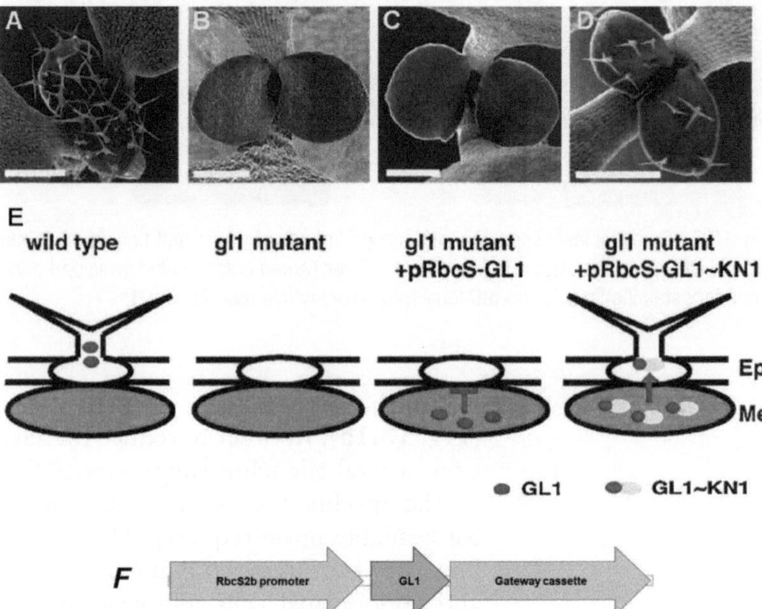

Fig. 13.3. (**a–d**) SEM images of wild-type seedling (**a**), *gl1* mutant seedling (**b**), *gl1* seedling carrying the control pRbcS::GL1 transgene (no trichome rescue) (c), and a *gl1* seedling expressing GL1–KN1 showing trichome rescue (**d**). (**e**) Schematic of the functional trafficking assay; in the wild type, GL1 functions in an epidermal precursor cell to initiate trichome formation. GL1 expressed in mesophyll cells is cell autonomous and cannot rescue trichomes in the *gl1* mutant. The KN1 fusion to GL1 can traffic into epidermal cells and rescue trichome formation in the *gl1* mutant. Ep, Epidermal cells; Me, mesophyll cells. Bars: (**b–d**), 50 μm; (**e–h**), 100 μm. Reprinted with permission from (6). (f) pCCL702 construct.

2. Materials

2.1. Plant Materials

1. A mutant with the glabrous phenotype, *gl1-1* which has a deletion in the GL1 locus (15), is used for trichome rescue. The *gl1-1* mutation was developed in Landsberg (Ler) background and then introgressed into Wassilewskija (WS) background.
2. GAL4-GFPer enhancer trap lines J0571 with *Arabidopsis* C24 ecotype (16) (Arabidopsis Biological Research Center).
3. Wild-type Col-0 is used as control.

2.2. Cloning of Gateway Destination Vector Fusion Constructs

1. Restriction enzymes: New England Biolabs (NEB).
2. Gateway cloning kits: BP Clonase II and LR Clonase II (Invitrogen).
3. Gateway donor vector pDonr207 (Invitrogen) and modified Gateway destination vectors pCCL293 and pCCL702 as described in **Section 3.1**.

2.3. Bacterial and Plant Culture Media

1. Luria broth (LB): Dissolve 5 g of yeast extract, 10 g of Bacto tryptone (Becto, Dickinson and Co.), and 10 g of NaCl in 1 L of double distilled water.
2. Yeast extract peptone (YEP): Dissolve 10 g of yeast extract (Becto, Dickinson and Co.), 10 g of peptone from meat (Merck & Co., Whitehouse Station, NJ, USA), and 5 g of NaCl in 1 L of double distilled water.
3. Murashige and Skoog (MS) medium: Dissolve 4.4 g of MS salt powder including vitamins (Duchefa Biochemie, Haarlem, Netherland) in 1 L of double distilled water, then add 0.5 g MES (Biopure, Canada) and dissolve well. Adjust the pH to 5.7 with 1 M KOH. Add 6 g of plant agar (Duchefa Biochemie, Haarlem, The Netherlands). Autoclave at 121°C, 15 psi for 20 min.
4. Peat mix is used for plant growth.

2.4. Antibiotics and Screening

1. Gentamycin sulfate salt (Sigma-Aldrich, Steinheim, Germany) is used at 25 mg/L for bacterial selection both in broth and agar solidified media.
2. Kanamycin sulfate (Biopure, Canada) is used at 50 mg/L for bacterial selection both in broth and agar solidified media.
3. Hygromycin B (Calbiochem, Darmstadt, Germany) is used at 50 mg/L in MS medium plates for selection of transgenic plants.

2.5. Propidium Iodide (PI) Staining Solution

1. Prepare a stock solution of PI in distilled water at a concentration of 10 mg/mL. Keep the stock solution at 4°C in dark.
2. Prepare a working solution by diluting the stock 1000 times in distilled water. The final working concentration will be 10 μg/mL.

3. Methods

3.1. Cloning of Gateway Destination Vector Fusion Constructs

1. The modified Gateway destination vector pCCL293 (**Fig. 13.1a**) was cloned by replacing GFPer with mCherry using *Sal*I/*Sac*I sites in pZY375 (generous gift from Dr. David Jackson) to get pCCL276. Subsequently a fragment corresponding to Gateway recombination sites obtained from pEarlyGate103 as a *Xho*I fragment was inserted into the *Sal*I site in pCCL276 to clone pCCL292 (*see* **Note 4**). Finally the fragment consisting of 5XUAS-Tev leader-p19-35S Ter, obtained by digesting pCCL1170 with *Acc*I, was inserted in *Cla*I site on pCCL292 to get pCCL293 constructs (*see* **Note 4**).
2. Another modified Gateway destination vector pCCL702 (**Fig. 13.3f**) was cloned for high-throughput cloning of different proteins or peptides as a C-terminal fusion with GL1 protein. The fragment corresponding to Gateway recombination sites was amplified from pDest22 (Invitrogen) with primer set attR-Bsgl-d1/attR-BKpn-r1 and inserted in the *Bgl*II/*Kpn*I sites of the pK1401 vector. Primer sequences are as follows:
   (a) attR-Bsgl-d1 – ggcgttcgaagatctcACAAGTTTGTACA-AAAAAGC
   (b) attR-BKpn-r1 – ggcgggtaccttatccttcgaacACCACTTTG-TACAAGAAAG
3. Target proteins or peptides are first cloned in pDONR207 according to the instructions of standard Gateway technology (Gateway technology, Invitrogen) and subsequently cloned in the required destination vectors to obtain the target expression vectors.
4. Expression vectors are transformed into *Agrobacterium* strain GV3101 by electroporation.
5. Positive clones are selected on LB plates supplemented with gentamycin and kanamycin double antibiotics (*see* **Note 5**).

3.2. Growth Conditions

1. Seed germination is synchronized by moist chilling at 4°C in darkness for at least 4 days. In particular, *Arabidopsis* C24

ecotype needs longer stratification to improve germination rate and synchrony.

2. *Arabidopsis* plants are grown at 22°C under 3-wave fluorescent light (Philips Co., Seoul, Korea) with cycles of 16 h light/8 h dark regimes either on soil or on MS plates in a growth chamber.
3. Four-week-old *Arabidopsis* plants are used for *Agrobacterium*-mediated transformation.

3.3. Transformation of Plants

1. A single colony of *Agrobacterium* harboring the target binary vector is inoculated in 3 mL of LB medium containing proper antibiotics (kanamycin and gentamycin). The culture is grown at 28°C for 48 h or until the turbidity of $OD_{600\ nm} = 1.0$.
2. This 3 mL culture is added to 500 mL of YEP medium containing proper antibiotics (kanamycin and gentamycin) and acetosyringone 200 μM. This culture is grown at 28°C for 24 h or until the turbidity at $OD_{600\ nm} = 1.0$.
3. 100 mL of 30% sucrose and 100 μL of Silwet L-77 (Pharmco Products Inc., USA) are added to the 500 mL culture.
4. Before floral dipping the already formed siliques from the flowering shoots are removed. The shoots are dipped for 15–20 s in culture and the plants are then put back in the culture room.
5. After maturity and proper drying the seeds are harvested and kept at 30°C for 3 days.
6. Transgenic seedlings are screened on MS medium containing 50 μg/mL hygromycin.

3.4. Scoring of mCherry Fluorescence

1. About 7- to 10-day-old transgenic plants expressing each transcription factor are used for observation with confocal laser scanning microscope. To observe fluorescence, a FluoView FV1000 confocal microscope (Olympus, Tokyo, Japan) is used. Excitation and emission spectra of GFP peak are at 488 and 510–540 nm, whereas the excitation and emission spectra of mCherry are at 543 and 587–625 nm, respectively.
2. The GFP-expressing lines are stained with fluorescent dye PI (Sigma-Aldrich, USA) which stains the cell walls in living cells (*see* **Note 6**). For PI staining, plant tissues are dipped in staining solution (10 μg/mL in water) for 5 min in the dark and then washed briefly in water. After washing, samples are mounted with water under a coverslip and observed under a confocal microscope using excitation and emission spectra at 543 and 587–625 nm, respectively.

Fig. 13.4. Trichome rescue in *gl1* mutants as observed with dissecting microscope in 2-week-old seedlings growing on MS plates. (**a**) Columbia wild type, (**b**) trichome rescue in *gl1* mutant, (**c**) *gl1* mutant control, (**d–f**) trichome observation at late stage in trichome rescued (**d** and inset showing closeup) and *gl1* control plants (**e** and inset showing closeup) on soil.

3.5. Scoring of Trichome Rescue

1. Trichome rescue is observed with a dissecting microscope (Olympus SZX12, Japan) directly in bright light. In this case trichomes are observed on the first two pairs of leaves of 2-week-old seedlings (**Fig. 13.3**) growing on MS plates (*see* **Note 3**).
2. Recently the authors have found delayed trichome appearance with a different cell-to-cell trafficking transcription factor (**Fig. 13.4**). In this case, trichomes are observed in 4-week-old plants growing on soil using a magnifying glass (hand lens). The plants are photographed with dissecting microscope (Olympus SZX12, Japan). Independent transgenic lines showed variation in trichome number per leaf as well as the leaf pair showing trichomes. The number and leaf pair varied with fusion peptide/protein (*see* **Note 3**).

3.6. Scanning Electron Microscopy (SEM)

1. Alternatively, individual samples are observed with SEM (**Fig. 13.3a–d**).
2. The shoot region from 1-week-old seedlings showing the first pair of leaves (**Fig. 13.3a–d**) or second pair of leaves from 4-week-old plants (data not shown) are fixed with 2.5%

glutaraldehyde in 0.1 M sodium cacodylate (pH 7.4) (*see* **Note 7**) overnight at 4°C. The samples are dehydrated in graded ethanol series (30, 50, 70, 80, 95 and 100% each 1 h). Then the tissues are dried in an Autosamdri-815 critical point dryer (CPD) (Tousimis Co., USA). Subsequently the samples are mounted on carbon tape stubs, coated with gold, and observed with JSM-6380LV (JEOL Co., Japan) and the observations recorded by using digital microscope camera.

4. Notes

1. There are some limitations of these assays. Intercellular protein trafficking is a tissue and developmentally dependent process; thus, trafficking of certain NCA-TFs may not be detected in either root or leaf trafficking assays. Both of the screening systems would be limited by the stability of certain TFs in expression domains mentioned here. In addition, fusion to mCherry or GL1 increases total molecule size by 28 kDa. This might reduce or abolish movement of NCA-TF that can move endogenously.
2. The endoplasmic reticulum retention signal ensures that the GFP does not move cell to cell and remains only in the cells where the gene is transcribed and translated. This also marks the cells where the TF–mCherry is produced.
3. The *gl1* mutant has some trichomes on the leaf edge at the late vegetative stage. Therefore, transgenic plants with only leaf margin trichomes should not be scored as trichome rescued. Only leaf surface trichomes should be counted to determine if trichome rescue has occurred or not. The trichome rescue may be variable in independent transgenic lines. In addition the transgenic approach may include ectopic gene expression depending on chromosome and T-DNA insertion site. Trichome number per leaf pair may also vary from zero to many but usually less than wild type. For all these reasons, it is not recommended to use a single transgenic line to assess trichome rescue, but rather a number of independent transgenic lines (at least 100) should be observed for trichome rescue. The results should be presented statistically as shown in **Fig. 13.5**. The length of bars along the *y*-axis represents the trichome rescue percentage with each construct; the average number of trichomes for each construct is shown with every bar.
4. *Xho*I and *Sal*I produce compatible cohesive ends. Likewise, *Acc*I and *Cla*I produce compatible cohesive ends.

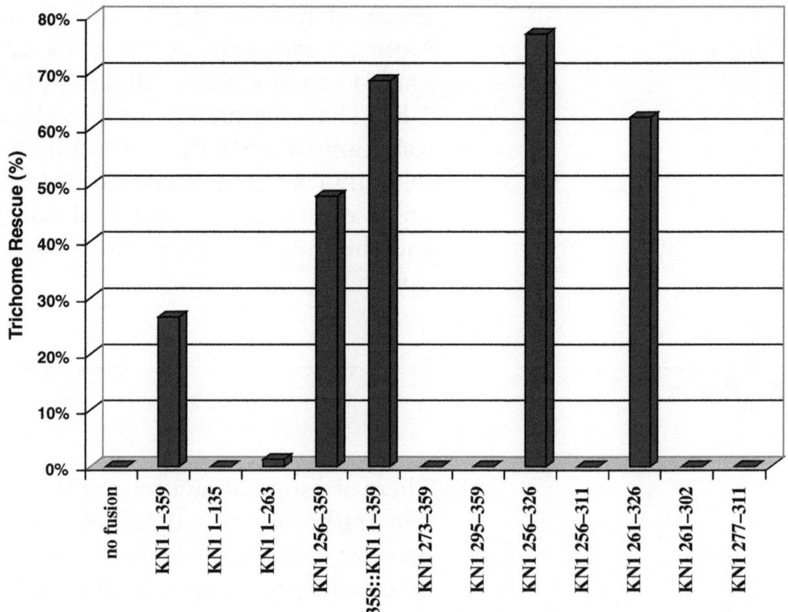

Fig. 13.5. Trichome rescue by different fusions of regions of KN1 to GL1. Trichome rescue rate (percent of T1 plants showing trichome rescue) is on the *Y*-axis. The number next to each bar is the average number of trichomes per leaf pair. Note that all constructs are fusions to GL1 driven by the pRbcS2b promoter, unless labeled "35S," in which case it is driven by the 35S promoter. Redrawn with permission from (6).

5. GV3101 cells are resistant to gentamycin. Transformation with the construct provides the kanamycin resistance.
6. While propidium iodide (PI) will stain nuclei in situations where it is able to cross the membrane, in living root cells, PI stains the cell wall.
7. Sodium cacodylate is harmful and should be treated as carcinogen while handling. It is recommended to wear gloves and handle in a fume hood.

Acknowledgments

This work was supported by World Class University Program (R33-10002), the National Research Lab Program (2009-0066339), and KRF-2008-314-C00362 through the National Research Foundation of Korea funded by the Ministry of Education, Science and Technology (to J.-Y. K.).

References

1. Lucas, W. J., Bouche-Pillon, S., Jackson, D. P., et al. (1995) Selective trafficking of KNOTTED1 homeodomain protein and its mRNA through plasmodesmata. *Science* **270**, 1980–1983.
2. Nakajima, K., Sena, G., Nawy, T., and Benfey, P. N. (2001) Intercellular movement of the putative transcription factor SHR in root patterning. *Nature* **413**, 307–311.
3. Kim, J. Y., Yuan, Z., Cilia, M., Khalfan-Jagani, Z., and Jackson, D. (2002) Intercellular trafficking of a KNOTTED1 green fluorescent protein fusion in the leaf and shoot meristem of Arabidopsis. *Proc. Natl. Acad. Sci. USA* **99**, 4103–4108.
4. Kim, J. Y., Yuan, Z., and Jackson, D. (2003) Developmental regulation and significance of KNOX protein trafficking in Arabidopsis. *Development* **130**, 4351–4362.
5. Lee, J. Y., Yoo, B. C., Rojas, M. R., Gomez-Ospina, N., Staehelin, L. A., and Lucas, W. J. (2003) Selective trafficking of non-cell-autonomous proteins mediated by NtNCAPP1. *Science* **299**, 392–396.
6. Kim, J. Y., Rim, Y., Wang, J., and Jackson, D. (2005) A novel cell-to-cell trafficking assay indicates that the KNOX homeodomain is necessary and sufficient for intercellular protein and mRNA trafficking. *Genes Dev.* **19**, 788–793.
7. Gallagher, K. L., Paquette, A. J., Nakajima, K., and Benfey, P. N. (2004) Mechanisms regulating SHORT-ROOT intercellular movement. *Curr. Biol.* **14**, 1847–1851.
8. Wu, X., Dinneny, J. R., Crawford, K. M., et al. (2003) Modes of intercellular transcription factor movement in the Arabidopsis apex. *Development* **130**, 3735–3745.
9. Riechmann, J. L., Heard, J., Martin, G., et al. (2000) Arabidopsis transcription factors: genome-wide comparative analysis among eukaryotes. *Science* **290**, 2105–2110.
10. Laplaze, L., Parizot, B., Baker, A., et al. (2005) GAL4-GFP enhancer trap lines for genetic manipulation of lateral root development in Arabidopsis thaliana. *J. Exp. Bot.* **56**, 2433–2442.
11. Gardner, M. J., Baker, A. J., Assie, J. M., Poethig, R. S., Haseloff, J. P., and Webb, A. A. (2009) GAL4 GFP enhancer trap lines for analysis of stomatal guard cell development and gene expression. *J. Exp. Bot.* **60**, 213–226.
12. Szymanski, D. B., Jilk, R. A., Pollock, S. M., and Marks, M. D. (1998) Control of GL2 expression in Arabidopsis leaves and trichomes. *Development* **125**, 1161–1171.
13. Esch, J. J., Chen, M. A., Hillestad, M., and Marks, M. D. (2004) Comparison of TRY and the closely related At1g01380 gene in controlling Arabidopsis trichome patterning. *Plant J.* **40**, 860–869.
14. Larkin, J. C., Oppenheimer, D. G., Pollock, S., and Marks, M. D. (1993) Arabidopsis GLABROUS1 gene requires downstream sequences for function. *Plant Cell* **5**, 1739–1748.
15. Oppenheimer, D. G., Herman, P. L., Sivakumaran, S., Esch, J., and Marks, M. D. (1991) A myb gene required for leaf trichome differentiation in Arabidopsis is expressed in stipules *Cell* **67**, 483–493.
16. Haseloff, J. and Hodge, S. (2001) Targeted gene expression in plants using GAL4. *US Patent* 6,255,558.

Section V

Protein–DNA Interaction

Chapter 14

SELEX (Systematic Evolution of Ligands by EXponential Enrichment), as a Powerful Tool for Deciphering the Protein–DNA Interaction Space

Chenglin Chai, Zidian Xie, and Erich Grotewold

Abstract

DNA-binding proteins, including transcription factors, play essential roles in many biological processes. The identification of the DNA sequences to which these proteins bind is a first, yet still challenging, step for determining their functions. SELEX provides an excellent tool for deciphering protein DNA-binding sequence specificity, and it has been widely adopted for addressing fundamental biological questions (1, 2). SELEX is an experimental procedure that involves the progressive selection, from a large combinatorial double-stranded oligonucleotide library, of DNA ligands with variable DNA-binding affinities and specificities by repeated rounds of partition and amplification. In this chapter, we describe a SELEX protocol that we have successfully applied to both plant and animal MYB transcription factors.

Key words: DNA-binding protein, protein–DNA interaction, SELEX, EMSA, consensus sequence, MYB.

1. Introduction

Protein–DNA interactions play vital roles in various cellular processes, such as transcription, chromatin packaging, genetic recombination, replication, and DNA repair (3). DNA-binding proteins are proteins which can bind to DNA sequences in a sequence-specific or non-sequence-specific fashion depending on their functions (4). Structural proteins, such as histones, bind to DNA in a non-sequence-specific manner, while some other proteins, for example, transcription factors, generally interact with the major or minor groove of the DNA double helix in a sequence-specific fashion (5). Although there is no simple universal recognition

L. Yuan, S.E. Perry (eds.), *Plant Transcription Factors*, Methods in Molecular Biology 754,
DOI 10.1007/978-1-61779-154-3_14, © Springer Science+Business Media, LLC 2011

code, like one-to-one correspondence between DNA bases and amino acid residues within protein–DNA complexes based on structural analyses (4, 6), a number of sequence motifs, such as the E-box (CANNTG) for bHLH regulators and the W-box (ttTGACC/T) for WRKY transcription factors, have been identified, highlighting how proteins with homologous DNA-binding domains recognize similar DNA sequences.

Various approaches have been adopted to explore the DNA–protein interaction space. These include computational prediction tools and structural methods (4, 6, 7), in vivo approaches such as ChIP (chromatin immunoprecipitation)-based methods (ChIP, ChIP-chip, and ChIP-Seq), TF–GR fusion methods (GR corresponding to the hormone-binding domain of the glucocorticoid receptor and this fusion is used to identify mRNAs induced/repressed in the presence of the GR ligand dexamethasone, for example, in the presence of an inhibitor of translation) (8–10), and in vitro methods that include SELEX, EMSA (electrophoretic mobility shift assay), and DNA footprinting (8, 9). In vivo techniques have been successfully used to identify direct targets of many transcription factors even on a genome-wide scale in complex organisms such as mammals and plants (8, 10). However, the technical challenges and requirements of known/candidate target sequences for checking in ChIP-based methods limit its application to well-characterized proteins.

In contrast, SELEX is easier to perform as it does not need many of these prerequisites (11). SELEX was initially described by two groups in 1990 and many variations have resulted from it (2, 11, 12). It is a screening technique that involves selecting specific targets, from a large combinatorial oligonucleotide pool, which could consist of a mixture of RNA and single-stranded (ss) or double-stranded (ds) DNA, by a reiterative process of partition and amplification (11). The molecules, which can bind to DNA molecules with variable affinities, can range from proteins to peptides, drugs, organic small molecules, or even metal ions (13). The products of SELEX using RNA or ssDNA include the aptamers that mimic properties of antibodies (14). In addition to serving as a technique to establish the in vitro DNA-binding specificity of a protein, SELEX also furnishes a powerful tool to determine whether a particular protein binds DNA in a sequence-specific fashion or not.

SELEX starts with a very large oligonucleotide library containing approximately 10^{12}–10^{18} different sequences. Each sequence in the pool is of the same length, containing a 20–30 base internal region in which positions along the polymer are randomly assigned to one of the four bases (A, G, C, and T/U). Non-random sequences of fixed length flank the random internal region and are used to design primers for PCR amplification (15). In the implementation of SELEX described here, the reverse primer is end-labeled by [γ-^{32}P]-ATP and T4 polynucleotide kinase (T4 PNK) and is then used as the primer (the oligonucleotide library as the template) to generate a double-stranded radioactively labeled DNA library. The radioactively labeled DNA pool is subsequently incubated with a putative DNA-binding protein, and sequences with different affinities and high specificities to the protein are selected by one of many possible techniques that include nitrocellulose membrane filtration, using affinity surfaces and affinity tags, cross-linking, antibody-based flow cytometry, and EMSA (16). The selected DNA molecules are PCR-amplified using a small number of cycles in order to avoid possible PCR bias issues (17). The amplified molecules are purified and used as substrates for the next round of the binding-site selection process. Initial rounds of SELEX can be performed under less stringent conditions, such as high protein/probe molar ratio and longer incubation time, whereas later rounds can involve more stringent conditions, which often include higher concentrations of non-specific competitor [e.g., bovine serum albumin (BSA), poly (dI/dC), salmon sperm DNA], monovalent or divalent cation, or shorter incubation times, in order to reduce non-specific protein binding (16). Attempts have been made to determine the number of rounds of SELEX sufficient for the enrichment of target sequences and, at the same time, elimination of non-specific binding (15). Usually 4–18 rounds are required to complete the whole SELEX process (2, 17–19). The enriched target molecules are then purified, PCR-amplified, cloned, and sequenced. The consensus DNA-binding sequence can then be derived by performing multiple alignments using specific computer algorithms and/or visual inspection, and candidates are finally verified by EMSA. In this chapter, we will describe a SELEX protocol that has been successfully applied to both plant and animal MYB transcription factors.

2. Materials

2.1. Labeling of Oligonucleotide Library and Purification

1. Oligonucleotide library:
   The oligonucleotide library for SELEX, described as oMC070 in a previous study (18), harbors the following sequences: 5′-ACTCGAGGAATTCGGTACCCCGGGT $(N)_{26}$TGGATCCGGAGAGCTCCCAACGCGT-3′, where N indicates an equimolecular distribution of A, C, G, and T. The oligonucleotide library is purified on a 12% denaturing PAGE (polyacrylamide gel electrophoresis) (*see* **Note 1**).

2. Protein:
   The protein used in this study, the R2R3 MYB domain of the maize P1 transcription factor (17), was expressed in *Escherichia coli* as an N-terminal histidine-tagged fusion (*see* **Note 2**) and purified using Ni^{2+}-NTA affinity chromatography. For proteins with GST (glutathione *S*-transferase) tags, GST·Bind Resin (Novagen, Inc) is used instead. The purified protein is dialyzed against a buffer suited for DNA binding ("binding buffer": e.g., we use 1X A-O buffer (10 mM Tris–HCl, pH 7.5, 50 mM NaCl, 1 mM EDTA, and 5% glycerol) for MYB proteins) before use.
3. Primers:
   Primers for generating double-stranded oMC070 (ds oMC070) and for PCR amplification are oMC068 (Forward primer): 5′-ACT CGA GGA ATT CGG TAC CCC GGG T-3′ and oMC069 (Reverse primer): 5′-ACG CGT TGG GAG CTC TCC GGA TCC A-3′. The primers were purified by 12% denatured PAGE (*see* **Note 1**).
4. γ-^{32}P-ATP: End-labeling grade, activity >6000 Ci/mmol (radiation hazard).
5. T4 PNK: 10,000 units/mL.
6. Glass microfiber filters (Whatman International Ltd.).
7. DNA Polymerase I, large (Klenow) fragment, 10,000 units/mL.
8. Vertical gel electrophoresis unit: Hoefer units, model number: SE400, glass plates, 18×16 cm (Hoefer®, Inc).
9. 5XTBE buffer: 450 mM Tris–HCl, pH 8.0, 450 mM Boric acid and 10 mM Na_2EDTA (*see* **Note 3**).
10. 40% acrylamide mix (acrylamide:bis-acrylamide = 80:1) was made by completely dissolving 80 g of acrylamide and 1 g of bis-acrylamide in ddH_2O with the final volume of 200 mL and stored at 4°C (this is a neurotoxin when unpolymerized and so care should be taken to minimize skin contact, *see* **Note 4**).
11. *N,N,N,N′*-Tetramethyl-ethylenediamine (TEMED).
12. 10% ammonium persulfate. Stored at –20°C and made up fresh each month.
13. 0.3 M sodium acetate, pH 5.2.
14. 95% ethanol.
15. Nanosep MF tube with pore size 0.45 μm (Pall Corporation).
16. 10% trichloroacetic acid in 10 mM sodium phosphate buffer, pH 7.5.

2.2. Site Selection and EMSA

1. N2+-NTA-agarose (Ni^{2+} beads, Qiagen) (*see* **Note 5**).
2. DTT, 100 mM, needs to be added just before use.
3. Poly (dI/dC) (Sigma), 1 μg/μL (*see* **Note 6**).

2.3. Cloning and Sequencing

1. DNA polymerase (*see* **Note 7**).
2. pCR-2.1-TOPO vector (Invitrogen) for TA cloning.

3. Methods

3.1. Labeling the Reverse Primer

1. Set up in a 1.5 mL Eppendorf tube the reaction (20 μL) including 50 pM of primer oMC069, 2 μL of T4 PNK buffer (10X), 2 μL of T4 PNK (20 U), and 8 μL of [γ-^{32}P]-ATP (23 μM, ≧6000 Ci/mmol). Mix well and spin down briefly.
2. Incubate at 37°C for 60 min to end-label the reverse primer.
3. Incubate at 65°C for 20 min to inactivate T4 PNK and cool on ice.
4. Take 1 μL from the reaction and dilute it in 500 μL of distilled water. Use 2 μL of diluted reaction to calculate the incorporation rate of [γ-^{32}P]-ATP into the primer and its radioactivity by scintillation assay (*see* **Note 8**).
5. Divide the reaction equally into two tubes, one of which is saved for subsequent PCR amplification after each round of site selection; the other one is for producing the dsDNA library (*see* **Section 3.2**).

3.2. Synthesis of the dsDNA Library (See Note 9)

1. Add 50 pmol (1 μL of 50 μM) of single-stranded (ss) DNA oMC070, 5 μL of Klenow buffer, and 27.5 μL of dH_2O to one of the two tubes from step 5 in **Section 3.1**, mix well and spin down.
2. Heat at 90°C for 5 min to denature the ssDNA, slowly cool down to 40°C.
3. Add 5 μL of dNTPs (10 mM) to a final concentration of 1 mM and 1 μL of Klenow, then spin down.
4. Incubate at 37°C for 60 min.
5. Heat at 65°C for 10 min to inactivate the enzyme.
6. Purify the probe by 8% native PAGE (*see* **Note 10**).
7. Quantify the incorporation rate by scintillation assay and calculate the total DNA concentration (including both radioactively labeled and non-labeled DNA).

3.3. First Round of SELEX

1. Set up a reaction (50 μL) including 1X binding buffer, 1 mM DTT, 400 ng of poly (dI/dC), 1 mg/mL BSA, and 5 ng of DNA-binding protein. Mix well by pipetting and then incubate on ice for 30 min.
2. Add 0.4 ng of ds oMC070 and mix well. Incubate on ice for 30 min.
3. Add 145 μL of 1X binding buffer and 5 μL of N^{2+}-NTA-agarose beads that have been equilibrated by 1X binding buffer (*see* **Note 11**).
4. Shake on rotator at 4°C for an hour.
5. Centrifuge at 800×*g* at 4°C for 1 min.
6. Remove the supernatant carefully to a new tube without disturbing the N^{2+}-NTA-agarose beads on the bottom.
7. Add 200 μL of 1X A-O binding buffer to the beads with gentle inversion and shake at 4°C for 5 min.
8. Centrifuge at 800×*g* for 1 min and remove the supernatant carefully without disturbing the beads.
9. Repeat steps 7 and 8 two more times.
10. Resuspend the N^{2+}-NTA-agarose beads in 20 μL of DNase-free ddH_2O.

3.4. PCR Amplification and DNA Gel Purification

1. Set up a PCR reaction in 50 μL containing 5 μL of resuspended beads as template and 1 pmol of radioactively labeled reverse primer oMC069, as well as 10 pmol of forward (oMC068) and reverse cold primers (oMC069). PCR reaction is performed using the following program: 5 min at 94°C; 20 cycles (round 1 of SELEX) or 15 cycles (other rounds) of 1 min denaturation at 94°C; 1 min at 50°C; 30 s at 72°C; and a final extension of 20 min at 72°C, followed by 4°C incubation.
2. The PCR product is purified and separated by PAGE (*see* **Notes 10** and **12**). The recovered DNA is quantified by scintillation counting and then used as probe for the next round of SELEX.

3.5. Subsequent Rounds of SELEX

Beginning with the second round of SELEX, EMSA is carried out in parallel with the procedure described in **Section 3.3**. EMSA is carried out as described below.

1. Set up a reaction (50 μL) including 1X A-O binding buffer, 1 mM DTT, 400 ng of poly (dI/dC), 1 mg/mL BSA, and 5 ng of DNA-binding protein (P1 in this study). Mix well by pipetting and incubate on ice for 30 min.

2. Add 0.2 ng of probe (1×10^5 cpm, approximately 0.01 μCi) to the above reaction, mix gently, and incubate on ice for 30 min.
3. Free probe and protein–DNA complex are resolved by 8% PAGE (80:1 acrylamide:bis-acrylamide) in 0.25X TBE buffer at about 40 V/cm for 60 min at 4°C.
4. After PAGE, gels are dried onto Whatman paper and then subjected to autoradiography at –70°C or directly subjected to autoradiography at 4°C.
5. After 6–12 h exposure to the X-ray film, the signal from shifted protein–DNA complex may be seen above the free probe.
6. Purify the DNA from shifted protein–DNA complex and resuspend DNA in 10 μL of DNase-free ddH_2O and quantify by scintillation.

If a shifted protein–DNA complex can be seen in EMSA, only EMSA is performed using recovered DNA (as probe) from the previous SELEX round. Otherwise, the procedure described in **Section 3.3** should be done in parallel with EMSA until a shifted protein–DNA complex can be seen in EMSA.

3.6. Cloning, Validation, and Sequencing

After 5–9 SELEX rounds, the protein-bound DNA may be enriched as reflected by increasing signal intensity from shifted protein–DNA complex (*see* **Note 13**).

1. The DNA recovered from the last round of SELEX is cloned into pCR-2.1-TOPO vector and transformed into TOPO 10 competent cells. Extract plasmids from 10 individual colonies.
2. Label probes by PCR using radioactively labeled oMC069 as one of the primer pairs and the plasmid DNA obtained from step 1 in this section as templates and perform EMSA with each probe.
3. If 8–10 out of 10 clones (from step 1 in this section) show shifted protein–DNA complexes in EMSA, then extract plasmid DNA from 40 or more colonies for sequencing (*see* **Note 14**).

3.7. Analysis of Sequence Data and Identifying the Consensus Binding Site

1. Retrieve the sequencing data and keep 3–5 flanking nucleotides in case of the possible contribution to the binding site from the flanking nucleotides.
2. Use online software The Gibbs Motif Sampler (http://bayesweb.wadsworth.org/gibbs/gibbs.html) to get the consensus sequence and WebLogo (http://weblogo.berkeley.edu/logo.cgi) to create the sequence logo.

4. Notes

1. Denaturing polyacrylamide gels contain 7 M urea, 12% acrylamide mix (80:1 acrylamide:bis-acrylamide). The single-stranded DNA library is resolved by denaturing PAGE in 1X TBE buffer at about 40 V/cm for 60 min at 4°C. Before loading, urea is added to the DNA samples to a final concentration around 7 M.
2. Since the fusion protein can have low molecular weight (as found with MYB domains which are ~13 kDa), the shifted protein–DNA complex could be very close to the free probe (1–2 cm separation) in EMSA.
3. 5X TBE buffer should be freshly made with no visible salt precipitation.
4. This step has to be done in a fume hood to prevent possible inhalation of the toxic powder.
5. N^{2+}-NTA-agarose needs to be pre-washed with 1X binding buffer (without DTT and poly dI/dC) before use.
6. Store at –20°C. To avoid repeated freeze-and-thaw cycles that might lead to degradation, aliquot the solution into small amounts when making this solution.
7. Use regular DNA polymerase (not high-fidelity) for PCR amplification to get products with a 3′-A overhang, which is suitable for subsequent TA cloning.
8. To check how much radioactivity is incorporated into the primer, 1 μL of 1:500 dilution radioactively labeled samples are added onto each of the two glass microfiber filters (Whatman International Ltd.) and dried at room temperature. One of the glass filters (precipitated) is washed with 10 mL of 10% trichloroacetic acid (TCA) in 10 mM sodium phosphate buffer, pH 7.5, for 10 min and washed with 10 mL of 95% ethanol for another 10 min. Finally the precipitated and unprecipitated glass microfiber filters are measured by scintillation counting. The incorporation rate of radioactivity is calculated by the following formula:

   Incorporation rate (%) = (scintillation score from precipitated glass filter)/(scintillation score from non-precipitated glass filter) × 100%. The radioactivity incorporation rate can be used for estimating the total DNA amount of the primer and PCR product.
9. Alternatively, ds oMC070 can be generated by PCR in which radioactively labeled oMC069 and non-radioactively labeled oMC068 are used as primers. The PCR product is PAGE-purified as described in step 6 of **Section 3.2** (*see* **Note 10**).

10. Cut the band exactly corresponding to the size of ds oMC070 as indicated by the appropriate DNA marker and elute in 400 μL of 0.3 M sodium acetate, pH 5.2, for 12 h using Nanosep MF tube. Spin down at maximum speed in a micro-centrifuge for 10 min and add 1 mL of 95% ethanol to the solution, mix well, and incubate at –20°C for 30 min. Spin down at maximum speed in a micro-centrifuge for 30 min to precipitate the DNA. Remove the supernatant and wash in 600 μL of cold 70% ethanol and dry on bench. Resuspend DNA in DNase-free ddH_2O.
11. Ni^{2+}-NTA-agarose needs to be completely resuspended and the tip of a pipette tip should be cut before use with a new razor blade to allow an aliquot of beads to be removed.
12. Be aware of possible contamination problem when dealing with multiple samples. At least one lane space between samples is preferred. Rinse the blade with DNase-free water after each sample.
13. The number of rounds of SELEX depends on the affinity of protein–DNA interaction. In general, the target sequences should be enriched with increasing rounds, which can be reflected by the increasing signal intensity of shifted protein–DNA complex. After observation of 3–5 consecutive increases in signal from the shifted protein–DNA complex, an attempt can be made to clone the selected sequences into a vector and test for adequate selection by EMSA.
14. If less than 50% of the clones tested by EMSA do not show shifted protein–DNA complex, non-specific protein–DNA binding can occur and more SELEX rounds might be needed.

Acknowledgments

Support in the Grotewold lab for projects involving regulation of gene expression is provided by NRI Grant 2007-35318-17805 from the USDA CSREES, DOE Grant DE-FG02-07ER15881, and NSF grant DBI-0701405. ZX was supported by a 1-year predoctoral Excellence in Plant Molecular Biology & Biotechnology fellowship.

References

1. Ng, E. W., Shima, D. T., Calias, P., Cunningham, E. T., Jr., Guyer, D. R., and Adamis, A. P. (2006) Pegaptanib, a targeted anti-VEGF aptamer for ocular vascular disease. *Nat. Rev. Drug Discov.* **5**, 123–132.
2. Tuerk, C., and Gold, L. (1990) Systematic evolution of ligands by exponential enrichment: RNA ligands to bacteriophage T4 DNA polymerase. *Science* **249**, 505–510.
3. Essers, J., Vermeulen, W., and Houtsmuller, A. B. (2006) DNA damage repair: anytime, anywhere? *Curr. Opin. Cell Biol.* **18**, 240–246.
4. Sarai, A., and Kono, H. (2005) Protein-DNA recognition patterns and predictions. *Annu. Rev. Biophys. Biomol. Struct.* **34**, 379–398.
5. Dervan, P. B. (1986) Design of sequence-specific DNA-binding molecules. *Science* **232**, 464–471.
6. Matthews, B. W. (1988) Protein-DNA interaction. No code for recognition. *Nature* **335**, 294–295.
7. Persikov, A. V., Osada, R., and Singh, M. (2009) Predicting DNA recognition by Cys2His2 zinc finger proteins. *Bioinformatics* **25**, 22–29.
8. Collas, P., and Dahl, J. A. (2008) Chop it, ChIP it, check it: the current status of chromatin immunoprecipitation. *Front. Biosci.* **13**, 929–943.
9. Grotewold, E., and Springer, N. (2009) Decoding the transcriptional hardwiring of the plant genome. In Annual Plant Reviews: Systems Biology. Coruzzi, G., and Guttierrez, R. (eds), Blackwell Publishing, Oxford, UK, vol. 35, pp. 196–227.
10. Haring, M., Offermann, S., Danker, T., Horst, I., Peterhansel, C., and Stam, M. (2007) Chromatin immunoprecipitation: optimization, quantitative analysis and data normalization. *Plant Methods* **3**, 11.
11. Yang, Y., Yang, D., Schluesener, H. J., and Zhang, Z. (2007) Advances in SELEX and application of aptamers in the central nervous system. *Biomol. Eng.* **24**, 583–592.
12. Ellington, A. D., and Szostak, J. W. (1990) In vitro selection of RNA molecules that bind specific ligands. *Nature* **346**, 818–822.
13. Stoltenburg, R., Reinemann, C., and Strehlitz, B. (2007) SELEX–a (r)evolutionary method to generate high-affinity nucleic acid ligands. *Biomol. Eng.* **24**, 381–403.
14. Jayasena, S. D. (1999) Aptamers: an emerging class of molecules that rival antibodies in diagnostics. *Clin. Chem.* **45**, 1628–1650.
15. Djordjevic, M. (2007) SELEX experiments: new prospects, applications and data analysis in inferring regulatory pathways. *Biomol. Eng.* **24**, 179–189.
16. Gopinath, S. C. (2007) Methods developed for SELEX. *Anal. Bioanal. Chem.* **387**, 171–182.
17. Grotewold, E., Drummond, B. J., Bowen, B., and Peterson, T. (1994) The myb-homologous P gene controls phlobaphene pigmentation in maize floral organs by directly activating a flavonoid biosynthetic gene subset. *Cell* **76**, 543–553.
18. Huang, H., Mizukami, Y., Hu, Y., and Ma, H. (1993) Isolation and characterization of the binding sequences for the product of the Arabidopsis floral homeotic gene AGAMOUS. *Nucleic Acids Res.* **21**, 4769–4776.
19. Nole-Wilson, S., and Krizek, B. A. (2000) DNA binding properties of the Arabidopsis floral development protein AINTEGUMENTA. *Nucleic Acids Res.* **28**, 4076–4082.

Chapter 15

Footprinting and Missing Nucleoside Analysis of Transcription Factor–DNA Complexes

Ivana L. Viola and Daniel H. Gonzalez

Abstract

In the following chapter we describe methods and protocols to analyze the interaction of proteins with DNA using footprinting and related techniques based on the modification of DNA with either hydroxyl radicals or methylating agents. Footprinting, based on the protection from chemical modification of DNA through the specific binding of a protein, gives information about the nucleotides that are in close contact with the protein upon binding. The derived missing nucleoside and interference techniques identify nucleotides that are energetically important for protein binding. These methods are highly valuable to study in detail the interaction of a transcription factor with nucleotides on both strands of its target DNA sequence.

Key words: Footprinting, hydroxyl radical, methylation interference, missing nucleoside, protein–DNA interaction, transcription factor.

1. Introduction

Footprinting methods allow the identification of sequences, within a DNA molecule, that are protected from modification by a defined chemical reagent upon binding of a protein (1). Basically, in footprinting, one of the DNA strands is labeled at its 3′- or 5′-end and, after protein binding, the DNA is subjected to a chemical modification reaction that produces hydrolysis of phosphodiester bonds at random positions. As a consequence, a population of labeled molecules whose size extends from the labeled end to each cleavage site is obtained. If the conditions are adjusted such that only one cleavage occurs per modified DNA molecule, then a range of sizes, ideally from one nucleotide to the size of

L. Yuan, S.E. Perry (eds.), *Plant Transcription Factors*, Methods in Molecular Biology 754,
DOI 10.1007/978-1-61779-154-3_15, © Springer Science+Business Media, LLC 2011

the entire fragment, will be observed when the DNA is analyzed in a denaturing polyacrylamide gel (**Fig. 15.1**). However, if modification is performed in the presence of a protein that recognizes a specific region of the DNA molecule under study, the modification of nucleotides present at positions where the protein is bound is reduced or even abolished, thus generating a "footprint" in the gel (**Fig. 15.1**). This footprint represents the nucleotides that are in close contact with the protein and are thus protected from modification by the reagent.

The nature of the contacts that are involved in generating the footprinting pattern depends on the nature of the cleavage reagent that is used (2). In DNase footprinting, for example, DNase I is used to hydrolyze phosphodiester bonds (3). Since DNase is a large molecule, the mere presence of a bound protein will preclude its action at nearby positions so that this method allows only the identification of the region of DNA where a protein binds. When small molecules are used as cleavage reagents, more intimate contacts between nucleotides and the protein of interest can be identified. One of the smallest reagents used in footprinting is the hydroxyl radical (4–6). This highly reactive species modifies nucleotides in a rather random fashion, irrespective of the DNA sequence, producing the cleavage of phosphodiester bonds and the elimination of nucleosides. This lack of specificity allows the generation of cleavages at every position in the DNA population so that a pattern that represents molecules differing in one nucleotide from each other can be obtained, producing results of high resolution. This method then allows the evaluation of the contacts established by a protein with each nucleotide of a DNA molecule in a single reaction (7). Since only one strand of the DNA is labeled at a time, the contacts established with each nucleotide of a complementary pair can be distinguished. It should be kept in mind that due to the nature of the modification (8), contacts with not only bases but also the DNA backbone will be identified.

A derivation of the footprinting technique is the missing nucleoside technique to study protein–DNA interactions (9, 10). Here, DNA is modified before protein binding. It can be envisaged that if a population of DNA molecules that are modified randomly at defined positions is used for binding of the protein, those molecules that have modifications at nucleosides that are required for binding will not interact with the protein, while those that are modified at unimportant positions will be bound. If the populations of bound and free (not bound) molecules are separated and analyzed, different patterns will be obtained in which sizes corresponding to positions important for binding will be underrepresented in the bound population and overrepresented in the free population (**Fig. 15.2**). Thus, this technique allows

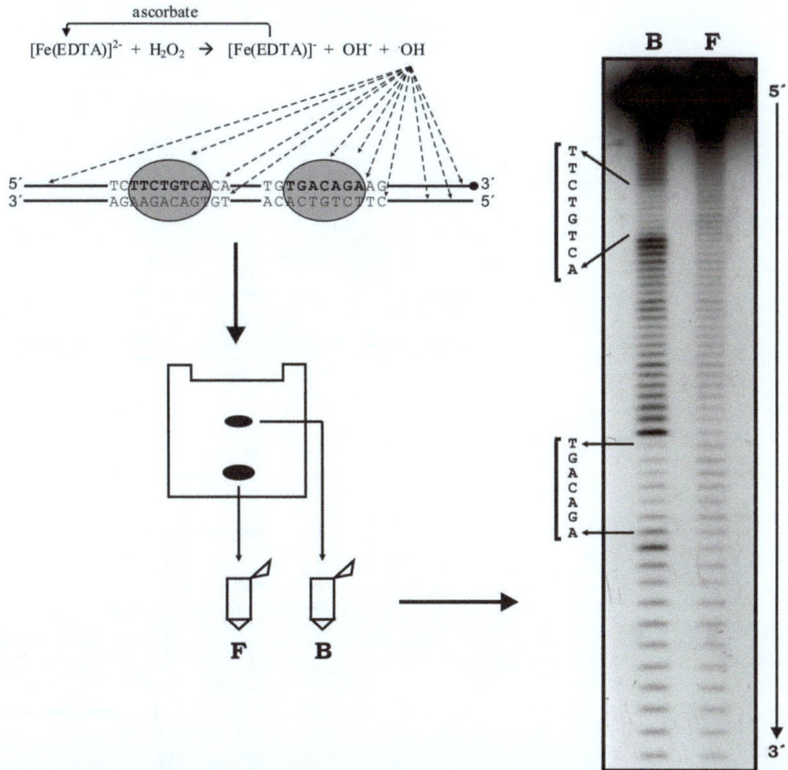

Fig. 15.1. The footprinting technique identifies nucleotides that are in close contact with a bound protein. In footprinting, a double-stranded DNA molecule that contains putative target sites for a transcription factor is labeled in one of its ends (*small circle* at the 3′-end of the upper strand in this case) and then used in a binding reaction with the protein of interest. After binding, the DNA is modified in a way that produces the hydrolysis of phosphodiester bonds adjacent to modified sites (the use of hydroxyl radical generated by a mix of Fe^{2+} and H_2O_2 is shown as an example). After modification, DNA that is not bound by the protein (F, free DNA) and bound DNA (B) are separated in an EMSA, eluted from the gel, and analyzed in a denaturing polyacrylamide gel. Typically, free DNA produces a uniform ladder of molecules of different sizes since all nucleotides are modified with similar efficiency. In bound DNA, however, certain nucleotides are protected from modification by the bound protein, resulting in an underrepresentation of molecules that correspond to these sites of attack (*brackets* in the bound DNA lane). Note: according to the scheme, the observed footprint corresponds to protected nucleotides of the upper (labeled) strand and the lower part of the gel shows positions that are closer to the 3′-end of this strand. Provided that a suitable molecular marker is used (*see* **Note 13** for details), the nucleotide positions that are protected from modification by the bound protein can be inferred. Also note that, to obtain a uniform ladder, most of the DNA remains unmodified and can be observed as a prominent band in the upper part of the gel.

the identification of nucleosides that are energetically important for the protein of interest to bind DNA.

In the next sections, we describe detailed protocols to perform an analysis of the interaction of transcription factors with DNA using hydroxyl radical footprinting and missing nucleoside

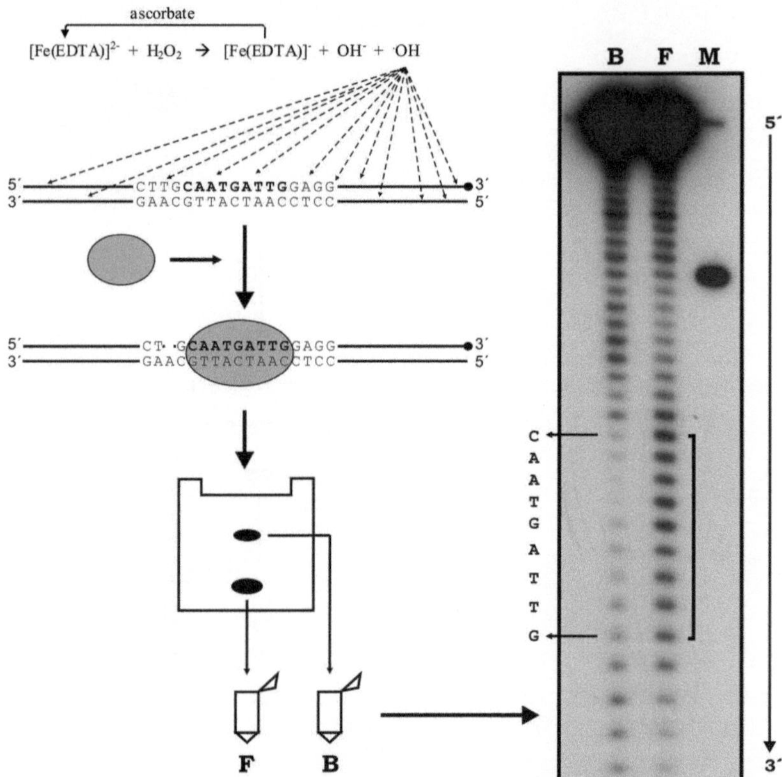

Fig. 15.2. The missing nucleoside/interference technique identifies nucleotides that are important for protein binding. The interference technique is derived from footprinting but, in this case, the free DNA, labeled at one of its ends, is modified randomly *before* protein binding. If modification of a given nucleotide interferes with protein binding, molecules in which this nucleotide has been modified will be poorly bound by the protein of interest and will be then underrepresented in the population of bound molecules and overrepresented in the population of unbound (free) molecules (*brackets* in the denaturing gel). *See* **Fig. 15.1** legend for further details. Note: the term missing nucleoside is usually used for modification with the hydroxyl radical, while interference is more generally employed for modifications with other reagents.

techniques. Briefly, the DNA is labeled, used in a binding reaction with the protein of interest, and subjected to hydroxyl radical cleavage. Since the hydroxyl radical is highly unstable, it is generated in situ from a mixture of Fe^{2+} complexed with EDTA and H_2O_2 (the Fenton reaction) (11). After cleavage, bound and free DNA are separated using an electrophoretic mobility shift assay (EMSA) and analyzed in a denaturing polyacrylamide gel. Ideally, a homogeneous pattern of molecules of different sizes will be obtained with the free DNA, while molecules of sizes that represent nucleotides that are protected by the protein will be underrepresented in the bound population. In missing nucleoside experiments, the steps of hydroxyl radical attack and protein binding are performed in a different order.

Footprinting and missing nucleoside (in this case named interference) experiments can also be performed using methylation of nucleotides (12, 13). Here, dimethyl sulfate is used to methylate guanines at N-7 (that faces the major groove) and adenines at N-3 (in the minor groove). DNA is then cleaved at modified positions using piperidine (only guanines) or NaOH (both guanines and adenines). Since the nature of the modification is different, the use of methylation can obviously detect interactions that are not observed when using hydroxyl radical and vice versa. Both methods are thus complementary. In addition, since methylation modifies bases, specific interactions may be detected in this case. A limitation of this technique is that only the interaction with purines can be monitored. We also outline basic protocols for the use of methylation in footprinting and interference in the sections below.

Finally, though less common, other modifications of DNA can also be used for similar purposes, such as ethylation and carbethoxylation (14, 15). The choice of each technique will depend on the specific purposes and on previously available information regarding the nature of the interaction of the transcription factor under study with DNA.

2. Materials

The quality of the water used in buffers and solutions for footprinting is very important because organic contaminants can interfere with the modification reaction. All solutions employed in these experiments must be prepared with water purified by a Milli-Q system.

2.1. Preparation of the Labeled DNA Fragment

1. Equipment and reagents for standard techniques in molecular biology: PCR, DNA purification, restriction enzyme cleavage.
2. Material for radiation protection (^{32}P) and space for manipulation of radioactive compounds.
3. Yeast tRNA (1 μg/μL). Store at −20°C.
4. [α-^{32}P]dATP (3000 Ci/mmol; 10 mCi/mL).
5. dCTP, dGTP, and dTTP mix (0.2 mM each). Store at −20°C.
6. Non-denaturing acrylamide gel:
   a. 38:2 Acrylamide:bisacrylamide solution (38 g of acrylamide and 2 g of *N*′,*N*′-methylenebisacrylamide in a final volume of 100 mL water) (*see* **Note 1**). Store at 4°C.

   b. 10× TBE: 0.9 M Tris base, 0.9 M boric acid, 0.2 M EDTA, pH 8.0. Store at room temperature.
   c. TEMED (*N′,N′,N′,N′*-tetramethylenediamine). Store at 4°C.
   d. 30% (w/v) Ammonium persulfate. Prepare a 10 mL stock solution in water and store at −20°C as 1 mL aliquots.
   e. 10% Acrylamide gel preparation: Mix 1.65 mL of 38:2 acrylamide:bisacrylamide solution and 0.6 mL of 10× TBE and bring to 6 mL with water. Before pouring the solution between the glass plates, add 12.6 μL of 30% ammonium persulfate and 3.6 μL of TEMED (*see* **Note 2**).
7. Sample buffer (for non-denaturing gel): 30% Glycerol, 0.25% (w/v) bromophenol blue, 0.25% (w/v) xylene cyanol. Store at 4°C.
8. X-ray film, exposure cassette, developing and fixing solutions.
9. Oligonucleotide elution buffer: 50 mM Tris–HCl, pH 8.0, 500 mM NaCl, 20 mM EDTA, pH 8.0.

2.2. Hydroxyl Radical Footprinting and Missing Nucleoside Assays

1. EMSA gel:
   a. 30% (w/v) Acrylamide solution (*see* **Note 1**). Store at 4°C.
   b. 2% (w/v) Bisacrylamide solution (*see* **Note 1**). Store at 4°C.
   c. 10× TBE: 0.9 M Tris base, 0.9 M boric acid, 0.2 M EDTA, pH 8.0. Store at room temperature.
   d. 50% Glycerol.
   e. TEMED (*N′,N′,N′,N′*-tetramethylenediamine). Store at 4°C.
   f. 30% (w/v) Ammonium persulfate. Prepare a 10 mL stock solution and store at −20°C as 1 mL aliquots.
   g. 5% EMSA gel preparation: Mix 975 μL of 30% acrylamide solution, 240 μL of 2% bisacrylamide solution, 0.6 mL of 50% glycerol, 0.6 mL of 10× TBE, and 4.2 mL of water. Before pouring the solution between the glass plates, add 22.5 μL of 30% ammonium persulfate and 6.6 μL of TEMED (*see* **Note 2**).
2. 3× Binding buffer: 60 mM HEPES–KOH, pH 7.5, 1.5% (w/v) Triton X-100, 1.5 mM EDTA, pH 8.0, 150 mM KCl, 6 mM $MgCl_2$, 0.7 mg/mL bovine serum albumin, and 3 mM dithiothreitol (DTT) (*see* **Note 3**). Store at −20°C.
3. Poly(dI-dC) 0.2 μg/μL. Store at −20°C.

4. 15% (w/v) Ficoll. Store at 4°C.
5. 20 mM Sodium ascorbate. Store at 4°C in a dark bottle.
6. Fe(II) solution: 2 mM $(NH_4)_2Fe(SO_4)_2{\cdot}6H_2O$. Prepare immediately before use.
7. Fe(II)–EDTA mix: Mix equal volumes of Fe(II) solution and 4 mM EDTA, pH 8.0 (*see* **Note 4**).
8. 0.6% (v/v) H_2O_2. Prepare from a 30% H_2O_2 solution immediately before use (*see* **Note 4**).
9. 87% Glycerol.
10. Sequencing gel sample buffer: 97.5% (v/v) Formamide, 10 mM EDTA, pH 8.0, 0.1% (w/v) bromophenol blue, 0.1% (w/v) xylene cyanol. Store in the dark at 4°C.
11. Sequencing gel:
    a. Ultrapure urea.
    b. 38:2 acrylamide:bisacrylamide solution (38 g of acrylamide and 2 g of *N′,N′*-methylenebisacrylamide in a final volume of 100 mL of water) (*see* **Note 1**). Store at 4°C.
    c. 10× TBE: 0.9 M Tris base, 0.9 M boric acid, 0.2 M EDTA, pH 8.0. Store at room temperature.
    d. TEMED (*N′,N′,N′,N′*-tetramethylenediamine). Store at 4°C.
    e. 30% (w/v) Ammonium persulfate. Prepare a 10 mL stock solution and store at −20°C as 1 mL aliquots.
    f. 12% Sequencing gel preparation: Add 16 mL of 38:2 acrylamide:bisacrylamide solution and 5 mL of 10× TBE to 24 g of urea and mix with mild heating until urea is completely dissolved. Bring to 50 mL with water and degas the solution for 5 min. Add 53 μL of 30% ammonium persulfate and 23 μL of TEMED just before pouring the solution between the glass plates.
12. Stop Mix: 0.1 M thiourea, 20 mM EDTA, pH 8.0. Store at 4°C.

2.3. Methylation Footprinting and Interference

1. Items 1–4 as described in **Section 2.2**.
2. Dimethyl sulfate (analytical grade).
3. 250 mM DTT. Store at −20°C.
4. 20 mM Potassium phosphate, pH 7.0, 1 mM EDTA.
5. 10 N NaOH.
6. 50 mM Sodium cacodylate, pH 8.0, 1 mM EDTA.
7. 1.5 M Sodium acetate, 1 M 2-mercaptoethanol. Store at −20°C.
8. Items 10 and 11 as described in **Section 2.2**.

2.4. Other Materials

1. Equipment for vertical slab gel electrophoresis and sequencing gel electrophoresis.
2. Scintillation counter.
3. X-ray film (e.g., preflashed Kodak XAR) and exposure cassette with intensifying screen.
4. Gel dryer.

3. Methods

3.1. Preparation of the Labeled DNA Fragment

In a footprinting assay, the optimum length of the employed DNA fragment is between 50 and 100 bp and about 100,000 to 250,000 cpm of singly end-labeled DNA is necessary (*see* **Note 5**). We usually clone fragments or synthetic double-stranded oligonucleotides containing the binding site of interest into the *Bam*HI and *Eco*RI sites of pBluescript and use adjacent *Hin*dIII and *Xba*I sites for labeling at the 3′-end (16–20) (*see* **Note 6**):

1. PCR amplify the fragment from the clone that contains the DNA to be analyzed using reverse and universal primers, purify the amplified DNA fragment from agarose gel, and perform a digestion of 0.2 μg of DNA with *Hin*dIII (from the pBluescript polylinker) in a 50 μL final volume (*see* **Note 7**). Add 2 μL of dCTP, dGTP, and dTTP mix, 0.5 μL (5 μCi) of [α-^{32}P]dATP (3000 Ci/mmol), and 2 U of the Klenow fragment of DNA polymerase and incubate at room temperature for 3 h.
2. Recover the labeled fragment by ethanol precipitation. For this, add 5 μL of 0.3 M NaAc, pH 5.2, 125 μL of ice-cold 100% ethanol, and 1 μL of 1 μg/μL tRNA. Keep at −80°C for at least 30 min.
3. Centrifuge in a microfuge at 12,000 rpm for 20 min at 4°C and remove the supernatant. Wash the pellet with 150 μL of ice-cold 70% ethanol. Spin again for 5 min at 12,000 rpm.
4. Remove the supernatant and dry the pellet at room temperature. Resuspend the DNA in water and digest with *Xba*I (50 μL final volume) (*see* **Note 8**).
5. Purify the labeled DNA fragment in a non-denaturing polyacrylamide gel (*see* **Section 2.1** and **Note 9**). For this, add 5 μL of sample buffer after digestion and apply onto a non-denaturing 10% polyacrylamide gel (*see* **Note 10**). Run in 0.5× TBE at 20 V/cm until the marker dye bromophenol blue is about 1.5–2 cm above the end of the gel.

6. Remove the minor glass plate and cover the gel with plastic wrap. Place an X-ray film on the gel and mark the position of the film. Put into an exposure cassette and expose for 5 min at room temperature.
7. Develop and mark the positions of the bands by placing the film on the gel. Cut the band corresponding to the fragment of interest with a scalpel and put the gel slice (without the plastic wrap) in a microcentrifuge tube containing 1 mL of oligonucleotide elution buffer. Elute overnight at 4°C (*see* **Note 11**).
8. Spin in a microcentrifuge for 1 min at maximum speed to separate the acrylamide. Transfer the supernatant to two new microcentrifuge tubes. Avoid transferring pieces of gel to the new tubes. Add 1 mL of ice-cold 100% ethanol and 1 μL of 1 μg/μL tRNA, mix, and keep at −80°C for at least 30 min (*see* **Note 12**).
9. Spin in a refrigerated microcentrifuge for 20 min at maximum speed. Remove the supernatant and wash once with 500 μL of ice-cold 70% ethanol. Dry the pellet at room temperature and resuspend the DNA in 5 μL of water (*see* **Note 8**). Transfer the DNA to a new tube. Determine the amount of radioactivity by Cherenkov counting. This is the labeled DNA to be used in the following sections.
10. Prepare the molecular weight marker for the sequencing gel by digestion of a 0.3 μL aliquot of the labeled DNA with enzyme(s) that cut at internal sites (*see* **Note 13**). Store at −20°C until use.
11. To obtain the DNA fragment labeled in the 3′-end of the other strand, follow a similar procedure but perform the first digestion with *Xba*I and the second one, after labeling, with *Hin*dIII.

3.2. Hydroxyl Radical Footprinting

1. Prepare a DNA-binding reaction by adding 5 μL of 3× binding buffer (*see* **Note 3**), 0.1–0.5 μg of poly(dI-dC) as a non-specific competitor, 100,000–250,000 cpm of labeled DNA, and the desired amount of protein in a final volume of 15 μL (*see* **Note 14**). Incubate for 30 min at the appropriate temperature to allow protein binding to DNA.
2. Place 3.5 μL each of previously prepared solutions of ascorbate and Fe(II)–EDTA as single drops on the inner wall of the tube. Rapidly add 3.5 μL of 0.6% H_2O_2, mix the three drops, and combine them with the sample (*see* **Note 15**).
3. Incubate at room temperature for 15 min.
4. Add 6 μL of 87% glycerol to stop the reaction.

5. During the binding and hydroxyl radical attack reactions, pre-run the EMSA gel in 0.5× TBE (*see* **Section 2.2**) at constant voltage.
6. Add 3 μL of 15% Ficoll to the sample and immediately load into the gel running at low voltage (*see* **Note 16**). In another well, load an aliquot of sample buffer (*see* **Section 2.1**) to monitor the progress of the run. Run the gel at 10 V/cm until the bromophenol blue is about 1.5–2 cm above the end of the gel (*see* **Note 17**).
7. Remove the minor glass plate and cover the gel with plastic wrap. Place an X-ray film on the gel and mark the position of the film. Put into an exposure cassette and expose for 40 min at −80°C.
8. Once the film is developed, mark the positions of the bound and free DNA bands in the gel. Cut each band with a scalpel and put the slices in microcentrifuge tubes (labeled bound DNA and free DNA) containing 1 mL of oligonucleotide elution buffer (*see* **Section 2.1**). Elute overnight at 4°C (*see* **Note 11**).
9. Precipitate the DNA as described in **Section 3.1**, steps 8 and 9, but resuspend each pellet in 5 μL of sequencing gel sample buffer (*see* **Section 2.2** and **Note 18**).
10. Transfer the sample to another tube. Ensure that no remaining gel pieces are transferred.
11. Adjust the amount of radioactivity and volume in each sample to about 5000–10,000 cpm in 4–5 μL (using Cherenkov radiation counting). Similar amounts of radioactivity of the bound and free DNA must be loaded in the sequencing gel.
12. Mix 5 μL of sequencing gel sample buffer with a small aliquot of the molecular weight marker prepared in **Section 3.1**, step 10 (200–300 cpm).
13. Denature the samples by heating to 94°C for no longer than 2 min and put on ice (*see* **Note 19**).
14. Load the samples on the 12% sequencing gel as quickly as possible. Run at constant power of 65–75 W for 1.75–2 h. The gel must be pre-run for at least 1 h at constant voltage until the gel temperature is between 45 and 50°C (*see* **Note 20**).
15. After electrophoresis, wait until the gel reaches room temperature, then remove one of the glass plates (*see* **Note 21**). Place a Whatman (Clifton, NJ) 3MM paper over the gel and lift it to transfer the gel onto the paper. Cover the gel with plastic wrap and dry in a gel dryer for 45 min at 80°C.

16. Expose the dried gel to preflashed Kodak (Rochester, MY) XAR film at −80°C for 16–24 h according to the amount of radioactivity loaded in each lane (*see* **Note 22**). Alternatively, a phosphorimager can be used if available.

3.3. Missing Nucleoside Assay

1. Prepare the 3′-end-labeled DNA fragment as described in **Section 3.1** but resuspend the DNA in 15 μL (instead of 5 μL) of water.
2. Perform the hydroxyl radical cleavage of DNA as described in **Section 3.2**, steps 2 and 3.
3. Add 7 μL of Stop Mix to stop the reaction.
4. Add 3.25 μL of 0.3 M NaAc, pH 5.2, 82 μL of ice-cold 100% ethanol, and 1 μL of 1 μg/μL tRNA. Keep at −80°C for at least 30 min.
5. Centrifuge in a microfuge at 12,000 rpm for 20 min at 4°C and remove the supernatant. Wash the pellet with 150 μL of ice-cold 70% ethanol. Spin again for 5 min at 12,000 rpm.
6. Remove the supernatant and dry the pellet at room temperature. Resuspend the DNA in 5 μL of water (*see* **Note 8**).
7. Perform a DNA-binding reaction as in **Section 3.2**, step 1.
8. During the DNA-binding reaction, pre-run the EMSA gel (*see* **Section 2.2**) at constant voltage.
9. Load samples in the EMSA gel and run as described in **Section 3.2**, step 6.
10. Expose the gel and elute the bound and free DNA as described in **Section 3.2**, steps 7 and 8.
11. Analyze the free and bound DNA in a sequencing gel as described in **Section 3.2**, steps 9–16.

3.4. Methylation Footprinting

1. Prepare the 3′-end-labeled DNA fragment as described in **Section 3.1**.
2. Perform a protein–DNA-binding reaction as described in **Section 3.2**, step 1 (*see* **Note 23**).
3. Add 0.5 μL of dimethyl sulfate and incubate for 10 min on ice (*see* **Note 24**).
4. Stop the reaction with 1/10 volume of 250 mM DTT.
5. Proceed to steps 5–9 of **Section 3.2** but resuspend the DNA in 100 μL of 20 mM potassium phosphate, pH 7.0, and 1 mM EDTA (*see* **Note 25**).
6. Incubate at 90°C for 5 min and then transfer to ice.
7. Add 1 μL of 10 N NaOH and incubate at 90°C for 5 min.

8. Transfer to ice and precipitate with ethanol.
9. Resuspend the DNA in 5 μL of sequencing gel sample buffer.
10. Proceed to steps 11–16 of **Section 3.2**.

3.5. Methylation Interference

1. Prepare the 3′-end-labeled DNA fragment as described in **Section 3.1** but resuspend the DNA in 200 μL of 50 mM sodium cacodylate, pH 8.0, and 1 mM EDTA (*see* **Note 26**).
2. Add 0.5 μL of dimethyl sulfate and incubate for 10 min on ice (*see* **Note 24**).
3. Stop the reaction with 50 μL of 1.5 M sodium acetate and 1 M 2-mercaptoethanol.
4. Precipitate with 850 μL of ice-cold ethanol and 1 μL of 1 μg/μL tRNA. Keep at −80°C for at least 30 min.
5. Proceed to steps 5–10, **Section 3.3**.
6. Precipitate the DNA as described in **Section 3.1**, steps 8 and 9, but resuspend the DNA in 100 μL of 20 mM potassium phosphate, pH 7.0, and 1 mM EDTA (*see* **Note 25**).
7. Proceed to steps 6–10, **Section 3.4**.

4. Notes

1. Acrylamide is a potent neurotoxin. Always wear gloves when working with acrylamide and, when handling the dry powder, a surgical mask is also essential.
2. Both the gel tank and the different components must be thoroughly cleaned and dried before use, especially if the equipment has been used previously for an SDS-PAGE. In addition, the gel plates must be cleaned with 95% ethanol.
3. The components of the binding buffer will depend on the protein that is analyzed. Sometimes, however, the reagents and buffers used in the protein–DNA-binding reaction may reduce the rate of DNA cleavage by the hydroxyl radical since they may act as scavengers. Glycerol, for example, is a potent scavenger and must be avoided. As a consequence, employ the most suitable binding buffer for your assay but without glycerol. Common buffers (such as Tris or HEPES) also reduce the rate of DNA cleavage, so it may be necessary to adjust the conditions of the DNA cleavage reaction accordingly to obtain a uniform footprinting pattern. In the missing nucleoside assay, there are no restrictions in the composition of the binding buffer since the

DNA is modified before binding. We usually add 30% glycerol to the 3× binding buffer for the interference assay.

4. The Fe(II) and H_2O_2 solutions must be freshly prepared before use, the first one from the solid reagent and the second one from a 30% stock solution. The Fe(II) and the EDTA solutions must be combined immediately before performing the DNA cleavage.

5. The length of the DNA fragment can also be between 100 and 200 bp, but we have found that elution from the polyacrylamide gels and separation of bound from free DNA are less efficient for fragments longer than 100 bp. In addition, the putative binding site must not be located too close to the ends of the fragment. We obtain good results when the label is 18–20 nucleotides away from the first nucleotide of the binding site.

6. 5′-Overhanging ends are required to label the fragment with the Klenow fragment of DNA polymerase. Through this strategy, the first nucleotide of the binding site is about 18 nucleotides away from the label. It is important to consider the length of the different fragments resulting after restriction enzyme digestion to achieve an efficient separation during polyacrylamide gel electrophoresis. The DNA fragment can also be labeled at its 5′-ends with T4 polynucleotide kinase and [γ-^{32}P]ATP (phosphatase treatment is required before labeling if a synthetic oligonucleotide is not used) followed by the removal of one label by restriction of the DNA with an appropriate enzyme. Note that in this case the fragments you will observe in the footprinting pattern will be in the 5′ to 3′ direction of the labeled strand from the lower to the upper part of the sequencing gel.

7. If vector primers are used for amplification, special care must be taken to avoid contamination, as many other clones that would serve as templates are routinely used in laboratories employing molecular biology techniques.

8. Take great care not to overdry the pellet to easily resuspend DNA and to prevent DNA damage. The quality of the labeled DNA is very important in this type of experiments. Nicks in the double strand will appear as additional bands in the sequencing gel and will obscure the footprinting results. In addition, we recommend performing the experiment immediately after labeling to prevent radiolysis of the DNA.

9. A standard system for polyacrylamide gel electrophoresis can be used. We have found that running the electrophoresis in vertical minigels (for example, the Mini-Protean III

system from Bio-Rad, Richmond, CA) enables a better elution of DNA and reduces assay times.

10. The percentage of the acrylamide gel will depend on the size of the fragments to be separated. *See* (21) for reference.
11. Elution can also be performed with agitation at room temperature.
12. We usually recover about 50% of the DNA from the gel through this procedure. To improve this, a second elution can be performed by adding 350 μL of elution buffer to the tubes containing the gel slices and incubating for 2–3 h.
13. This is necessary to exactly locate the position of the footprint in the DNA sequence. We usually digest the DNA in a restriction site (*Bgl*II) that we include in the cloned oligonucleotide adjacent to the putative binding site. Alternatively, other sites present in the fragment can be used, but it is preferable that the size of the labeled product lies close to the position of the binding site in the gel. It should be kept in mind that after hydroxyl radical cleavage, the nucleoside that is the site of attack is destroyed and as a consequence, the resulting DNA fragment is one nucleotide shorter than if it was terminated by the residue that was modified (8). This correction in size must be applied when using a marker generated by restriction enzyme cleavage. An alternative is to use a G or G+A Maxam and Gilbert chemical sequencing reaction of the same labeled fragment. The procedure for this is the same as outlined for methylation interference (**Section 3.5**), except that step 5 is omitted. Because in the sequencing reaction the modified nucleoside is also eliminated after cleavage, a direct comparison can be made between the size of the fragments obtained after hydroxyl radical attack and the sequence of the fragment as read in the gel. To avoid problems related to, for example, incomplete filling-in reactions or partial degradation of the fragment before labeling, we do not recommend the use of size markers or sequencing reactions prepared from other DNA fragments. With the use of the same fragment, these problems are avoided.
14. The amount of label necessary to obtain a discernable footprinting pattern is higher than that usually employed in an EMSA. This implies that a higher mass of DNA will be used in the binding reaction. Accordingly, the optimal amount of protein to obtain a good proportion of free and bound DNA must be determined experimentally for each system. In this sense, the limiting factor is usually the amount of bound DNA that is obtained. Usually, a band that can be

observed after a 40-min exposure is enough to obtain a footprinting pattern. It must also be considered that a high proportion of bound DNA (more than 30–50%) may be the consequence of non-specific interactions of the protein with DNA. We usually obtain good results using between 5 and 15 times more protein than in an EMSA.

15. These reagents must be freshly prepared during the protein–DNA-binding reaction.
16. Loading of the samples while the gel is running minimizes the time that the protein–DNA complex takes to enter the gel and, as a consequence, reduces the possibility of complex dissociation. Use plastic tips to avoid risks of electrocution. Layer the samples carefully into the wells without mixing with the gel running buffer.
17. The samples should not contain loading dyes to avoid interference with binding. The progress of loading can be visualized due to the differential refraction of the denser sample (due to the presence of Ficoll) compared to that of the running buffer. The best voltage and the time of run should be determined empirically. We pre-run the band shift mini gels at 100 V and perform the run at 130 V. In 5% gels, the bromophenol blue runs as a 65-bp DNA fragment and the xylene cyanol as a 250-bp DNA fragment.
18. Usually, there is a visible pellet caused by some remaining acrylamide. The resuspension volume depends on the radioactivity of the sample but it should be taken into account that a small volume must be loaded in the sequencing gel.
19. Longer heating or boiling creates additional cuts in the DNA.
20. These conditions are for a 50-cm-long, 21-cm-wide, and 0.4-mm-thick gel. They must be adjusted according to the system used. To avoid cross-contamination, we leave an empty lane between the samples in which we load 5 μL of sequencing gel sample buffer (this helps to obtain a uniform width of the bands in the entire ladder). During electrophoresis, the gel temperature should return to the original pre-running temperature and then remain constant. The percentage of acrylamide in the gel and the duration of electrophoresis depend on the distance between the region of interest and the labeled end of the DNA (i.e., on the size of the fragments that must be visualized). For separating oligonucleotides of 10–50 nucleotides, we use a 12% polyacrylamide gel in which the xylene cyanol and the bromophenol blue migrate as fragments of 40 and 10 nucleotides, respectively.

21. The gel must return to room temperature to facilitate its separation from the glass plate and the adherence to Whatman paper.
22. A first overnight exposure at −80°C can be useful to allow an early look at the gel and to determine the optimal exposure time. Usually a gel containing 5000–10,000 cpm/lane should be exposed at −80°C for 16–24 h, while 1000 cpm/lane takes about 5 days of exposure.
23. Sodium cacodylate (sodium dimethyl arsenate) is the preferred buffer for the methylation reaction because it is not modified by the methylating agent dimethyl sulfate. However, other buffers can be used if sodium cacodylate is not suitable for the DNA-binding reaction. Sodium cacodylate is highly toxic by ingestion, inhalation, and skin contact.
24. The amount of dimethyl sulfate and the incubation time necessary to produce a uniform ladder must be optimized according to the buffer that is used and the length of the DNA fragment. Because dimethyl sulfate is highly reactive, all steps including this agent must be performed in a fume hood.
25. The cleavage method described here (using NaOH) produces the hydrolysis of phosphodiester bonds at modified guanines and adenines. To cleave only in guanines, the DNA can be resuspended in 50 μL of piperidine (1:10 in water), incubated for 30 min at 90°C, dried under vacuum, and resuspended again in sequencing gel sample buffer.
26. Sodium cacodylate is highly toxic by ingestion, inhalation, and skin contact.

Acknowledgments

We thank Adriana Tron for the interference shown in **Fig. 15.2**. This work was supported by grants from Agencia Nacional de Promoción Científica y Tecnológica (ANPCyT) and Universidad Nacional del Litoral. ILV and DHG are members of CONICET.

References

1. Tullius, T. D. (1989) Physical studies of protein–DNA complexes by footprinting. *Annu. Rev. Biophys. Biophys. Chem.* **18**, 213–237.
2. Papavassiliou, A. (1995) Chemical nucleases as probes for studying DNA–protein interactions. *Biochem. J.* **305**, 345–357.

3. Galas, D. J., and Schmitz, A. (1978) DNase footprinting: a simple method for the detection of protein–DNA binding specificity. *Nucleic Acids Res.* **5**, 3157–3170.
4. Tullius, T. D. (1988) DNA footprinting with hydroxyl radical. *Nature* **332**, 663–664.
5. Dixon, W. J., Hayes, J. J., Levin, J. R., Weidner, M. F., Dombroski, B. A., and Tullius, T. D. (1991) Hydroxyl radical footprinting. *Methods Enzymol.* **208**, 380–413.
6. Jain, S. S., and Tullius, T. D. (2008) Footprinting protein–DNA complexes using the hydroxyl radical. *Nat. Protoc.* **3**, 1092–1100.
7. Tullius, T. D., and Dombroski, B. A. (1986) Hydroxyl radical "footprinting": high-resolution information about DNA–protein contacts and application to lambda repressor and Cro protein. *Proc. Natl. Acad. Sci. USA* **83**, 5469–5473.
8. Balasubramanian, B., Pogozelski, W. K., and Tullius, T. D. (1998) DNA strand breaking by the hydroxyl radical is governed by the accessible surface areas of the hydrogen atoms of the DNA backbone. *Proc. Natl. Acad. Sci. USA* **95**, 9738–9743.
9. Hayes, J. J., and Tullius, T. D. (1989) The missing nucleoside experiment: a new technique to study recognition of DNA by protein. *Biochemistry* **28**, 9521–9527.
10. Brunelle, A., and Schleif, R. F. (1987) Missing contact probing of DNA–protein interactions. *Proc. Natl. Acad. Sci. USA* **84**, 6673–6676.
11. Fenton, H. J. H. (1894) Oxidation of tartaric acid in the presence of iron. *J. Chem. Soc.* **65**, 899–910.
12. Shaw, P. E., and Stewart, A. F. (1994) Identification of protein–DNA contacts with dimethyl sulfate: methylation protection and methylation interference. *Methods Mol. Biol.* **30**, 79–87.
13. Shaw, P. E., and Stewart, A. F. (2009) Identification of protein/DNA contacts with dimethyl sulfate. Methylation protection and methylation interference. *Methods Mol. Biol.* **543**, 97–104.
14. Wissmann, A., and Hillen, W. (1991) DNA contacts probed by modification protection and interference studies. *Methods Enzymol.* **208**, 365–379.
15. Manfield, L. W., and Stockley, P. G. (2009) Ethylation interference footprinting of DNA–protein complexes. *Methods Mol. Biol.* **543**, 105–120.
16. Tioni, M. F., Viola, I. L., Chan, R. L., and Gonzalez, D. H. (2005) Site-directed mutagenesis and footprinting analysis of the interaction of the sunflower KNOX protein HAKN1 with DNA. *FEBS J.* **272**, 190–202.
17. Tron, A. E., Comelli, R. N., and Gonzalez, D. H. (2005) Structure of homeodomain–leucine zipper/DNA complexes studied using hydroxyl radical cleavage of DNA and methylation interference. *Biochemistry* **44**, 16796–16803.
18. Viola, I. L., and Gonzalez, D. H. (2006) Interaction of the BELL-like protein ATH1 with DNA: role of homeodomain residue 54 in specifying the different binding properties of BELL and KNOX proteins. *Biol. Chem.* **387**, 31–40.
19. Viola, I. L., and Gonzalez, D. H. (2007) Interaction of the PHD-finger homeodomain protein HAT3.1 from *Arabidopsis thaliana* with DNA. Specific DNA binding by a homeodomain with histidine at position 51. *Biochemistry* **46**, 7416–7425.
20. Viola, I. L., and Gonzalez, D. H. (2009) Binding properties of the complex formed by the Arabidopsis TALE homeodomain proteins STM and BLH3 to DNA containing single and double target sites. *Biochimie* **91**, 974–981.
21. Sambrook, J., Fritsch, E. F., and Maniatis, T. (1989) Molecular cloning. A laboratory manual. Cold Spring Harbor Laboratory Press, New York, NY.

Chapter 16

Chromatin Immunoprecipitation to Verify or to Identify In Vivo Protein–DNA Interactions

Yumei Zheng and Sharyn E. Perry

Abstract

Chromatin immunoprecipitation (ChIP) is a valuable tool to detect the interaction in vivo between a DNA-associated protein and DNA fragments. Combined with approaches to assess gene expression in response to accumulation of a transcription factor, it is possible to identify direct responsive targets from targets that are indirectly responsive to accumulation of the transcription factor. ChIP may be used to confirm in vivo association of a transcriptional regulator with suspected target DNA fragments. ChIP may also be used to discover new targets, and when combined with high-throughput approaches to identify DNA fragments associated with a transcription factor, it may provide a tool to study the gene regulatory networks active during plant development and/or response to the environment. Furthermore, ChIP is also a powerful means to map epigenetic modifications within a genome.

Key words: Chromatin immunoprecipitation, protein–DNA interaction, gene regulation, transcription, transcription factor, epigenetics, DNA modification.

1. Introduction

Protein–DNA interactions regulate transcription as well as DNA replication, recombination, and repair. Histone and DNA modifications provide an additional layer of control of gene expression. Proteins involved in these processes include transcription factors that specifically bind to generally short motifs of 5–10 bp. Key to understanding transcriptional regulators is identification of direct targets of these factors. Thus, it is important to identify specific DNA fragments bound by the transcription factor of interest. There are several methods to investigate protein–DNA interactions in vivo, including chromatin immunoprecipitation (ChIP),

L. Yuan, S.E. Perry (eds.), *Plant Transcription Factors*, Methods in Molecular Biology 754,
DOI 10.1007/978-1-61779-154-3_16, © Springer Science+Business Media, LLC 2011

Dam methylase identification (DamID) [(1, 2); *see* Chapter 18], footprinting [(3, 4); *see* Chapter 15], and yeast one-hybrid assay (*see* Chapter 3). Induction of expression or nuclear localization of a transcription factor and measurement of response of gene expression, either at short timepoints after induction and/or in the presence of cycloheximide, have also been used to identify putative direct targets [(5–8); *see* Chapter 7]. Each method has advantages as well as limitations.

ChIP is a powerful and increasingly used technique to isolate DNA fragments associated with a specific protein or complex in vivo (9–14). When combined with whole genome (tiling) arrays or high-throughput sequencing, ChIP may be used to globally map the targets of a protein (15–20). Recently published protocols for ChIP in plants can be found in (13, 21–24). The procedure of ChIP involves seven main steps: (1) stabilize protein–DNA complexes in vivo by using a cross-linking reagent, most commonly formaldehyde; (2) isolate nuclei and lyse to obtain chromatin; (3) shear the isolated chromatin to reasonable size (~1000 bp or smaller); (4) immunoprecipitate the desired protein–DNA complexes; (5) recover the DNA fragments; (6) reverse the protein–DNA cross-links with heat and remove proteins with proteinase K followed by extraction and precipitation to purify the DNA fragments; and (7) check for association with DNA fragments of interest with semi-quantitative PCR or quantitative PCR.

Formaldehyde is most commonly used to cross-link DNA-associated proteins to DNA for ChIP. It is a short (approximately 2 Å) cross-linker that reacts with primary amines on amino acids and DNA/RNA bases (25). The cross-links are reversible by heat. Formaldehyde easily enters cells and cross-linking times can range from minutes to hours depending on the protocol. Because formaldehyde rapidly inactivates enzymes, it provides a "snapshot" of the interactions occurring within the cell. However, formaldehyde can disrupt epitopes needed for immunoprecipitation, either by modifying the antigen (e.g., epitopes with lysine(s) that are a main target of formaldehyde) (14) or simply by denaturing the protein, and too long of fixation times lead to poor solubilization of chromatin. Unlike cross-links generated by UV irradiation, formaldehyde cross-links proteins to DNA and proteins to proteins so that even if a protein is associated with DNA via another DNA-binding protein, in theory it is still possible to isolate DNA fragments indirectly associated with the protein of interest. There have been reports of longer cross-linking agents increasing the efficiency of this type of ChIP (26, 27).

Highly specific antibody suitable for ChIP is essential for the protocol and in some cases may include antibodies that recognize particular posttranslational modifications. Alternatively, epitope tags may be engineered into the protein (e.g., HA, c-myc, and GFP have been used for ChIP in plants) (17, 20, 28) and commercially available antibody specific to a tag and suitable for ChIP purchased. Commercially available antibodies are also available for epigenetic modifications. Some tags (e.g., tandem affinity purification and TAP tag) (29) can eliminate the need for an antibody. Protein A (or G depending on the primary antibody) Sepharose or magnetic beads allow separation of the desired protein–DNA complexes from the rest of the chromatin. Depending on the approach, different controls may be used including precipitation with preimmune or no serum, precipitation with immune serum from chromatin prepared from tissue lacking the particular DNA-binding protein (e.g., knockout mutant tissue), and precipitation from tissue expressing an untagged form of the protein or from tissue expressing an unrelated protein that does not associate with DNA but that includes the epitope tag. Total (input) DNA may also be used as a reference. It is possible to assess and troubleshoot the immunoprecipitation by following the protein if the protein accumulates to high enough amounts.

Downstream analysis depends on the goal of the investigation. In some cases, the aim is to test whether a suspected target gene is associated with a particular DNA-binding protein or an epigenetic mark. In these cases, a directed test for enrichment of the target fragment in the immune precipitation compared to controls may be performed with semi-quantitative (*see* **Section 3.6**) or quantitative approaches. The fold enrichment of the target in the immune precipitation may be compared to the preimmune precipitation, or the differential site occupancy of a target site and non-bound control can be compared in the same immune precipitation. In cases where the desire is to map occupation by a factor globally, high-throughput approaches such as ChIP-(on)-chip or ChIP-SEQ may be utilized. *See* Chapter 17 for a discussion of analysis for ChIP-SEQ data. If no targets are known, it may be helpful to use a low-throughput clone and sequence approach to identify some targets to use to test populations before proceeding with the expensive high-throughput analysis.

AGL15 is a MADS-domain regulatory factor that accumulates primarily, although not exclusively, during embryogenesis and promotes somatic embryo development (30, 31). Here we present a ChIP protocol that we have used to identify and assess direct targets of AGL15.

2. Materials

2.1. Tissue Preparation

1. MC buffer: 10 mM Potassium phosphate, pH 7, 50 mM NaCl, 0.1 M sucrose.
2. 37% Formaldehyde (Sigma), store for up to 1 year at room temperature.

3. 1.25 M Glycine, store at −20°C.
4. Liquid nitrogen.
5. Vacuum desiccator and vacuum line/pump.

2.2. Nuclei Isolation and Chromatin Solubilization (See Note 1)

1. 100 mM PMSF in isopropanol, store at 4°C. Use care as PMSF is toxic. Add PMSF to buffers immediately before use because PMSF has a short half-life in aqueous solutions.
2. Extraction buffer 1 (EB1): 0.4 M Sucrose, 10 mM Tris–HCl, pH 8.0, 5 mM β-mercaptoethanol, 0.1 mM PMSF, store at −20°C; EB1, 2, and 3 are from (13).
3. Miracloth (Calbiochem).
4. Extraction buffer 2 (EB2): 0.25 M Sucrose, 10 mM Tris–HCl, pH 8.0, 5 mM β-mercaptoethanol, 0.1 mM PMSF, 10 mM $MgCl_2$, 1% Triton X-100, store at −20°C.
5. Extraction buffer 3 (EB3): 1.7 M Sucrose, 10 mM Tris–HCl, pH 8.0, 5 mM β-mercaptoethanol, 0.1 mM PMSF, 2 mM $MgCl_2$, 0.15% Triton X-100, store at −20°C.
6. M3 Buffer: 10 mM Potassium phosphate, pH 7.0, 0.1 M NaCl, 10 mM β-mercaptoethanol, 0.1 mM PMSF, store at −20°C.
7. Sonication buffer: 10 mM Potassium phosphate, pH 7.0, 0.1 M NaCl, 0.5% sarkosyl, 10 mM EDTA, 0.1 mM PMSF, store at −20°C.
8. Probe sonicator with small tip.
9. Mortar and pestle.
10. Acetone.
11. Small paintbrush.
12. Centrifuge and microcentrifuge.

2.3. Immunoprecipitation of Protein–DNA Complexes

1. Antibody (for our experiments, AGL15 antiserum and preimmune serum).
2. Immunoprecipitation buffer (IP bf): 50 mM Hepes, pH 7.5, 150 mM KCl, 5 mM $MgCl_2$, 10 μM $ZnSO_4$, 1% Triton X-100, 0.05% SDS, store at −20°C.
3. Protein A-Sepharose 4B (Invitrogen), store at 4°C. Depending on the source, you may need to wash the beads with IP buffer to remove ethanol or other preservatives before use.
4. Pelco R2 rotary mixer or other mixer that will gently mix end-over-end.
5. Drawn-out Pasteur pipettes or fine (gel-loading) pipette tips.

2.4. Elution and Recovery of DNA Fragments

1. Glycine elution buffer: 0.1 M Glycine, 0.5 M NaCl, 0.05% Tween 20, pH 2.8, store at 4°C.

2. 1 M Tris, pH 9.0, autoclave and store at room temperature.
3. 10 mg/mL DNase-free RNase A in TE buffer. RNase A must be boiled for 30 min to inactivate any DNase.
4. Proteinase K, 20 mg/mL in phosphate buffered saline, store in small aliquots at −20°C.
5. Phenol (buffered to pH 8):chloroform:isoamyl alcohol, 25:24:1.
6. Molecular biology-grade glycogen, 20 mg/mL, store at −20°C.
7. 3 M Sodium acetate, pH 5.2, store at room temperature.
8. Ethanol.

2.5. Western Blot Analysis

1. 4× SDS-PAGE sample buffer: 40% Glycerol, 8% sodium dodecyl sulfate (SDS), 50% 4× upper (stacking) gel buffer, 0.05 mg/mL bromophenol blue, 6% (w/v) dithiothreitol (DTT).
2. 4× Upper (stacking) gel buffer: 0.5 M Tris, pH 6.8, 0.4% SDS.
3. Standard equipment and solutions for running SDS-PAGE protein gels and for blotting to membrane.
4. Primary antibody to detect the protein of interest or an epitope tag on the protein of interest.
5. Detection system including an enzyme-labeled secondary antibody and appropriate detection reagents.
6. Heat block.

2.6. Semi-quantitative Enrichment Test to Verify a Suspected Target

1. Oligonucleotide primers to amplify ~300–400-bp target fragment.
2. Oligonucleotide primers to amplify ~200-bp control fragment.
3. Taq enzyme (KlenTaq from Ab Peptides, Inc. is an economical choice) and buffer for the enzyme.
4. 2 mM dNTPs (dATP, dCTP, dTTP, and dGTP).
5. Thermal cycler.
6. 2% Agarose gel with 1 μg/mL ethidium bromide.
7. Standard equipment for running and documenting agarose gels.

3. Methods

3.1. Tissue Preparation

1. Choose a tissue source where the transcription factor of interest is accumulating within the nuclei, preferably to normal endogenous levels. Slice the tissue into small pieces in cold MC buffer on ice and place into a beaker on ice. We use 8–12 g of somatic embryo tissue generated from cultured developing zygotic embryos (30) per experiment in 40 mL of MC buffer but have used as little as 1 g of flower buds from *35S:AGL15* plants (*see* **Note 2**).
2. Add formaldehyde to a 1% final concentration from a 37% formaldehyde stock.
3. Put the beaker onto ice within a vacuum desiccator. Pull the vacuum until the first bubbles are observed at the top of the buffer and release. Repull the vacuum and seal for 1 h on ice (*see* **Note 3**).
4. After 1 h, release the vacuum and quench the unreacted formaldehyde by addition of glycine to 0.125 M final concentration. Incubate on ice for about 10–30 min.
5. Wash the tissue thoroughly with MC buffer (three times), pull off as much buffer as possible, and flash freeze in a seal-a-meal bag in liquid nitrogen. Store at −80°C for future use.

3.2. Nuclei Isolation and Chromatin Solubilization

1. Powder the frozen tissue in liquid nitrogen with a mortar and pestle (10 g tissue).
2. Transfer the pulverized tissue into a 50-mL Falcon tube and resuspend in 20 mL of EB1 with freshly added PMSF to 0.1 mM final concentration (amount of EB1 buffer needed is approximately twice the weight of the tissue, *see* **Notes 4** and **5**). Mix and shake to thaw the suspension and add more EB1 buffer as needed to generate a thick but pourable slurry. Remove large tissue pieces by filtering through miracloth into centrifuge tubes. Save 75 μL for a Western blot if desired and label this "total protein."
3. Pellet the crude nuclei by centrifuging at 2500×*g* in a fixed angle rotor (e.g., Sorvall SS-34) for 20 min at 4°C.
4. Save 750 μL of supernatant and concentrate the proteins by adding 450 μL of acetone. Place on ice or at −20°C for at least 2 h, and then pellet precipitated protein in a microcentrifuge at top speed for 2 min. Remove the remaining supernatant (decant carefully) and resuspend pellet in 100 μL of 1× SDS-PAGE sample buffer (*see* **Section 2.5** for a 4× stock recipe). This is a "nuclear-depleted" sample for any desired troubleshooting on a Western blot.

5. Remove the rest of the supernatant from the crude nuclear pellet and resuspend the pellet in EB2 plus freshly added PMSF by first using a few milliliters and a paintbrush to gently resuspend the pellet and then adding 10–20 mL of EB2. Allow to sit on ice for 10 min and then centrifuge at 15,000×*g* in a fixed angle rotor (Sorvall SS-34) for 10 min at 4°C. Remove the supernatant by careful decanting (*see* **Note 6**).
6. Resuspend the pellet in 1 mL of EB3 plus freshly added PMSF using a paintbrush. Layer onto 0.5 mL EB3 cushions in 1.5-mL Eppendorf microcentrifuge tubes. Centrifuge at top speed for 30–60 min at 4°C.
7. Remove the supernatant and wash the pellet in 1 mL of M3 buffer in a 1.5-mL microcentrifuge tube. If desired, a 10 μL aliquot may be saved for DAPI staining to check nuclei using fluorescence microscopy. Centrifuge at top speed for 5 min at 4°C.
8. Remove the supernatant and resuspend the nuclei pellet in 1 mL of sonication buffer and add PMSF to a 0.1 mM final concentration. Save 50 μL for Western analysis labeled as "nuclei."
9. Shear the chromatin with a probe sonicator by four pulses of 10–15 s each, setting 50 (on a Fisher brand sonicator, 50 is one-half the maximum). Keep the samples on ice during sonication and cool on ice in-between pulses for at least 4 min (*see* **Note 7**).
10. Pellet unsolubilized material in a microcentrifuge at top speed for 5 min at 4°C. Remove the supernatant to a new tube. This is the solubilized chromatin (also called sonication supernatant). Save the pellet (sonication pellet) for troubleshooting should the protein of interest be missing from the solubilized chromatin. The solubilized chromatin may be flash frozen in liquid nitrogen for later use. If frozen, upon proceeding, first centrifuge at top speed for 2 min in a microcentrifuge to remove any precipitate. Save 50 μL of solubilized chromatin as total (input) DNA for DNA workup and another 75 μL for Western blot analysis to verify solubilization of the protein of interest. The DNA sample is needed for the enrichment test by semi-quantitative and quantitative PCR. It is also useful to check the size of the fragmented chromatin by running 10 μL of undiluted total (input) DNA on an agarose gel with a DNA ladder (*see* **Note 3**).

3.3. Immunoprecipitation of the Protein–DNA Complexes

1. Divide the remaining solubilized chromatin into two equal aliquots to new 1.5-mL microcentrifuge tubes. Add an equal amount of IP buffer. Add immune serum to one tube

and preimmune serum (or no serum if necessary) to the other tube. The amount added will depend on the antibody/antiserum and must be empirically determined. Typically 2.5 μL of immune or preimmune serum works well for our tissue and antibody. Incubate for at least 1 h at 4°C on a mixer that will gently mix the solution end-over-end such as a rotating wheel (Pelco R2 rotary mixer). Depending on the antibody, this incubation may go overnight (*see* **Notes 8** and **9**).

2. Microcentrifuge at top speed for 2 min at 4°C.
3. Remove the top 90% of the volume to a new tube. Add 40 μL of 50% protein A-Sepharose slurry (Invitrogen) and incubate for 1 h at 4°C with rotation. The remaining 10% of the sample left in the tube may be saved for Western analysis if desired.
4. Pellet the beads by microcentrifuging at top speed for 1–2 min at 4°C. Remove the supernatant and save 75 μL of solution as "post-bind" immune or preimmune for Western analysis. If all goes well, the protein of interest should be at least somewhat depleted from the immune precipitation but not the preimmune.
5. Wash the beads by addition of cold IP buffer (1 mL each tube). Invert to mix and place on a rotating wheel for 10 min at room temperature. Depending on the antibody and/or the stability of the protein of interest, this step may be performed at 4°C. Beads are pelleted by microcentrifuging at top speed for 1 min at room temperature (or 4°C if needed) and the wash removed as thoroughly as possible by using a drawn-out Pasteur pipette. The wash is repeated for a total of 3–4 times, moving the suspended beads to a fresh microcentrifuge tube at the last wash. After removal of the last wash, the tube is subjected to centrifugation for 1 min at top speed and any remaining buffer is removed with a drawn-out Pasteur pipette.

3.4. Elution and Recovery of DNA Fragments

1. Add 100 μL of cold glycine elution buffer to the beads, vortex for 30 s to 1 min and pellet the beads at top speed for 1 min at room temperature. Transfer the supernatant to a new tube with 50 μL of 1 M Tris, pH 9.0, to neutralize. Repeat the elution and neutralization twice more for approximately 450 μL total volume.
2. Microcentrifuge the eluted samples at top speed for 2 min at room temperature. Transfer the top 375 μL to a new tube for DNA workup, taking care to avoid any beads. Save the remaining sample as "eluted immune or preimmune" for Western analysis. If all goes well, the protein will be

detected in the eluted immune sample but not the eluted preimmune sample. Beads after elution may be neutralized with 10–20 μL of 1 M Tris, pH 9, and boiled in 1× SDS-PAGE sample buffer to test degree of elution from beads.

3. To the input (total) DNA that was saved from the sonicated chromatin (step 10 in **Section 3.2**), add 108 μL of 1 M Tris (pH 9.0) and 217 μL of cold glycine elution buffer.
4. Add 1 μL of 10 mg/mL DNase-free RNase A to each sample for DNA workup (immune, preimmune, and total) and incubate at 37°C for 15–30 min.
5. Add proteinase K to 0.4 mg/mL (7.5 μL of 20 mg/mL stock in PBS) and incubate overnight at 37°C.
6. The next day, add another 5.0 μL of 20 mg/mL proteinase K stock in PBS and incubate for at least 6 h at 65°C. Let cool to room temperature and chill on ice (*see* **Note 10**).
7. Extract the DNA by adding an equal volume of phenol:chloroform:isoamyl alcohol (25:24:1). Vortex for 15–30 s and microcentrifuge at top speed for 5 min. Remove the top aqueous phase to a new tube and precipitate the DNA by adding 1 μL of 10 mg/mL glycogen, followed by 0.1 volume of 3 M sodium acetate, pH 5.2, and finally 2.5 volumes of 100% ethanol. Incubate overnight at −20°C.
8. The next day, pellet the DNA at top speed in a microcentrifuge for 30 min at 4°C. Remove the supernatant, taking care to not disturb the pellet. Wash the pellet with 1 mL of 70% ethanol and spin at top speed for 5 min. Dry the samples in a SpeedVac or by leaving the tubes open on the benchtop.
9. Resuspend the pellet in 20–30 μL of MilliQ sterile water.

3.5. Western Blot Analysis

1. Samples reserved for Western blot analysis during the protocol can help diagnose problems and test the efficiency of precipitation. SDS-PAGE sample buffer is added to a final 1× concentration and samples are heated at 95–100°C for 5 min in a heat block. Heating will reverse formaldehyde cross-links.
2. Use standard SDS-PAGE protocols to separate proteins in the reserved samples and standard Western blotting protocols to transfer the separated proteins to membrane such as Immobilon PVDF (Millipore). Be aware of the need to separate the protein of interest from the IgG because the heavy chain of IgG (55–60 kDa) may also be detected by Western analysis. Because AGL15 is a low-abundance protein, even when expression is driven with the 35S CaMV promoter, we use a sensitive chemiluminescent detection system.

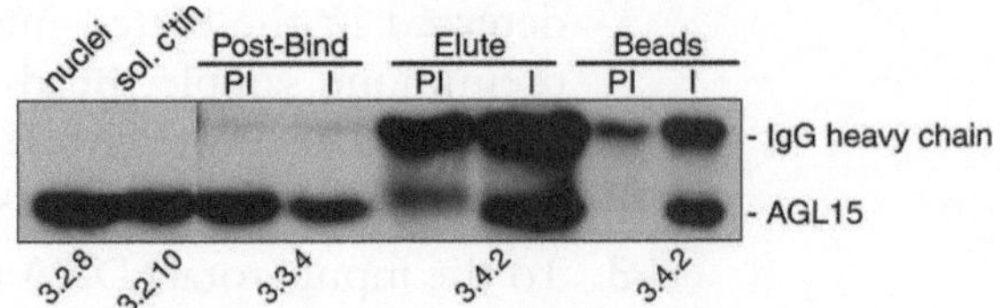

Fig. 16.1. Western blot probed with anti-AGL15 serum to detect AGL15 at different steps of the ChIP protocol. Section and step numbers are shown under the lanes. Total protein (**Section 3.2**, step 2), nuclear-depleted sample (**Section 3.2**, step 4), the insoluble pellet after sonication (**Section 3.2**, step 10), and the samples after addition of immune or preimmune serum but before protein A-Sepharose beads (**Section 3.3**, step 3) are not shown.

3. As shown in **Fig. 16.1**, AGL15 is detected in the crude nuclei and is solubilized in the solubilized chromatin (sonicated supernatant) samples. After the protein A-Sepharose beads are pelleted, the remaining supernatant shows at least some depletion of AGL15 in the immune sample as compared to the preimmune samples (compare post-bind I and PI). Also notable is the fact that nearly all IgG has been bound in these samples, leaving very little in the supernatant. This result indicates that the amount of antibody added could be increased to attempt to precipitate more AGL15 in the immune sample. With the amount of antibody and conditions used, the volume of protein A-Sepharose was sufficient to bind nearly all of the IgG. In the eluted sample, AGL15 is recovered in the immune but not the preimmune sample and some AGL15 was left associated with the beads.

3.6. Semi-quantitative Enrichment Test to Verify a Suspected Target

1. To perform a relatively simple semi-quantitative enrichment test, design primers to amplify the target DNA fragment of interest (target) and primers to amplify a DNA fragment that is not bound by the protein of interest and is not near (i.e., within 1000 bp) to a region that is bound (control). If multiplex PCR is desired, as shown in **Fig. 16.2**, design the target primers to amplify a fragment of 300–400 bp and the control primers to amplify a fragment of about 200 bp. Designing the target to be larger than the control fragment will prevent

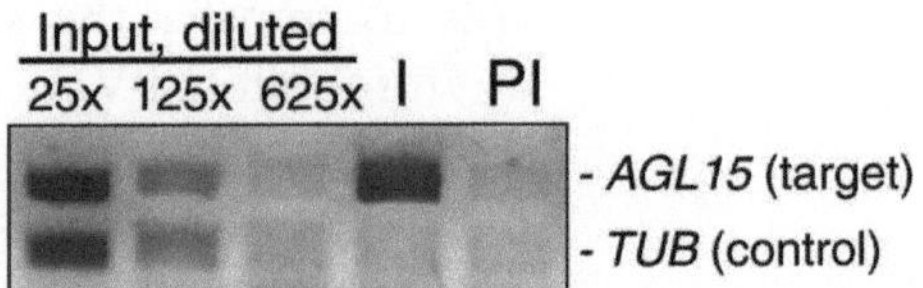

Fig. 16.2. Semi-quantitative multiplex enrichment test to show enrichment of a target DNA fragment (*AGL15*) compared to a non-bound DNA fragment (*TUB2*) after ChIP with immune (I) antibody compared to input (total) DNA and to the preimmune (PI) control.

enrichment of target being observed simply because of DNA fragmentation during the ChIP procedure. Primers typically are 20–24 bp in length with a predicted annealing temperature of approximately 55°C.

2. Total (input) DNA is diluted 25-, 125-, and 625-fold and 0.5 μL is used per 15 μL PCR reaction. Immune and preimmune precipitations are used undiluted (0.5 μL per reaction).
3. Multiplex PCR reactions are set up using 1× PC2 buffer, 0.2 mM dNTPs, 1 mM target primer pair, 1 mM control primer pair, and 1 U KlenTaq (Ab Peptides). PCR conditions are typically 94°C for 5 min, followed by 32–35 cycles of 94°C for 30 s, 52–55°C for 30 s, 72°C for 30 s, followed by 72°C for 5 min.
4. PCR products are resolved on a 2% agarose gel containing ethidium bromide.
5. As shown in **Fig. 16.2**, in the input DNA, the control band (*TUBULIN*) is roughly of equal intensity as the target band (*AGL15*, AGL15 binds to its own regulatory regions) (32). However, in the immune precipitation, the target band is much more intense than the control band, and in the preimmune control precipitation, very little target or control band is present. This is a good ChIP with very strongly bound target (AGL15).
6. qPCR should be performed to quantitate enrichment as in (33).

4. Notes

1. The nuclear preparation protocol presented in (12) works equally well but is somewhat more time consuming than the protocol presented in this chapter that follows the procedure in (13).
2. Use enough tissue for ChIP such that you can detect the protein on a Western blot. While not necessary, the ability to follow the protein by Western blot analysis will aid in any needed troubleshooting.
3. The optimal time for formaldehyde treatment of tissue depends on the particular tissue. For yeast, the optimal treatment time is around 15–20 min, whereas for *Caenorhabditis elegans,* it is about 20 min. For larger plant pieces, longer time is needed. Also the time for reaction depends on temperature. Vacuum helps the fixative to

infiltrate into plant tissue. Too long of cross-linking will lead to difficulty obtaining DNA fragments of suitable size and will lead to decreased yield of chromatin. The size and amount of DNA obtained may be checked by running some total (input) DNA (from step 10, **Section 3.2** and after DNA workup described in **Section 3.4**) on a 1% agarose gel with 1 μg/mL ethidium bromide and a DNA ladder. The majority of the smear should be between 200 and 1000 bp.

4. Some proteins are more unstable than others. In our protocol, we have used PMSF as the sole protease inhibitor and even this is not essential. PMSF is added to the buffer immediately before use because it has a relatively short half-life in aqueous solutions. Depending on the protein of interest, more extensive protection of the protein may be needed throughout the experiment including isolation of nuclei, sonication, and immunoprecipitation steps. Additional protease inhibitors that may be used include aprotinin, pepstatin A, leupeptin, antipain, TPCK, and benzamidine (13). Alternatively, protease inhibitor cocktails may be commercially purchased such as from Sigma (P9599, protease inhibitor cocktail for plant cell and tissue extracts). Additionally, while AGL15 is stable even with incubations at room temperature, for other proteins, keeping all steps cold may be more important.
5. Some plant tissues are especially challenging, including those with high accumulation of phenolic compounds and/or polysaccharides such as *Arabidopsis* developing seed. We have found that increasing the amount of EB1 buffer twofold in step 2 of **Section 3.2** and addition of 1% PEG (34) helps, but the ChIP is still not as robust as from the culture tissue we usually use.
6. If the pellet is still very green after the EB2 step (**Section 3.2**, step 5), then the EB2 wash may be repeated.
7. Samples should be kept on ice during sonication because sonication can heat the sample, leading to cross-link reversal. Foaming during sonication should be avoided. Sonication may need to be empirically determined and may vary depending on fixation conditions. DNA fragment size can be checked as in **Note 3**. If DNA fragments are too long, reduce fixation time/temperature and/or increase sonication. Recovery of the protein of interest in the solubilized chromatin sample can also be diagnostic. Generally it is difficult to get too short of DNA fragments by sonication and if fragments are excessively small, cleanliness of solutions and the probe should be considered. As an example, when DNA fragments were too small, we found the shared

probe sonicator was being used with DNase by another group. Cleaning the sonication probe before use solved the problem. Methods to inactivate DNase include heating (immersion of the probe in water at 65–70°C), 75% ethanol, and/or EDTA treatment.

8. Successful immunoprecipitation depends to a large extent on the specificity of the antibody and the ability of the antibody to immunoprecipitate. Not all antibodies that work well on a Western blot are suitable for immunoprecipitation and the suggested Western analysis should help diagnose any problem. Many commercial sources of antibodies to epitope tags will indicate if they are suitable for ChIP but the tag must be accessible in the particular complex. The degree of immunoprecipitation will also depend on the affinity of the antibody, the amount of antibody and protein A-Sepharose used, and the stringency of the binding and wash conditions, and conditions may need to be adjusted. For example, some antibody-antigen interactions may be disrupted by too high amounts (>0.1%) of ionic detergents such as SDS and sarkosyl.
9. Non-specific background can also be a problem in ChIP. One should make sure that the antibody is specific to the target protein by Western blot and that any DNA co-precipitated is specific by performing ChIP with an independent tag and/or by using tissue from a knockout mutant of the protein of interest. The rotation time for the protein A-Sepharose step should be limited to 1–2 h as increased time will increase non-specific DNA association. If too much non-specific binding is observed, it may be helpful to preclear the sonicated supernatant by incubation with protein A-Sepharose. Also the stringency of the wash solutions and the number of washes may be adjusted.
10. Depending on the timing of the experiment, it is possible to perform the step at 65°C for 6 h and follow this step with the incubation at 37°C overnight.

Acknowledgments

The authors would like to thank Jeanne Hartman for comments on the manuscript. This work was supported by grants from the National Science Foundation (IBN-9984274 and IOS-0922845).

References

1. Germann, S., Juul-Jensen, T., Letarnec, B., and Gaudin, V. (2006) DamID, a new tool for studying plant chromatin profiling in vivo, and its use to identify putative LHP1 target loci. *Plant J.* **48**, 153–163.
2. Greil, F., Moorman, C., and van Steensel, B. (2006) DamID: mapping of in vivo protein–genome interactions using tethered DNA adenine methyltransferase, in Methods in Enzymology. *DNA Microarrays Part A: Array Platforms and Wet-Bench Protoc.* **410**, 342–359.
3. Ikuta, T., and Kan, Y. W. (1991) In vivo protein DNA interactions at the beta-globin gene locus. *Proc. Natl. Acad. Sci. USA* **88**, 10188–10192.
4. Kato, K., Nomoto, M., Izumi, H., Ise, T., Nakano, S., Niho, Y., and Kohno, K. (2000) Structure and functional analysis of the human STAT3 gene promoter: alteration of chromatin structure as a possible mechanism for the upregulation in cisplatin-resistant cells. *BBA-Gene Struct. Expr.* **1493**, 91–100.
5. Braybrook, S. A., Stone, S. L., Park, S., Bui, A. Q., Lee, B. H., Fischer, R. L., Goldberg, R. B., and Harada, J. J. (2006) Genes directly regulated by LEAFY COTYLEDON2 provide insight into the control of embryo maturation and somatic embryogenesis. *Proc. Natl. Acad. Sci. USA* **103**, 3468–3473.
6. Ko, J. H., Kim, W. C., and Han, K. H. (2009) Ectopic expression of MYB46 identifies transcriptional regulatory genes involved in secondary wall biosynthesis in *Arabidopsis. Plant J.* **60**, 649–665.
7. Passarinho, P., Ketelaar, T., Xing, M. Q., van Arkel, J., Maliepaard, C., Hendriks, M. W., Joosen, R., Lammers, M., Herdies, L., den Boer, B., van der Geest, L., and Boutilier, K. (2008) BABY BOOM target genes provide diverse entry points into cell proliferation and cell growth pathways. *Plant Mol. Biol.* **68**, 225–237.
8. Sablowski, R. W. M., and Meyerowitz, E. M. (1998) A homolog of *NO APICAL MERISTEM* is an immediate target of the floral homeotic genes *APETALA3/PISTILLATA*. *Cell* **92**, 93–103.
9. Grandori, C., Mac, J., Siëbelt, F., Ayer, D. E., and Eisenman, R. N. (1996) Myc-Max heterodimers activate a DEAD box gene and interact with multiple E box-related sites in vivo. *EMBO J.* **15**, 4344–4357.
10. Orlando, V., Strutt, H., and Paro, R. (1997) Analysis of chromatin structure by in vivo formaldehyde cross-linking. *Methods* **11**, 205–214.
11. Orlando, V. (2000) Mapping chromosomal proteins in vivo by formaldehyde-crosslinked-chromatin immunoprecipitation. *TIBS* **25**, 99–104.
12. Wang, H., Tang, W., Zhu, C., and Perry, S. (2002) A chromatin immunoprecipitation (ChIP) approach to isolate genes regulated by AGL15, a MADS-domain protein that preferentially accumulates in embryos. *Plant J.* **32**, 831–843.
13. Bowler, C., Benvenuto, G., Laflamme, P., Molino, D., Probst, A. V., Tariq, M., and Paszkowski, J. (2004) Chromatin techniques for plant cells. *Plant J.* **39**, 776–789.
14. Das, P. M., Ramachandran, K., vanWert, J., and Singal, R. (2004) Chromatin immunoprecipitation assay. *Biotechniques* **37**, 961–969.
15. Buck, M. J., and Lieb, J. D. (2004) ChIP-chip: considerations for the design, analysis, and application of genome-wide chromatin immunoprecipitation experiments. *Genomics* **83**, 349–360.
16. Johnson, C., Boden, E., Desai, M., Pascuzzi, P., and Arias, J. (2001) In vivo target promoter-binding activities of a xenobiotic stress-activated TGA factor. *Plant J.* **28**, 237–243.
17. Lee, J., He, K., Stolc, V., Lee, H., Figueroa, P., Gao, Y., Tongprasit, W., Zhao, H., Lee, I., and Deng, X. W. (2007) Analysis of transcription factor HY5 genomic binding sites revealed its hierarchical role in light regulation of development. *Plant Cell* **19**, 731–749.
18. Zheng, Y., Ren, N., Wang, H., Stromberg, A. J., and Perry, S. E. (2009) Global identification of targets of the Arabidopsis MADS domain protein AGAMOUS-Like 15. *Plant Cell* **21**, 2563–2577.
19. Kaufmann, K., Muiño, J. M., Jauregui, R., Airoldi, C. A., Smaczniak, C., Krajewski, P., and Angenent, G. C. (2009) Target genes of the MADS transcription factor SEPALLATA3: Integration of developmental and hormonal pathways in the *Arabidopsis* flower. *PLoS Biol.* 7, 0854–0875.
20. Oh, E., Kang, H., Yamaguchi, S., Park, J., Lee, D., Kamiya, Y., and Choi, G. (2009) Genome-wide analysis of genes targeted by PHYTOCHROME INTERACTING FACTOR 3-LIKE5 during seed germination in *Arabidopsis. Plant Cell* **21**, 403–419.
21. Kaufmann, K., Muino, J. M., Osteras, M., Farinelli, L., Krajewski, P., and Angenent, G.

C. (2010) Chromatin immunoprecipitation (ChIP) of plant transcription factors followed by sequencing (ChIP-SEQ) or hybridization to whole genome arrays (ChIP-CHIP). *Nat. Protoc.* **5**, 457–472.
22. Morohashi, K., Xie, Z., and Grotewold, E. (2009) Gene-specific and genome-wide ChIP approaches to study plant transcriptional networks. In *Methods in Molecular Biology: Plant Systems Biology* (Belostotsky, D. A., Ed.), pp 3–12, Springer, Clifton, NJ.
23. Ricardi, M. M., Gonzalez, R. M., and Iusem, N. D. (2010) Protocol: fine-tuning of a chromatin immunoprecipitation (ChIP) protocol in tomato. *Plant Methods* **6**, article 11. doi: 10.1186/1746-4811-6-11.
24. Xie, Z. D., and Grotewold, E.. (2008) Serial ChIP as a tool to investigate the co-localization or exclusion of proteins on plant genes. *Plant Methods* **4**, article 25. doi: 10.1186/1746-4811-4-25.
25. Aparicio, O., Geisberg, J. V., and Struhl, K. (2004) Chromatin immunoprecipitation for determining the association of proteins with specific genomic sequences in vivo. *Curr. Protoc. Cell Biol. Macromol. Interact. Cell Suppl.* **23**, 1–23. John Wiley & Sons, Inc.
26. Kim, T. H., and Ren, B. (2006) Genome-wide analysis of protein–DNA interactions. *Annu. Rev. Genomics Hum. Genet.* **7**, 81–102.
27. Nowak, D. E., Tian, B., and Brasier, A. R. (2005) Two-step cross-linking method for identification of NF-KB gene network by chromatin immunoprecipitation. *Biotechniques* **39**, 715–725.
28. de Folter, S., Urbanus, S. L., van Zuijlen, L. G., Kaufmann, K., and Angenent, G. C. (2007) Tagging of MADS domain proteins for chromatin immunoprecipitation. *BMC Plant Biol.* 7, article 47. doi: 10.1186/1471-2229-7-47.
29. Puig, O., Caspary, F., Rigaut, G., Rutz, B., Bouveret, E., Bragado-Nilsson, E., Wilm, M., and Séraphin, B. (2001) The tandem affinity purification (TAP) method: A general procedure of protein complex purification. *Methods* **24**, 218–229.
30. Harding, E. W., Tang, W., Nichols, K. W., Fernandez, D. E., and Perry, S. E. (2003) Expression and maintenance of embryogenic potential is enhanced through constitutive expression of *AGAMOUS-Like 15*. *Plant Physiol.* **133**, 653–663.
31. Thakare, D., Tang, W., Hill, K., and Perry, S. E. (2008) The MADS-domain transcriptional regulator AGAMOUS-Like 15 promotes somatic embryo development in *Arabidopsis* and soybean. *Plant Physiol.* **146**, 1663–1672.
32. Zhu, C., and Perry, S. E. (2005) Control of expression and autoregulation of *AGL15*, a member of the MADS-box family. *Plant J.* **41**, 583–594.
33. Mukhopadhyay, A., Deplancke, B., Walhout, A. J. M., and Tissenbaum, H. A. (2008) Chromatin immunoprecipitation (ChIP) coupled to detection by quantitative real-time PCR to study transcription factor binding to DNA in *Caenorhabditis elegans*. *Nat. Protoc.* **3**, 698–709.
34. Gehrig, H. H., Winter, K., Cushman, J., Borland, A., and Taybi, T. (2000) An improved RNA isolation method for succulent plant species rich in polyphenols and polysaccharides. *Plant Mol. Biol. Rep.* **18**, 369–376.

Chapter 17

Visualizing and Characterizing In Vivo DNA-Binding Events and Direct Target Genes of Plant Transcription Factors

Jose M. Muiño, Gerco C. Angenent, and Kerstin Kaufmann

Abstract

Physical interactions between transcription factors and specific DNA sites are essential for gene regulation. Recent progress in genome-wide in vivo techniques, like chromatin immunoprecipitation followed by high-throughput sequencing (ChIP-SEQ), enables plant researchers to generate genome-wide, high-resolution DNA-binding maps of transcription factors. These new types of data require the use of advanced bioinformatic tools in order to understand the molecular mechanisms of functional specificity and target gene regulation by transcription factors. Here, we will review the use of a genome browser to visualize genome-wide DNA-binding maps of plant transcription factors along with other publicly available data and the program MEME to determine DNA sequence motifs in the bound regions. We also describe a tool for functional classification of target genes using GO annotations. Analysis of transcriptional regulatory networks requires the integration of multiple types of data, and this chapter aims at giving an overview about different bioinformatic approaches for meta-analysis and data integration.

Key words: Transcription factor, target genes, chromatin immunoprecipitation followed by high-throughput sequencing (ChIP-SEQ), genome browser, sequence motifs, gene ontology (GO).

1. Introduction

Molecular interactions between transcription factors (TFs) and specific DNA sequences are at the heart of gene-regulatory cascades and networks. Recent technological progress enables us to characterize in vivo protein–DNA interactions at unprecedented resolution at a genome-wide scale. In chromatin immunoprecipitation (ChIP) experiments, protein–DNA interactions in intact plant tissues are stabilized by cross-linking; the DNA is then sheared and DNA fragments that are bound by the protein

L. Yuan, S.E. Perry (eds.), *Plant Transcription Factors*, Methods in Molecular Biology 754,
DOI 10.1007/978-1-61779-154-3_17, © Springer Science+Business Media, LLC 2011

of interest are purified by immunoprecipitation using protein-specific antibodies. To identify TF-binding sites (TFBSs) at a genome-wide scale, the protein-bound DNA fragments can be recovered and then subjected to high-throughput sequencing (ChIP-SEQ) or hybridized to whole-genome tiling arrays (ChIP-CHIP) (for review, see (1)).

ChIP-SEQ and ChIP-CHIP have been successfully used to characterize DNA-binding sites of plant TFs (see, e.g. (2–6)). Protocols of the occasionally challenging, sample preparation procedures for these approaches have been described elsewhere (7, 8). Bioinformatic analysis of these data and combination with other types of genomic information can help to understand how TFs recognize their binding sites and what are the regulatory effects of DNA binding. ChIP-SEQ data analysis software, like CisGenome (9) or CSAR (10), usually generates several types of output data, among them TF DNA-binding maps in wig format files for visualization genome browsers (*see* **Section 3.1**) and lists of genes in the vicinity of the TF-binding sites (potential "direct" target genes).

TFs bind DNA in a sequence-specific manner. One important issue in the analysis of ChIP-SEQ or ChIP-CHIP data is to identify common DNA sequence elements ("motifs") that are required for binding of TFs to specific genomic sites. There are two main bioinformatic approaches to tackle this problem (1): look for enrichment of known DNA-binding sites in the genomic regions bound by a certain TF or (2) de novo DNA sequence motif discovery. For both approaches, the DNA sequences covering all genomic regions that are bound by the TF of interest need to be extracted from the ChIP-SEQ/ChIP-CHIP data. Typically, 200 to 1000 bp surrounding the peak maximum score positions are used depending on the exact purpose.

The DNA-binding preferences of only a very limited number of plant TFs have been studied using in vitro methods. DNA-binding characteristics of TFs are determined using SELEX (Systematic Evolution of Ligands by Exponential Enrichment) experiments (*see*, e.g. (11)) or by protein-binding microarrays (PBMs; (12)). The combination of SELEX and modern high-throughput sequencing techniques can improve the quality and efficacy of the approach (13). The DNA sequence motifs that are generated by SELEX or PBM experiments can be visualized in the form of consensus sequences or position weight matrices (PWMs). Information of DNA-binding preferences of TFs is accessible via databases. Commonly used databases are AGRIS (12, 13), TRANSFAC (commercial and public versions) (14), and JASPAR (15). Related TFs sharing a conserved DNA-binding domain and belonging to the same TF family often bind related or even nearly identical DNA sequences. Recent findings of PBMs also showed that a single TF can bind to multiple unrelated

DNA sequence elements in vitro (14). Analysis of overrepresentation of known TF-binding motifs regarding a control set of sequences can be done using web-based or other tools (16–19), although not all tools support plant datasets yet. Positional information of binding sites can be used to identify composite models of TF-binding motifs in genomic regions, since most regulatory processes require combinatorial binding of several TFs to *cis*-regulatory elements in promoters of target genes. The PEAKS positional footprinting web server (20) detects motifs that cluster at a particular distance from a reference element, for instance a ChIP-SEQ peak maximum position (*see* **Section 3.3**).

De novo DNA motif discovery is an alternative approach to characterize genomic regions bound by a TF of interest. Commonly used tools for de novo motif discovery are MEME (21) (*see* **Section 3.2**) and MatrixReduce (22). Both are available as web server implementations, however, with limitations concerning sequence dataset size. Here, we will introduce the use of the MEME web server.

Another line of research in analyzing TF-binding data is the bioinformatic characterization of genes associated with TF-binding events (potential direct targets). Usually the gene that is closest to the TF-binding site is taken for these analyses, but it cannot be excluded that genes more distant from the binding site are (also) targets of the TF. Gene products can be classified according to their molecular function, the biological process in which they are involved, or their localization. The Gene Ontology (GO) terminology aims at comprehensive and standardized functional classification of gene products. Currently, approximately 75% of all *Arabidopsis* gene products have annotated biological processes, and 70% have predicted or described molecular functions and localization (23). These data provide valuable information on the functions of specific sets of TF target genes. A variety of tools are available to test for overrepresentation of genes with specific localization or function within sets of genes, for instance BiNGO (24) (*see* **Section 3.4**) or AmiGO! (25).

One of the challenges of TF DNA-binding data analysis is to identify the genes that are regulated by individual DNA-binding events, making it necessary to link TF-binding data with gene expression datasets which measure changes in gene expression in vivo upon induction or depletion of the TF. TF-specific time-series gene expression data are most valuable for the analysis (*see*, e.g. (5)). Combining DNA motif discovery and information on the up- or downregulation of target genes can lead to a deeper understanding of molecular modes of action and specificity of particular TFs. A large amount of gene expression data have been published particularly for *Arabidopsis thaliana*. There are many web resources to access and analyze these datasets (26).

Finally, it is important to keep in mind that not all DNA-binding events may have a measurable effect on the expression of nearby genes, and at the same time also physically distant TF-binding sites, e.g., at enhancer elements, may be essential for gene regulation. Eukaryotic genes often require the combinatorial assembly of several TFs along their promoters in order to induce/repress transcription in a spatially and temporally correct manner, so the regulatory effect may not be immediately measurable after DNA binding of a TF. In order to prove the regulatory effect of specific DNA-binding events, validation by wetlab experiments is essential, like site-directed mutagenesis to abolish specific TF-binding sites or to change their genomic position as well as reporter gene assays to determine the necessity of a specific TF.

2. Materials

A computer which has the Java Runtime Environment (JRE) version 1.5 or higher and a web browser (e.g., FireFox) installed is needed. The computer should have a fast Internet connection available when required. Java is available for many operating systems, including Windows, Linux, and Macintosh. Java software is free and can be downloaded from http://java.com/. Most computers have a preinstalled web browser – if this is not the case, download one (e.g., Firefox from http://www.mozilla.com). The amount of RAM memory required depends on the amount of data intended to view on the genome browser; a minimum of 1 GB RAM is recommended.

Download and install Cytoscape and BiNGO on your computer. Cytoscape can be downloaded from the webpage http://cytoscape.org/download_list.php and installed locally using standard settings. After installing Cytoscape, the BiNGO.jar file can be downloaded and saved in the /plugins directory under your Cytoscape installation directory (http://www.psb.ugent.be/cbd/papers/BiNGO/).

3. Methods

3.1. ChIP-SEQ Data Visualization Using the IGB Genome Browser

Genome browsers can be used to visualize and explore many types of genome-scale datasets, independent of how they were produced (e.g., tiling array data, next-generation sequencing results, or in silico predictions), along with genome coordinates and corresponding DNA sequence as well as gene annotation. In such a way, it is possible to simultaneously visualize binding maps from ChIP-SEQ experiments, microarray expression data, nucleotide conservation obtained through DNA sequence comparison of different species or ecotypes, and positions of specific DNA motifs.

Currently available genome browsers are mostly implemented as part of web-based or stand-alone browsers (27). Stand-alone software allows the users to visualize data on their own computers without the need of data transfer to a web server, which allows a more flexible and faster analysis. The Integrated Genome Browser (IGB) is a well-documented stand-alone browser that incorporates useful tools for visualization and analysis. The most recent *Arabidopsis* gene annotations are supported (TAIR7, 8, and 9). It also supports the visualization of *Arabidopsis* TF–DNA binding maps provided by the web server PRI-CAT (10). Here we will describe the most common features that the user usually will need; for more detailed information, a complete software manual can be found at http://genoviz.sourceforge.net/IGB_User_Guide.pdf.

3.1.1. Launching and Setting IGB to Work with TAIR9 and PRI-CAT

1. Execute IGB URL http://igb.bioviz.org/download.shtml;
2. In the bottom panel, "Data access" tab. Select "Arabidopsis thaliana" and "A_thaliana_Jun_2009" in the "Choose" field.
3. In case that the DAS servers "Bioviz (Quickload)" and "PRI-CAT" are not shown on the tree located in the bottom-left corner of the "Data access" tab, click on "Configure…" button. In the new window, click "Add" and insert the next information for each server.

   Bioviz

   Name: Bioviz

   Type: QuickLoad

   URL:http://bioviz.org/quickload/

   PRI-CAT

   Name: PRI-CAT

   Type: QuickLoad

   URL:http://www.ab.wur.nl/pricat/quickload/
4. Open the tree for Bioviz (QuickLoad) server and mark the desired annotations to be visualized (e.g., TAIR9 mRNA). Be sure that in the right panel the "Choose Load Mode" is set to "Whole Genome" for the desired features.
5. Open the tree for PRI-CAT (QuickLoad) server and mark the desired annotations to be visualized (e.g., AP1-BindingSites). Be sure that in the right panel the "Choose Load Mode" is set to "Whole Genome" for the desired

features. *Warning*: Be always sure that all the features/annotations that you visualize correspond to the same genome version. Mixing TAIR8 and TAIR9 can create positional errors due to changes in the chromosome lengths, and the gene annotation can differ.

6. Load the DNA sequence for the genomic region shown on the screen: click "Load Sequence in View." To load all DNA sequences of the genome, click "Load All Sequence." The latter option is computationally more expensive.
7. Load your own data for visualization: File→Open File. In the new window browse the file that you want to load. IGB is able to visualize a variety of file formats, among them .wig files as a standard format for visualization of genomic data. For the visualization of ChIP-SEQ data, wig files are generated by the Bioconductor package CSAR (*C*hIP-*S*EQ *a*nalysis using *R*; (10)). Other ChIP-SEQ data analysis packages, like CisGenome (9) produce .bar files for visualization. Genomic annotations are usually represented in gff format, and FASTA files are used to represent DNA sequences in the IGB.

3.1.2. Example of Use

In this example we will try to identify a DNA sequence within the *SOC1* promoter to which the transcription factor APETALA1 (AP1) binds (*see* **Note 1**):

1. Open IGB as described before. Load the "TAIR9 mRNA" annotation from the Bioviz server and the AP1 binding map (AP1_Replicate1) and the DNA sequence variation from the 80 *Arabidopsis* ecotypes (Athaliana ecotypes DNA variation) from the PRI-CAT server.
2. Search for the *SOC1* locus: Click the "Search" tab on the bottom panel. Select "Matching IDs" and "genome" as options and write in the blank box the TAIR locus identifier of the gene of interest (AT2G45660.1 in this example). Click "Search." If needed, select the annotation of interest between the search results provided in the bottom panel of the genome browser by double-clicking on the locus search result. You will see the locus in the browser window, and you can modify position and zoom of the visible region. An overview of the genomic region with ChIP-SEQ peaks and variation between ecotypes is shown in **Fig. 17.1a**.
3. Load the DNA sequence for the genomic region in the screen as explained in step 6.
4. Search for the CArG box consensus. On the Search tab, select this time "Matching residues" and "chr2" as options. Write in the blank box the regular expression describing

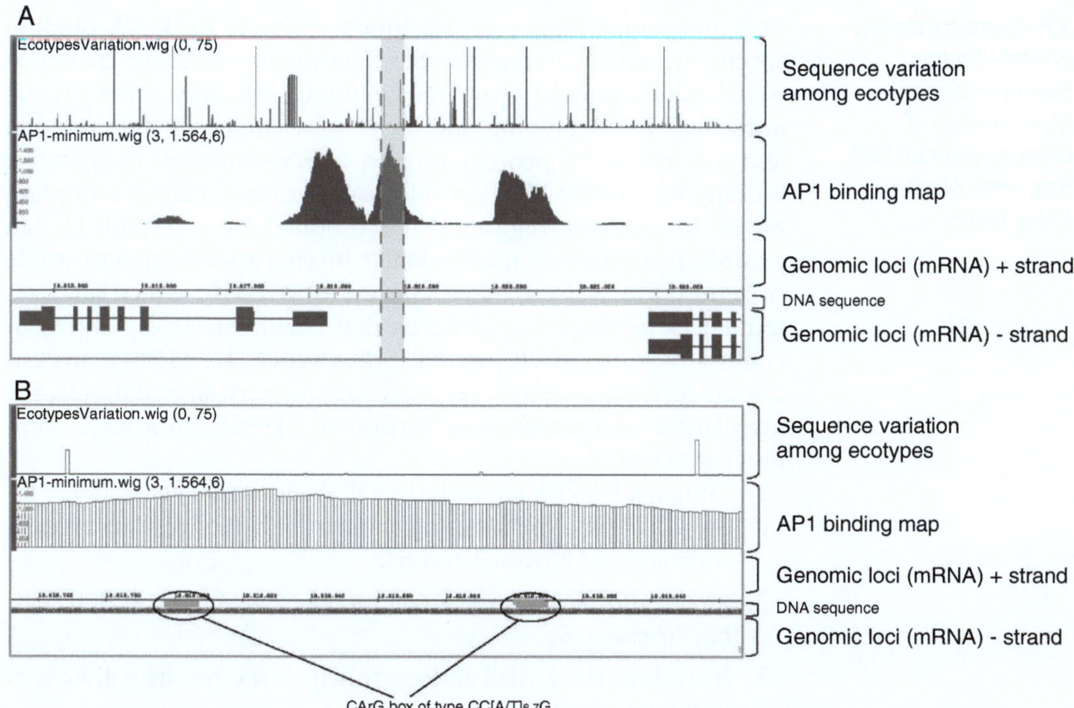

Fig. 17.1. IGB genome browser images at different spatial resolution. (**a**) AP1 ChIP-SEQ peaks and ecotype sequence variation in the *SOC1* promoter. The *shaded part* corresponds to the genomic region enlarged in (**b**). (**b**) Zoom-in of a region comprising a single ChIP-SEQ peak. Two CArG boxes of the consensus CC[AT] {6,7} G are present in the peak core, indicated above the DNA sequence as *grey boxes*.

the DNA consensus sequence of interest, for example, the expression CC[AT] {6,7}G will search for DNA sequences starting by two Cs, ending by G with a tract of length between 6 and 7 adenines and/or thymines in both strands of the loaded DNA sequence. For more information about regular expressions of DNA bases supported by IGB refer to (28). In **Fig. 17.1b**, you can see that two CArG boxes of the indicated consensus sequence are present in a selected ChIP-SEQ peak in the *SOC1* promoter.

5. The nucleotide sequence variability can be used to give more evidence on the functionality of the predicted binding sites. In our example, the complete genomic region comprising both CArG boxes is largely conserved among different ecotypes (**Fig. 17.1b**).
6. Select the DNA sequence of interest in the genome browser and click "Copy Selected Residues to Clipboard" from the "Edit" menu. Now you can paste your sequence in any other software that supports the "copy and paste" function for further analyses.

3.2. Characterization of DNA-Binding Events In Vivo I: Identification of Overrepresented DNA Sequence Motifs Using MEME

One important aspect of TF functionality is its DNA-binding specificity, which determines the "regulatory code" of transcriptional regulation. In vivo DNA binding not only relies on the individual DNA-binding specificity of a single TF but is often also influenced by protein–protein interactions and cooperative binding with other TFs. For this reason, analysis of DNA-binding motifs in genomic regions that are bound by a certain TF can provide insights into its molecular in vivo specificity and mode of action. We will introduce the use of MEME, which analyzes sequences for sequence patterns that occur repeatedly and generates a motif for each pattern it discovers (21). MEME analysis can be done via a web server; however, for very large datasets, the installation of a stand-alone version of MEME on a local computer is required.

1. Obtain a FASTA file with the DNA sequences associated with the ChIP-SEQ peaks using standard ChIP-SEQ data analysis software (*see* **Notes 2** and **3**).
2. Open the URL http://meme.sdsc.edu/meme4_3_0/cgi-bin/meme.cgi.
3. Introduce the e-mail address to which the results will be sent in the required box.
4. Upload your (masked) DNA sequence dataset in FASTA format (*see* **Note 4**). The dataset should contain less than 300 sequences and 60,000 nucleotides in total. For instance, use the top 300 peaks with 200 nucleotides surrounding the peak maximum score positions.
5. Mark the option "Zero or one per sequence" if you expect the motif to occur not more than once in each sequence (assuming one DNA-binding event per peak).
6. Click "Start search."
7. As a control, repeat steps 1–5 and mark the option "Shuffle sequence letters."

3.3. Characterization of DNA-Binding Events In Vivo II: PEAKS Analysis of TF-binding Sites and Promoter Structure

A next step in the analysis of genomic regions bound by a TF of interest is to characterize the spatial distribution and co-occurrence of individual DNA sequence motifs. For this purpose, we will introduce the PEAKS positional footprinting server (20). PEAKS can be used to study the overrepresentation of TF motifs, as well as their spatial distribution along the sequence. We will exemplify the motif characterization of ChIP-SEQ data with this program.

1. Obtain a FASTA file with the DNA sequences associated with the ChIP-SEQ peaks. Alternatively, if you aim at characterizing promoter sequences of a set of TF target genes, you can download them from TAIR (29) (*see* **Note 5**).

2. Open Peaks positional footprinting server in your web browser (http://evolutionarygenomics.imim.es/pf/pf_input.php) and upload the FASTA file.
3. Enter the "reference element," e.g., the peak score maximum position, which is usually in the center of the sequences in case of ChIP-SEQ peak data (e.g., position 500 in a 1000 bp FASTA sequence). If a set of promoter sequences is used, you can use the transcriptional start site (TSS) as reference element (e.g., position 1000 in a 1000 bp long promoter sequence extracted from TAIR) (*see* **Note 6**).
4. Choose the matrices that you want to use (e.g., Transfac 7 library-plant), mention your e-mail, and click "LOAD DATASET." In the next window, click "SUBMIT to ANALYSIS," and you will get your results by e-mail. An example output is given in **Fig. 17.2**.

3.4. Functional Classification of TF Target Genes Using BiNGO

The Gene Ontology (GO) project gathers knowledge on gene function using a controlled terminology that is applicable to all organisms (27). The current classification consists of three hierarchically structured ontologies that describe gene products in terms of their associated biological processes, molecular functions, and cellular components.

A variety of tools for analyzing GO data have been developed (30), and we will exemplify the use of GO classification in the functional characterization of specific gene sets, e.g., TF target genes, with BiNGO, which is an open-source Java tool implemented as a plug-in to the Cytoscape software environment. We will use typical settings, which, however, can be varied according to specific aspects of individual datasets (e.g., dataset size) (*see* **Note 7**).

1. *Open BiNGO*: Start the Cytoscape software and select BiNGO from the "Plugins" menu.
2. *Analysis*: In the BiNGO settings window, choose and enter a name of the data file, select "Paste genes from text." Then, copy a list of TAIR locus identifiers of the gene set to be analyzed into the box below. The list should be formatted so that each identifier is in a separate row. Select "Hypogeometric test," "Benjamini & Hochberg False Discovery Rate (FDR) correction," and a significance level of 0.05. Select to show "Overrepresented categories after correction" and "Test cluster versus whole annotation." Select "GO_slim plants" or alternatively any of the standard complete GO categories (e.g., "GO_Biological_Process") (*see* **Note 8**). Then, select "Arabidopsis thaliana" as organism. Check the box for saving data and select a folder in which the results files should be saved.

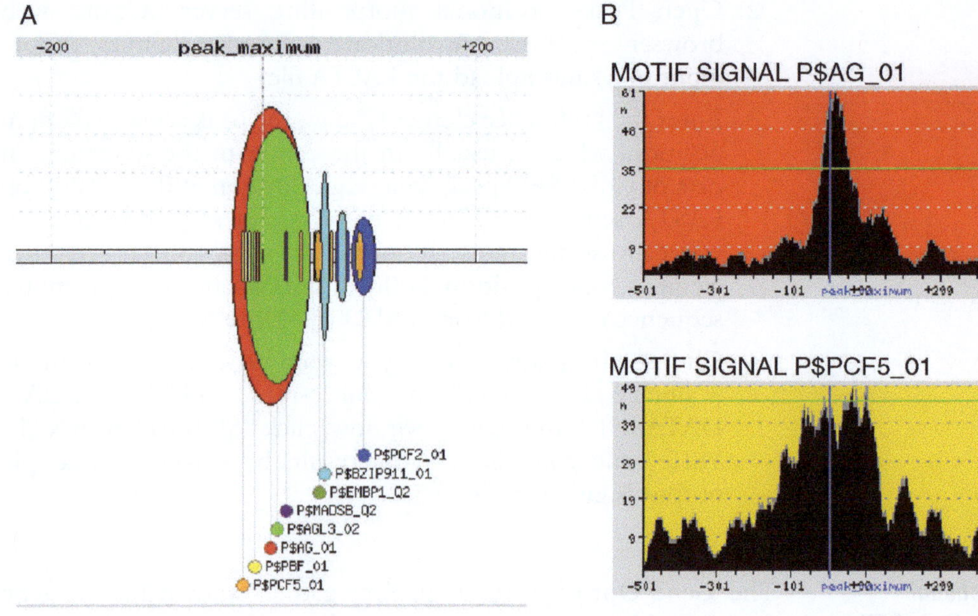

C

RANK	MOTIF	SCORE	PVAL	MAX PEAK		RANGES		
				n seq.	position	n	size	positions
1	P$AG_01	68.2	1.063e-36	61	+8	1	75	-28..+46
2	P$AGL3_02	44.4	4.007e-31	62	+19	1	65	-17..+47
3	P$PCF2_01	14.6	5.905e-11	49	+98	3	27	+71..+108
4	P$PCF5_01	11.9	2.573e-9	49	+57	6	23	-18..+96
5	P$PBF_01	24.6	1.781e-8	360	-16	3	7	-16..-6
6	P$MADSB_Q2	16.1	1.964e-8	72	+26	1	4	+23..+26
7	P$EMBP1_Q2	10.5	2.930e-8	78	+55	2	5	+51..+56
8	P$BZIP911_01	2.9	3.240e-8	7	+58	2	20	+56..+81

Fig. 17.2. Typical PEAKS output. The dataset comprised the 1001 bp surrounding the ChIP-SEQ peak score maximum position of the top 1000 SEP3 peaks in wild-type inflorescences taken from (4). (**a**) Significant motifs and ranges (score *P*-value <1e-7), (**b**) examples of profiles of individual motifs, and (**c**) detailed table of motif positions represented in (**a**).

3. *Interpretation of the results*: The output of the analysis will appear in a new window in the form of a table as well as a network visualization which shows the dependencies of different GO categories. In the table, you will see a list of overrepresented GO categories, along with results of the statistical analysis and the loci corresponding to each category. In some cases, the number of loci that can be identified is very low. We usually consider only categories that are identified as overrepresented with a minimum number of five loci in the dataset. If the output window is empty, no GO categories are overrepresented using the given settings. If this is the case, you may try to modify the input gene lists and/or the analysis settings.

4. Notes

1. MADS domain containing transcription factors, such as AP1, binds to *cis* motifs with a consensus sequence consisting of (C)C-A/T rich stretch-G(G) and these motifs are called "CArG boxes."
2. FASTA files of ChIP-SEQ peak regions need to be generated for MEME and PEAKS: these are the standard output of CisGenome (9). Alternatively, they can be generated by simple scripts using positional information that is the output of other kinds of ChIP-SEQ data analysis software, like CSAR.
3. You can change the vertical position of individual panels in the genome browser window by "drag and drop" using your computer mouse cursor.
4. For de novo motif discovery using MEME or other software, it is usually recommended to filter the DNA sequence dataset for repeated and low-complexity regions. This can be done using REPEATMASKER (31).
5. In the PEAKS software, you have the option to change the sliding window size – test the effect of changes in this parameter to identify motifs that have more/less strong positional enrichment. The larger the window size, the more likely you will identify motifs that have a less precise location.
6. In our experience, PEAKS performs better when using larger sequence datasets (e.g., using the sequences of several hundred top ChIP-SEQ peaks).
7. In BiNGO, you can create a custom annotation file in order to show enrichment GO categories in specific subsets of target gene lists. For instance, use all genes with a TF-binding event to create a new customs annotation file and then test for overrepresentation of GO categories in a sub-dataset, e.g., genes that are identified as differentially expressed in a microarray or are common targets with another TF. A description of how to create custom annotation files can be found at http://www.psb.ugent.be/cbd/papers/BiNGO/annotations.htm.
8. In BiNGO, you can use different types of annotation. GO_slim categories contain a subset of the terms in the whole GO, thus giving an overview of the ontology content.

References

1. Park, P. J. (2009) ChIP-seq: advantages and challenges of a maturing technology. *Nat. Rev.* **10**, 669–680.
2. Morohashi, K., and Grotewold, E. (2009) A systems approach reveals regulatory circuitry for Arabidopsis trichome initiation by the GL3 and GL1 selectors. *PLoS Genet.* **5**, e1000396.
3. Zheng, Y., Ren, N., Wang, H., Stromberg, A. J., and Perry, S. E. (2009) Global identification of targets of the Arabidopsis MADS domain protein AGAMOUS-Like15. *Plant Cell* **21**, 2563–2577.
4. Kaufmann, K., Muiño, J. M., Jauregui, R., Airoldi, C. A., Smaczniak, C., Krajewski, P., and Angenent, G. C. (2009) Target genes of the MADS transcription factor SEPALLATA3: integration of developmental and hormonal pathways in the Arabidopsis flower. *PloS Biol.* 7, e1000090.
5. Kaufmann, K., Wellmer, F., Muiño, J. M., Ferrier, T., Wuest, S. E., Kumar, V., Serrano-Mislata, A., Madueno, F., Krajewski, P., Meyerowitz, E. M., Angenent, G. C., and Riechmann, J. L. (2010) Orchestration of floral initiation by APETALA1. *Science (New York, NY)* **328**, 85–89.
6. Mathieu, J., Yant, L. J., Murdter, F., Kuttner, F., and Schmid, M. (2009) Repression of flowering by the miR172 target SMZ. *PLoS Biol.* 7, e1000148.
7. Morohashi, K., Xie, Z., and Grotewold, E. (2009) Gene-specific and genome-wide ChIP approaches to study plant transcriptional networks. *Methods Mol. Biol. (Clifton, NJ)* **553**, 3–12.
8. Kaufmann, K., Muiño, J. M., Osteras, M., Farinelli, L., Krajewski, P., and Angenent, G. C. (2010) Chromatin immunoprecipitation (ChIP) of plant transcription factors followed by sequencing (ChIP-SEQ) or hybridization to whole genome arrays (ChIP-CHIP). *Nat. Protoc.* **5**, 457–472.
9. Ji, H., Jiang, H., Ma, W., Johnson, D. S., Myers, R. M., and Wong, W. H. (2008) An integrated software system for analyzing ChIP-chip and ChIP-seq data. *Nat. Biotechnol.* **26**, 1293–1300.
10. Muiño, J. M., Hoogstraat, M., van Ham, R. C. H. J., and van Dijk, A. D. J. (2011) PRI-CAT: A web-tool for the analysis, storage and visualization of plant ChIP-seq experiments. *Nucleic Acids Res.* (In press)
11. Bouvet, P. (2001) Determination of nucleic acid recognition sequences by SELEX. *Methods Mol. Biol. (Clifton, NJ)* **148**, 603–610.
12. Berger, M. F., and Bulyk, M. L. (2009) Universal protein-binding microarrays for the comprehensive characterization of the DNA-binding specificities of transcription factors. *Nat. Protoc.* **4**, 393–411.
13. Jolma, A., Kivioja, T., Toivonen, J., Cheng, L., Wei, G., Enge, M., Taipale, M., Vaquerizas, J. M., Yan, J., Sillanpaa, M. J., Bonke, M., Palin, K., Talukder, S., Hughes, T. R., Luscombe, N. M., Ukkonen, E., and Taipale, J. (2010) Multiplexed massively parallel SELEX for characterization of human transcription factor binding specificities. *Genome Res.* **20**, 861–873.
14. Matys, V., Kel-Margoulis, O. V., Fricke, E., Liebich, I., Land, S., Barre-Dirrie, A., Reuter, I., Chekmenev, D., Krull, M., Hornischer, K., Voss, N., Stegmaier, P., Lewicki-Potapov, B., Saxel, H., Kel, A. E., and Wingender, E. (2006) TRANSFAC and its module TRANSCompel: transcriptional gene regulation in eukaryotes. *Nucleic Acids Res.* **34**, D108–D110.
15. Portales-Casamar, E., Thongjuea, S., Kwon, A. T., Arenillas, D., Zhao, X., Valen, E., Yusuf, D., Lenhard, B., Wasserman, W. W., and Sandelin, A. (2010) JASPAR 2010: the greatly expanded open-access database of transcription factor binding profiles. *Nucleic Acids Res.* **38**, D105–D110.
16. Hestand, M. S., van Galen, M., Villerius, M. P., van Ommen, G. J., den Dunnen, J. T., and 't Hoen, P. A. (2008) CORE_TF: a user-friendly interface to identify evolutionary conserved transcription factor binding sites in sets of co-regulated genes. *BMC Bioinformatics* **9**, 495.
17. Frith, M. C., Fu, Y., Yu, L., Chen, J. F., Hansen, U., and Weng, Z. (2004) Detection of functional DNA motifs via statistical over-representation. *Nucleic Acids Res.* **32**, 1372–1381.
18. Chekmenev, D. S., Haid, C., and Kel, A. E. (2005) P-Match: transcription factor binding site search by combining patterns and weight matrices. *Nucleic Acids Res.* **33**, W432–W437.
19. Haudry, Y., Ramialison, M., Paten, B., Wittbrodt, J., and Ettwiller, L. (2010) Using Trawler_standalone to discover overrepresented motifs in DNA and RNA sequences derived from various experiments including chromatin immunoprecipitation. *Nat. Protoc.* **5**, 323–334.
20. Bellora, N., Farre, D., and Mar Alba, M. (2007) PEAKS: identification of regulatory motifs by their position in DNA

sequences. *Bioinformatics (Oxford, England)* **23**, 243–244.
21. Bailey, T. L., Williams, N., Misleh, C., and Li, W. W. (2006) MEME: discovering and analyzing DNA and protein sequence motifs. *Nucleic Acids Res.* **34**, W369–W373.
22. Foat, B. C., Morozov, A. V., and Bussemaker, H. J. (2006) Statistical mechanical modeling of genome-wide transcription factor occupancy data by MatrixREDUCE. *Bioinformatics (Oxford, England)* **22**, e141–e149.
23. www.arabidopsis.org
24. Maere, S., Heymans, K., and Kuiper, M. (2005) BiNGO: a Cytoscape plugin to assess overrepresentation of gene ontology categories in biological networks. *Bioinformatics (Oxford, England)* **21**, 3448–3449.
25. Carbon, S., Ireland, A., Mungall, C. J., Shu, S., Marshall, B., and Lewis, S. (2009) AmiGO: online access to ontology and annotation data. *Bioinformatics (Oxford, England)* **25**, 288–289.
26. Brady, S. M., and Provart, N. J. (2009) Web-queryable large-scale data sets for hypothesis generation in plant biology. *Plant Cell.* **21**, 1034–1051.
27. Nielsen, C. B., Cantor, M., Dubchak, I., Gordon, D., and Wang, T. (2010) Visualizing genomes: techniques and challenges. *Nat. Methods* 7, S5–S15.
28. http://java.sum.com/j2se/1.5.0/docs/api/java/util/regex/Pattern.html.
29. http://www.arabidopsis.org/tools/bulk/sequences.
30. http://www.geneontology.org.
31. Tarailo-Graovac, M., and Chen, N. (2009) Using RepeatMasker to identify repetitive elements in genomic sequences. *Current Protocols in Bioinformatics* 4.10.1–4.10.14.

Chapter 18

Mapping In Vivo Protein–DNA Interactions in Plants by DamID, a DNA Adenine Methylation-Based Method

Sophie Germann and Valérie Gaudin

Abstract

DamID (DNA adenine methylation identification) is an adenine methylation-based tagging method designed to map protein–DNA interactions in vivo. DamID, an alternative method to chromatin immunoprecipitation (ChIP), is based on the covalent linking of a "fingerprint" in the vicinity of the DNA-binding sites of the protein of interest. The fingerprints can be further mapped by simple molecular approaches. First developed by van Steensel's group in *Drosophila melanogaster* (1), DamID was successfully adapted to *Arabidopsis thaliana,* and its feasibility demonstrated by using the well-known yeast GAL4 transcription factor (2). The method was further used to establish a genome-wide map of the target sites of LHP1, a regulatory chromatin protein in *A. thaliana* (3).

Key words: In vivo protein–DNA interactions, DamID, adenine methylation, chromatin profiling, *Arabidopsis thaliana.*

1. Introduction

Interactions between proteins and DNA are essential in nuclear processes, such as chromosome segregation, DNA replication and recombination, and in transcriptional control mechanisms. To study protein–DNA interactions within dynamic chromatin environments, the Dam identification (DamID) method was successfully used in several model organisms in both the animal and plant kingdoms, such as human, mouse, fly, yeast, and *Arabidopsis thaliana* (4, 5). DamID allowed the mapping of genomic targets of several proteins (2, 3, 6–22).

In the DamID technique, the protein of interest, which can be either a specific transcription factor, a co-factor, or a

L. Yuan, S.E. Perry (eds.), *Plant Transcription Factors*, Methods in Molecular Biology 754,
DOI 10.1007/978-1-61779-154-3_18, © Springer Science+Business Media, LLC 2011

chromatin-associated protein, is fused to the prokaryotic adenine methyltransferase protein (Dam), which methylates adenine at the N^6-position within the GATC DNA sequence (23). Produced transiently or in transgenic organisms, the Dam fusion protein is tethered to the native genomic binding sites of the protein of interest and leaves specific $G^{6m}ATC$ methylation fingerprints in the vicinity, which can then be analyzed by methylation-sensitive restriction enzymes (i.e., *Dpn*I and *Dpn*II, which cut and do not cut the $G^{6m}ATC$ sites, respectively). This method takes advantage of the low level or the absence of adenine methylation in eukaryotes and of a suitable GATC motif distribution in the eukaryotic genomes (one site every 200–400 bp on average), thus allowing probing almost every region of the genome. Finally,

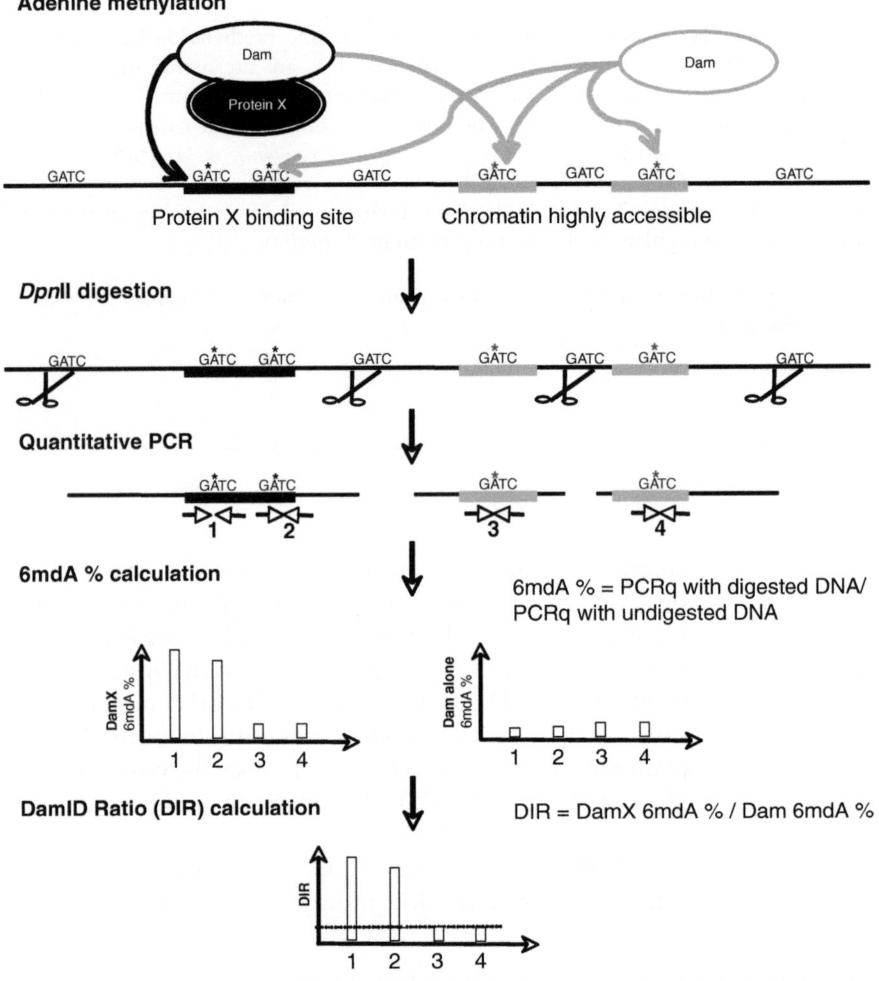

Fig. 18.1. Principle for quantification of the methylation level by qPCR. Methylation at target GATC sites (*black box*) and non-target GATC sites (*grey box*). * Methylated adenine.

Dam methylation has minor effects on the overall topology of the DNA (24).

A few DamID and ChIP experiments, performed side by side to study the same proteins, revealed that both methods provide extensively overlapping data (6, 13, 25). Furthermore, statistical methods have been developed to score overlaps between ChIP and DamID profiles, which encourage direct comparisons of profiles obtained with both methods (26). The appropriate technique can be chosen according to the DamID- and ChIP-specific advantages and constraints, the model system, and the protein of interest. DamID has the great advantage of allowing detection of the direct genomic targets of any given factor without the

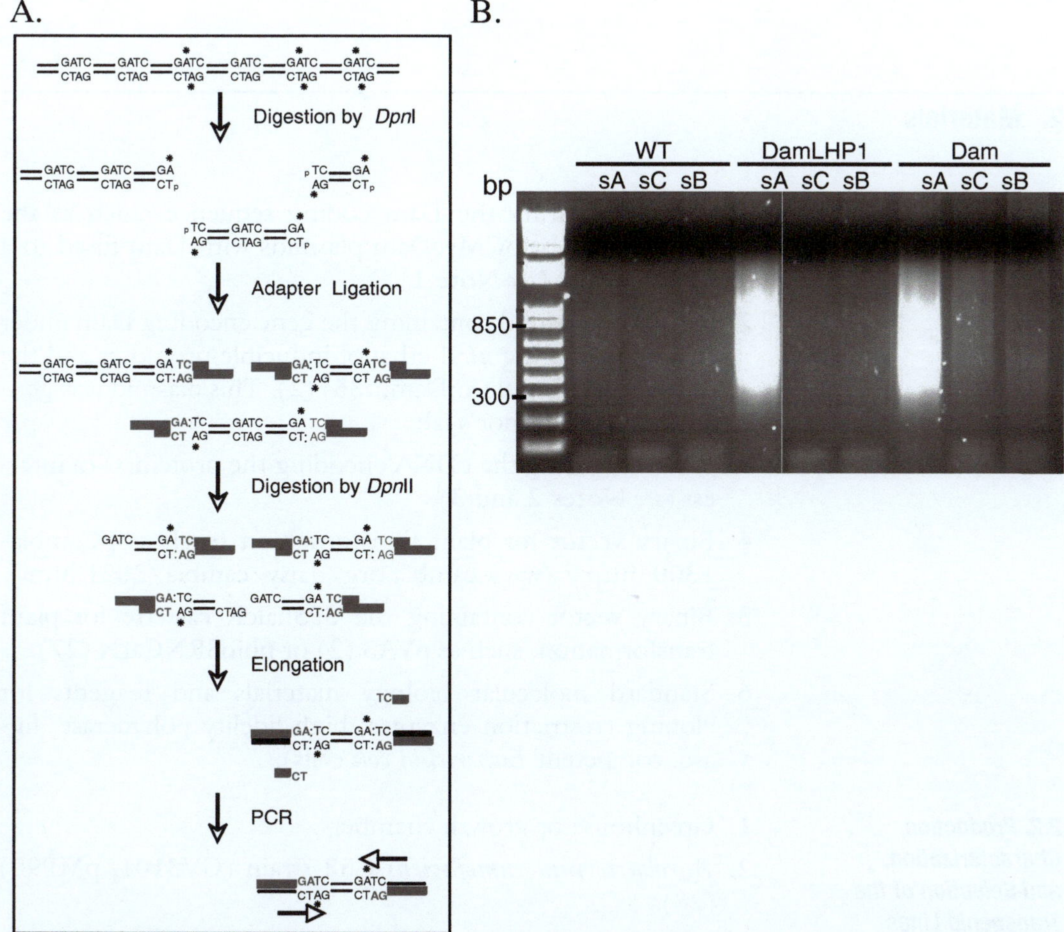

Fig. 18.2. Selective amplification of Dam-methylated DNA for microarray hybridization. **a**. Schematic representation of the method. Double-stranded adapter in gray; * Methylated adenine; $_p$ phosphorylated 5′-end; : covalent link. **b**. PCR products after selective amplification starting from DNA extracted from wild-type plants (WT) and DamLHP1 or Dam plants. No PCR product is obtained in control experiments without *Dpn*I (sC) or without ligase (sB), whereas a smear is obtained in sample A (sA).

need for DNA cross-linking (which can lead to artifactual interactions) or the use of an antibody. Furthermore, as methylation is via a covalent bond, interaction events can be recorded by the deposition of this stable fingerprint. Maps of proteins that associate transiently, indirectly, or loosely with the genome can be easily established. Finally, DamID requires only simple molecular biology techniques (DNA extraction, enzymatic digestion, and ligation) and it is used to analyze protein–DNA interaction not only at specific loci by coupling with selective real-time quantitative PCR (qPCR) (**Fig. 18.1**) but also at a genome-wide scale by coupling with microarray hybridization (**Fig. 18.2**). The development of high-throughput sequencing technologies might facilitate genome-wide DamID profiling.

2. Materials

2.1. Plasmid Constructions

1. Plasmids bearing the Dam coding sequence (such as the pNDamMyc or pCMycDam plasmids with Dam fused to a Myc epitope) (*see* **Note 1**).
2. pL4-Dam plasmid containing the gene encoding Dam under the control of the *alcA* ethanol-inducible promoter and the 35S terminator (alcA::Dam::t35) (2). This plasmid was generated in the author's lab.
3. Plasmid bearing the cDNA encoding the protein(s) of interest (*see* **Notes 2** and **3**).
4. Binary vector for plant transformation (such as pCambia-1300, http://www.cambia.org/daisy/cambia/2021.html).
5. Binary vector containing the 35S::alcR cassette for plant transformation, such as pVA3 (2) or pbinSRNCatN (27).
6. Standard molecular biology materials and reagents for cloning (restriction enzymes, high-fidelity polymerase, ligase, competent *Escherichia coli* cells).

2.2. Production, Characterization, and Selection of the Transgenic Lines Expressing Dam Fusion Proteins

1. Greenhouse or growth chambers.
2. *Agrobacterium tumefaciens* C58 strain (GV3101, pMP90) (28).
3. 1.2% (w/v) *Arabidopsis* medium (DU0742.0025; Duchefa, Haarlem, The Netherlands) supplemented with 2 mM $Ca(NO_3)_2$, pH 5.8, and 0.7% (w/v) agarose.
4. Antibiotics (hygromycin B, kanamycin; Duchefa, Haarlem, The Netherlands).

5. RNA isolation kit: GenElute Mammalian Total RNA Miniprep Kit (Sigma-Aldrich, Saint-Louis, MO, USA) or RNeasy Mini Kit (Qiagen).
6. RNase-free DNase I.
7. RNaseOUT^{TM} ribonuclease inhibitor.
8. SuperScript II reverse transcriptase (Invitrogen, Carlsbad, CA, USA).
9. Primers to assess the expression of the Dam fusions and the elongation factor 1-alpha gene (*EF-1α*, At5g60390), respectively: Dam05 (5′-AAACAAATTGCGCCGAGGTTT-3′)/Dam06 (5′-CCGAACGGCACGTTAAACTCA-3′); EF1F (5′-CTGGAGGTTTTGAGGCTGGTAT-3′)/EF1R 5′-CCAAGGGTGAAAGCAAGAAGA-3′).

2.3. DNA Preparation

1. Mortar, pestle, and liquid nitrogen or a rotative plastic pestle and sand.
2. CTAB (cetyltrimethylammonium bromide) buffer: 2% CTAB, 1.4 M NaCl, 20 mM EDTA, 100 mM Tris–HCl, pH 8, 0.2% β-mercaptoethanol (to add immediately before use).
3. TE buffer: 10 mM Tris–HCl, pH 8, 1 mM EDTA.
4. Chloroform (Sigma-Aldrich, Saint-Louis, MO, USA).
5. Isopropanol (Sigma-Aldrich, Saint-Louis, MO, USA).
6. Isoamyl alcohol (Sigma-Aldrich, Saint-Louis, MO, USA).
7. Lambda DNA (Invitrogen, Carlsbad, CA, USA).
8. 70% Ethanol.
9. Standard materials and equipment to run and visualize DNA agarose gels.

2.4. Quantification of Dam Methylation by Real-Time Quantitative PCR for Analysis at Selected Genomic Sites

1. *Dpn*II restriction enzyme (New England Biolabs, Beverly, MA, USA).
2. Primers surrounding the GATC sites, designed with the online Primer3 software (http://frodo.wi.mit.edu/primer3) (29) by using the following settings: amplicon size of 50–200 bp, primer length of 21 bases, optimal melting temperature of 65°C, and GC percentage between 40 and 60% (*see* **Note 4**). The chosen regions should be unique in the *A. thaliana* genome (confirmation by Blast analysis).
3. LightCycler instrument (Roche Molecular Biochemicals, Mannheim, Germany) (*see* **Note 4**).
4. Ready-to-use "Hot Start" reaction mix (LightCycler FastStart DNA Master SYBR Green I; Roche Molecular Biochemicals, Mannheim, Germany) (*see* **Note 4**).

2.5. PCR Amplification of Dam-Methylated DNA for Genome-Wide Analysis

1. *Dpn*I and *Dpn*II restriction enzymes (New England Biolabs, Beverly, MA, USA).
2. T4 DNA ligase (Roche Molecular Biochemicals, Mannheim, Germany).
3. Primers for the preparation of the AdR double-stranded adaptor (50 μM): AdRt (5′-CTAATACGACTC ACTATAGGGCAGCGTGGTCGCGGCCGAGGA-3′) and AdRb (5′-TCCTCGGCCG-3′). The two primers are not phosphorylated at the 5′-end. In an Eppendorf tube, mix 50 μL of AdRt primer (100 μM) and 50 μL of AdRb primer (100 μM), vortex, and place the tube in a beaker with boiling water. Let the beaker cool down to room temperature to anneal slowly the two primers into the AdR double-stranded adaptor.
4. AdPCR primer (5′-GGTCGCGGCCGAGGATC-3′).
5. BD Advantage^{TM} 2 PCR Enzyme System (Clontech, Palo Alto, CA, USA).
6. QIAquick PCR purification kit (Qiagen).
7. 100 and 70% Ethanol.
8. 3 M sodium acetate, pH 5.5.

3. Methods

For an unbiased and successful DamID mapping of protein–DNA interactions in vivo, it is crucial to take into account chromatin accessibility and to express Dam transgenes at very low levels. For the first requirement, adenine methylation percentages (^{6m}dA%) are measured for each genomic region of interest, both in materials producing the Dam protein alone (accessibility control) and the Dam protein fused to the protein of interest. The two percentages are then compared and a DamID ratio (DIR) is calculated, which thus takes into account chromatin structure and accessibility at each region. DIRs obtained for different genomic regions can then be compared (**Fig. 18.1**). The second requirement, a very low level of expression of the Dam transgenes, prevents saturation of the system (*see* **Note 5**), and thus target regions of the protein of interest can be distinguished from non-target regions. This condition is also particularly important in plants, since plants seem to be sensitive to adenine methylation (30). We indeed observed that plants expressing Dam at high level (^{6m}dA%>15%) present some phenotypic defects such as altered and dwarf architectures, smaller rosette leaves, decreased fertility, and premature senescence (Germann, Gaudin, unpublished). Using appropriate transgenic lines without any morphological phenotype is thus important. In *A. thaliana*, the *alcA*/AlcR ethanol-inducible system from *Aspergillus nidulans* (31, 32) or the leakiness of the *alcA* ethanol-inducible promoter without ethanol induction gave satisfactory results with DamID (2).

The step-by-step protocol presented here is based on the use of stable transgenic plants (*see* **Note 6**) expressing Dam fusion transgenes using the *alcA*/AlcR system. Transgenes encoding Dam or Dam-X proteins are placed under the control of the *alcA* promoter, which contains AlcR binding sites. These constructs are introduced into *A. thaliana*, and *alcA*::Dam and *alcA*::Dam-X transgenic lines are crossed with 35S::AlcR lines. Upon ethanol induction of the F1 transgenic lines, the AlcR factor binds to the *alcA* promoter, driving expression of the Dam transgenes. We showed that the *alcA* promoter, in the absence of the AlcR factor or in the absence of ethanol induction, was sufficiently leaky to allow DamID analyses. The presence of AlcR and short ethanol induction (up to 24 h) was also exploited to help in discriminating target versus non-target sites (2). This strategy provides great flexibility to generate appropriate methylation levels. Different methylation levels are indeed expected in heterozygous plants versus homozygous plants, in plants with and without AlcR, and finally in plants before or after induction (*see* **Note 7**). According to methylation levels on control sites, one or the other materials will be preferred.

3.1. Plasmid Constructions

1. Subclone in frame the cDNA encoding the protein of interest (X) in a plasmid containing the cDNA encoding Dam and place the transgene under the control of an inducible promoter by using standard cloning techniques (33). For example, the cDNA can be cloned in the pL4-Dam vector at the *Nco*I site (2) (*see* **Notes 1, 2,** and **3**) to generate an alcA::DamMyc-X::t35 cassette (*see* **Note 8**). Verify constructs by restriction analysis, PCR, and DNA sequencing.
2. Introduce the alcA::Dam-X::t35 (or/and alcA::X-Dam::t35) and alcA::Dam::t35 cassettes into a binary vector such as pCambia1300.

3.2. Production, Characterization, and Selection of the Transgenic Lines Expressing Dam Fusion Proteins

1. Introduce binary vectors by electroporation into *A. tumefaciens* and transform *A. thaliana* plants (34). Select transformants in vitro on *Arabidopsis* medium in the presence of the appropriate antibiotic (10 mg/L hygromycin B or 50 mg/L kanamycin) and subsequently transfer the resistant plants to soil. Perform segregation analyses to select homozygous transgenic lines for the different constructs (*see* **Note 9**).
2. Characterize the morphological phenotypes of the homozygous lines (*alcA*::Dam; alcA::Dam-X; or/and alcA::X-Dam),

on soil and in in vitro conditions, and select 4–10 lines with wild-type phenotypes for each construct.

3. Analyze the expression of the Dam and Dam fusion transgenes by quantitative RT-PCR in the selected lines. Collect material and isolate total RNA from seedlings or other plant material using RNA extraction kits. Treat 1 μg of total RNA with RNase-free DNase I and perform reverse transcription reactions with SuperScript II reverse transcriptase, according to the manufacturer's instructions. Finally, assess the expression of the Dam fusions and of the elongation factor *EF-1α* by qPCR with primers Dam05/Dam06 and EF1F/EF1R, respectively, and perform normalization using the *EF-1α* expression (*see* **Notes 5** and **10**).
4. On three to five selected lines with a normal phenotype and expressing Dam alone and Dam fusion transgenes at low levels, quantify the Dam methylation at several control GATC sites. Control GATC sites are defined as sites located in genomic regions which are not supposed to be target sites of the protein of interest or located at distance from the potential binding sites (>5–10 kbp) (proceed to **Sections 3.3** and **3.4**) (*see* **Note 11**). Compare the different lines and choose the lines with low methylation at the control sites for further analyses (*see* **Notes 12** and **13**).
5. Cross two to three homozygous transgenic lines with low methylation levels at control GATC sites for each construct with at least two AlcR homozygous transgenic lines (*see* **Note 7**). Check the Dam methylation in the F1 transgenic lines at the control GATC sites. Select F1 transgenic lines with a methylation percentage lower than 15%.
6. Proceed to the analyses of the adenine methylation at the regions of interest by qPCR or genome-wide analyses. Induction with ethanol can be performed by exposing plants to ethanol vapor, using 96.2% (v/v) ethanol, for 0–72 h, as previously described (32). Briefly, place open 500-μL microfuge tubes filled with ethanol into each pot with a plant and cover groups of four pots with transparent bag (12 cm × 12 cm × 30 cm). Upon induction, the methylation levels at target sites are expected to increase more rapidly than the methylation levels at control GATC sites.

3.3. DNA Preparation

1. Collect plant material from the transgenic lines, at precise developmental stages, with or without ethanol induction (*see* **Note 14**).
2. Extract DNA as described (35). Briefly, grind plant material in a 1.5-mL Eppendorf tube with a rotative plastic pestle in

the presence of 400 μL of CTAB buffer and sand (or in a mortar in the presence of liquid nitrogen).

3. Incubate at 60°C for 30 min. Add 400 μL of chloroform/isoamyl alcohol (24:1), vortex, and spin for 10 min at high speed. Recover the aqueous phase in a 1.5-mL Eppendorf tube, add 350 μL of isopropanol, mix and spin for 10 min at maximum speed. Remove the supernatant and wash the pellet with 600 μL of 70% ethanol, air-dry, and resuspend the pellet in water or TE buffer.
4. Load a DNA aliquot on an agarose gel and quantify by comparing to a commercial DNA solution with standardized concentrations such as lambda DNA (*see* **Note 15**).

3.4. Quantification of Dam Methylation by Real-Time Quantitative PCR for Analysis at Selected Genomic Sites

1. Extract and quantify DNA from the transgenic lines (*see* **Section 3.3**).
2. Prepare two tubes for each DNA sample. Incubate 300 ng of DNA for 1 h at 37°C without (undigested sample) and with 10 units of the adenine methylation-sensitive *Dpn*II enzyme (digested sample). After digestion, incubate the two tubes for 20 min at 80°C to inactivate the enzyme (**Fig. 18.1**).
3. Adjust the DNA concentration to 8 ng/μL for each sample.
4. Design primers surrounding the control GATC sites and the GATC sites in regions of interest and check their efficiency in qPCR with a standard dilution series of genomic DNA (at least five points of dilution covering the range of expected DNA quantity (from 15 pg to 50 ng). A LightCycler instrument can be used with the ready-to-use "Hot Start" reaction mix (LightCycler FastStart DNA Master SYBR Green I), according to the manufacturer's recommendations (*see* **Note 4**). The three steps for real-time PCR are: an initial denaturation step 95°C, 8 min; then 45 cycles as follows: 95°C 10 sec, 60°C 7 sec, 72°C 10 sec with a temperature transition rate of 20°C s^{-1}; and a final cycle as follows: 95°C 1 sec, 65°C 30 sec and an increase to 95°C with a temperature transition rate of 0.1°C s^{-1} for determination of the fusion curve. Determine the slope of the curve (p) by using the "second derivative maximum" method. Calculate the primer efficiency (E), given by the following equation: $(1 + E) = 10^{(-1/\text{slope})}$. Retain only primers with a PCR efficiency of 1.8–2.2 (i.e., 80–120%) and that do not form dimers.
5. Perform qPCR on 40 ng of DNA of the undigested and digested samples. For each selected genomic region to analyze, perform qPCR with two to three pairs of primers, according to its size and the desired resolution. Mix 5 μL DNA (40 ng), 0.5 μM of each primer (10 μM), 2 μL of

LightCycler Fast Start DNA Master SYBR Green I (10×), and 2.4 μL $MgCl_2$ (25 mM) to reach a 4 mM $MgCl_2$ final concentration and adjust the final volume to 20 μL with water.

6. Determine slope (*p*) and intercept (*i*) by using the "second derivative maximum" method and calculate the PCR product quantity (PCRq) given by the following formula: $PCRq = 10^{(Ct-i)/p}$. Calculate the adenine methylation percentage (^{6m}dA%) at a specific GATC site. ^{6m}dA% is the fraction of DNA resistant to *Dpn*II digestion and is given by the formula ^{6m}dA% = [(PCRq obtained with digested sample)/(PCRq obtained with undigested sample)]×100.
7. Calculate the DamID ratio (DIR) at a specific GATC site. The DIR corresponds to the ratio of ^{6m}dA% of the Dam fusion line to ^{6m}dA% of the Dam alone line. The DIR takes into account chromatin accessibility differences between genomic regions and allows a correct representation of the binding profile. Comparisons of DIRs obtained for different GATC sites are thus possible.

3.5. PCR Amplification of Dam-Methylated DNA for Genome-Wide Analysis

1. Extract, quantify DNA from selected transgenic lines (**Section 3.3**), and perform preliminary analyses by qPCR at control GATC sites on the selected transgenic lines (**Section 3.4**).
2. Precipitate DNA of the selected lines with 2.5 volumes of 100% ethanol and 0.1 volume of sodium acetate (3 M, pH 5.5). Place the sample for 1 h at −80°C or 2–3 h at −20°C. Spin at 4°C for 30 min, at maximum speed, remove the supernatant, and wash the pellet with 1 volume of 70% ethanol. Spin at 4°C for 5 min, at maximum speed. Remove the supernatant and air-dry the pellet. Dissolve the pellet in water or TE buffer.
3. In three Eppendorf tubes [samples A, B, and C (sA, sB, and sC)], add 2.5 μg of DNA and 1× digestion buffer. Only in samples A and B, add 10 U of *Dpn*I (which cuts only G^{6m}ATC sites). Sample C is the negative digestion control. Incubate the three tubes for 16 h at 37°C (*see* **Note 16**).
4. Incubate the three tubes at 80°C for 20 min to inactivate *Dpn*I. In all tubes, add 40 pmol of the double-stranded adaptor AdR (50 μM) and ligation buffer. In samples A and C, add 5 U of T4 DNA ligase (5 U/μL). Incubate the three tubes for 2 h at 16°C. Incubate all tubes for 10 min at 65°C to inactivate T4 DNA ligase. Sample B is a ligase-negative control (**Fig. 18.2**).
5. To prevent amplification of DNA fragments bearing unmethylated GATC sites, incubate the three samples for

1 h at 37°C with 10 U of *Dpn*II (which cuts only unmethylated GATC sites).

6. For each sample, perform PCR amplifications using 0.5 μg of the treated DNA using the BD Advantage 2 PCR Enzyme System and the AdPCR primer. The PCR conditions are as follows: 10 min at 68°C; one cycle of 1 min at 94°C, 5 min at 65°C and 15 min at 68°C; three cycles of 1 min at 94°C, 1 min at 65°C, and 10 min at 68°C; and 17 cycles of 1 min at 94°C, 1 min at 65°C, and 2 min at 68°C.
7. Purify the PCR products using QIAquick PCR purification columns according to the manufacturer's recommendations and load 5 μL of the PCR products on a 1% agarose gel. A smear between 300 and 800 bp is expected with sample A, whereas no product should be obtained in samples B and C (**Fig. 18.2**).
8. Selectively amplified DNA samples (samples A) derived from lines producing the protein of interest fused to Dam and Dam alone (control) are then used for hybridization on microarrays (*see* **Note 17**).

4. Notes

1. The following website (http://research.nki.nl/vansteensellab/damid.htm) provides useful information on Dam sequences and bibliography. Step-by-step protocols developed for animal systems can also be valuable for adaptation of the DamID method to plant systems (36–38).
2. It may be useful to have control proteins such as the protein of interest bearing mutations affecting its DNA-binding properties or a protein with already characterized binding sites.
3. It is possible to introduce a nuclear localization signal (NLS) in the fusion proteins to increase the efficiency of the targeting of the Dam fusion proteins to the nucleus (2).
4. Parameters for designing primers are indicative. In fact, different real-time quantitative PCR systems can be used, provided the selection of primers and the analyses are performed on the same machine, according to the manufacturer's recommendations.
5. *Escherichia coli* Dam methylation is a highly processive mechanism (39). It may explain why a very low quantity of the enzyme is enough to generate a detectable signal

in DamID, and why methylation levels can be observed at both target and non-target sites when Dam is expressed at high levels. In animal studies, the Dam protein was undetectable by western blot or immunolocalization. In *A. thaliana*, we observed that a sixfold difference in expression between two lines can lead to a 26-fold difference in methylation (2).

6. DamID could also probably be adapted to use transient expression in plant cells or seedlings, as already used with animal cell models, provided the same rigorous controls and quantification of methylation levels are performed.
7. It was reported in both plants and animals that the leakiness of an inducible system is sufficient to map adenine methylation. Other inducible systems or leaky promoters may be used. Although the cross with AlcR line is not necessary, this strategy provides great flexibility to generate appropriate methylation levels by performing or not ethanol induction. Furthermore, homozygous transgenic lines may accumulate methylation over generations. A cross with an AlcR transgenic line or a wild-type plant allows decreasing methylation. Therefore, working with heterozygous or homozygous plants for the transgene may be critical. However, homozygous lines before crosses may also be good material if the methylation is low enough.
8. Dam retains its methyltransferase activity whether fused to other proteins at its C- or N terminus (1). However, according to the protein of interest, one or the other orientation may be preferred.
9. Getting a large set of transgenic plants bearing only one site of T-DNA insertion for each construct is critical for further selection of transgenic lines.
10. Myc tag is present in the Dam fusion protein. The quantification of the protein levels can be performed by western blot using a commercial anti-Myc antibody. However, even if there is no detection of the protein by western blot, lines should not be discarded.
11. According to the protein of interest, control sites defined in (2) might be reused.
12. Plants with methylation rate higher than 15% should be discarded. We estimated that lines with Dam expression levels lower than those of the *EF-1α* gene by a factor of 40 and with methylation percentages at control sites of 0.5 to 3–4% were suitable for analysis.

13. If a few binding sites of the protein of interest are already known, they can be analyzed simultaneously with control sites to help in the selection of the lines.
14. To avoid variability in methylation percentages between experiments, it is important to grow plants in well-controlled conditions and to collect material on plants at the same developmental stages in biological and technical repeats and from the same parts of the organs. The use of two to three different transgenic lines expressing Dam fusion can also help to identify targets and obtain robust data.
15. New sensitive technologies (i.e., Nanodrop, QIAxcel), more widely available nowadays, may also be used for DNA quantification.
16. DNA control extracted from wild-type plants can also be used to set the digestion, ligation, and PCR conditions (**Fig. 18.2**).
17. Due to the rapid development of microarray hybridization techniques, the labeling and hybridization protocols are not described in this chapter.

Acknowledgments

The authors thank Bas van Steensel for supplying the plasmids with the Dam coding sequence, as well as for his constant assistance, advice, and support for establishment of the DamID technique in plants. We thank Nicole Houba-Hérin and Mark Tepfer for critical reading of the manuscript.

References

1. van Steensel, B., and Henikoff, S. (2000) Identification of in vivo DNA targets of chromatin proteins using tethered dam methyltransferase. *Nat. Biotechnol.* **18**, 424–428.
2. Germann, S., Juul-Jensen, T., Letarnec, B., and Gaudin, V. (2006) DamID, a new tool for studying plant chromatin profiling in vivo, and its use to identify putative LHP1 target loci. *Plant J.* **48**, 153–163.
3. Zhang, X., Germann, S., Blus, B. J., Khorasanizadeh, S., Gaudin, V., and Jacobsen, S. E. (2007) The *Arabidopsis* LHP1 protein colocalizes with histone H3 Lys27 trimethylation. *Nat. Struct. Mol. Biol.* **14**, 869–871.
4. van Steensel, B., and Henikoff, S. (2003) Epigenomic profiling using microarrays. *Biotechniques* **35**, 346–350, 352–354, 356–357.
5. Southall, T. D., and Brand, A. H. (2007) Chromatin profiling in model organisms. *Brief Funct. Genomics Proteomics* **6**, 133–140.
6. Moorman, C., Sun, L. V., Wang, J., de Wit, E., Talhout, W., Ward, L. D., Greil, F., Lu, X. J., White, K. P., Bussemaker, H. J., and van Steensel, B. (2006) Hotspots of transcription factor colocalization in the genome of *Drosophila melanogaster*. *Proc. Natl. Acad. Sci. USA* **103**, 12027–12032.

7. van Steensel, B., Delrow, J., and Henikoff, S. (2001) Chromatin profiling using targeted DNA adenine methyltransferase. *Nat. Genet.* **27**, 304–308.
8. Sun, L. V., Chen, L., Greil, F., Negre, N., Li, T. R., Cavalli, G., Zhao, H., Van Steensel, B., and White, K. P. (2003) Protein–DNA interaction mapping using genomic tiling path microarrays in *Drosophila*. *Proc. Natl. Acad. Sci. USA* **100**, 9428–9433.
9. Greil, F., van der Kraan, I., Delrow, J., Smothers, J. F., de Wit, E., Bussemaker, H. J., van Driel, R., Henikoff, S., and van Steensel, B. (2003) Distinct HP1 and Su(var)3–9 complexes bind to sets of developmentally coexpressed genes depending on chromosomal location. *Genes Dev.* **17**, 2825–2838.
10. de Wit, E., Greil, F., and van Steensel, B. (2007) High-resolution mapping reveals links of HP1 with active and inactive chromatin components. *PLoS Genet.* **3**, e38.
11. Bianchi-Frias, D., Orian, A., Delrow, J. J., Vazquez, J., Rosales-Nieves, A. E., and Parkhurst, S. M. (2004) Hairy transcriptional repression targets and cofactor recruitment in *Drosophila*. *PLoS Biol.* **2**, E178.
12. Orian, A., van Steensel, B., Delrow, J., Bussemaker, H. J., Li, L., Sawado, T., Williams, E., Loo, L. W., Cowley, S. M., Yost, C., Pierce, S., Edgar, B. A., Parkhurst, S. M., and Eisenman, R. N. (2003) Genomic binding by the *Drosophila* Myc, Max, Mad/Mnt transcription factor network. *Genes Dev.* **17**, 1101–1114.
13. Tolhuis, B., de Wit, E., Muijrers, I., Teunissen, H., Talhout, W., van Steensel, B., and van Lohuizen, M. (2006) Genome-wide profiling of PRC1 and PRC2 Polycomb chromatin binding in *Drosophila melanogaster*. *Nat. Genet.* **38**, 694–699.
14. Pickersgill, H., Kalverda, B., de Wit, E., Talhout, W., Fornerod, M., and van Steensel, B. (2006) Characterization of the *Drosophila melanogaster* genome at the nuclear lamina. *Nat. Genet.* **38**, 1005–1014.
15. Greil, F., de Wit, E., Bussemaker, H. J., and van Steensel, B. (2007) HP1 controls genomic targeting of four novel heterochromatin proteins in *Drosophila*. *EMBO J.* **26**, 741–751.
16. Song, S., Cooperman, J., Letting, D. L., Blobel, G. A., and Choi, J. K. (2004) Identification of cyclin D3 as a direct target of E2A using DamID. *Mol. Cell Biol.* **24**, 8790–8802.
17. Vogel, M. J., Guelen, L., de Wit, E., Peric-Hupkes, D., Loden, M., Talhout, W., Feenstra, M., Abbas, B., Classen, A. K., and van Steensel, B. (2006) Human heterochromatin proteins form large domains containing KRAB-ZNF genes. *Genome Res.* **16**, 1493–1504.
18. Lebrun, E., Fourel, G., Defossez, P. A., and Gilson, E. (2003) A methyltransferase targeting assay reveals silencer–telomere interactions in budding yeast. *Mol. Cell Biol.* **23**, 1498–1508.
19. Venkatasubrahmanyam, S., Hwang, W. W., Meneghini, M. D., Tong, A. H., and Madhani, H. D. (2007) Genome-wide, as opposed to local, antisilencing is mediated redundantly by the euchromatic factors Set1 and H2A.Z. *Proc. Natl. Acad. Sci. USA* **104**, 16609–16614.
20. Reddy, K. L., Zullo, J. M., Bertolino, E., and Singh, H. (2008) Transcriptional repression mediated by repositioning of genes to the nuclear lamina. *Nature* **452**, 243–247.
21. Pindyurin, A. V., Moorman, C., de Wit, E., Belyakin, S. N., Belyaeva, E. S., Christophides, G. K., Kafatos, F. C., van Steensel, B., and Zhimulev, I. F. (2007) SUUR joins separate subsets of PcG, HP1 and B-type lamin targets in Drosophila. *J. Cell. Sci.* **120**, 2344–2351.
22. Braunschweig, U., Hogan, G. J., Pagie, L., and van Steensel, B. (2009) Histone H1 binding is inhibited by histone variant H3.3. *EMBO J.* **28**, 3635–3645.
23. Brooks, J. E., Blumenthal, R. M., and Gingeras, T. R. (1983) The isolation and characterization of the *Escherichia coli* DNA adenine methylase (dam) gene. *Nucleic Acids Res.* **11**, 837–851.
24. Barras, F., and Marinus, M. G. (1989) The great GATC: DNA methylation in *E. coli*. *Trends Genet.* **5**, 139–143.
25. Negre, N., Hennetin, J., Sun, L. V., Lavrov, S., Bellis, M., White, K. P., and Cavalli, G. (2006) Chromosomal distribution of PcG proteins during *Drosophila* development. *PLoS Biol.* **4**, e170.
26. Fu, A. Q., and Adryan, B. (2009) Scoring overlapping and adjacent signals from genome-wide ChIP and DamID assays. *Mol. Biosyst.* **5**, 1429–1438.
27. Caddick, M. X., Greenland, A. J., Jepson, I., Krause, K. P., Qu, N., Riddell, K. V., Salter, M. G., Schuch, W., Sonnewald, U., and Tomsett, A. B. (1998) An ethanol inducible gene switch for plants used to manipulate carbon metabolism. *Nat. Biotechnol.* **16**, 177–180.
28. Koncz, C., and Schell, J. (1986) The promoter of the TL-DNA gene 5 controls the tissue-specific expression of chimeric genes carried by a novel type of *Agrobacterium*

binary vector. *Mol. Gen. Genet.* **204**, 383–396.

29. Rozen, S., and Skaletsky, H. J. (2000) Primer3 on the WWW for general users and for biologist programmers. in "Bioinformatics Methods and Protocols: Methods in Molecular Biology" (Krawetz, S., and Misener, S., Eds.), pp. 365–386, Humana Press, Totowa, NJ.
30. van Blokland, R., Ross, S., Corrado, G., Scollan, C., and Meyer, P. (1998) Developmental abnormalities associated with deoxyadenosine methylation in transgenic tobacco. *Plant J.* **15**, 543–551.
31. Roslan, H. A., Salter, M. G., Wood, C. D., White, M. R., Croft, K. P., Robson, F., Coupland, G., Doonan, J., Laufs, P., Tomsett, A. B., and Caddick, M. X. (2001) Characterization of the ethanol-inducible alc gene-expression system in *Arabidopsis thaliana*. *Plant J.* **28**, 225–235.
32. Deveaux, Y., Peaucelle, A., Roberts, G. R., Coen, E., Simon, R., Mizukami, Y., Traas, J., Murray, J. A., Doonan, J. H., and Laufs, P. (2003) The ethanol switch: a tool for tissue-specific gene induction during plant development. *Plant J.* **36**, 918–930.
33. Sambrook, J., Fritsch, E. F., and Maniatis, T. (1989) Molecular Cloning: A Laboratory Manual, Cold Spring Harbor Laboratory Press, Cold Spring Harbor, NY.
34. Bechtold, N., Ellis, J., and Pelletier, G. (1993) *In planta Agrobacterium* mediated gene transfer by infiltration of adult *Arabidopsis thaliana* plants. *C. R. Acad. Sci. Paris Life Sci.* **316**, 1194–1199.
35. Doyle, J. J., and Doyle, D. J. (1990) Isolation of plant DNA from fresh tissues. *Focus* **12**, 13–15.
36. Vogel, M. J., Peric-Hupkes, D., and van Steensel, B. (2007) Detection of in vivo protein–DNA interactions using DamID in mammalian cells. *Nat. Protoc.* **2**, 1467–1478.
37. Greil, F., Moorman, C., and van Steensel, B. (2006) DamID: mapping of in vivo protein–genome interactions using tethered DNA adenine methyltransferase. *Methods Enzymol.* **410**, 342–359.
38. Orian, A., Abed, M., Kenyagin-Karsenti, D., and Boico, O. (2009) DamID: a methylation-based chromatin profiling approach. *Methods Mol. Biol.* **567**, 155–169.
39. Urig, S., Gowher, H., Hermann, A., Beck, C., Fatemi, M., Humeny, A., and Jeltsch, A. (2002) The *Escherichia coli* dam DNA methyltransferase modifies DNA in a highly processive reaction. *J. Mol. Biol.* **319**, 1085–1096.

Section VI

Engineering Metabolic Pathways Using Transcription Factors

Chapter 19

Directed Evolution Through DNA Shuffling for the Improvement and Understanding of Genes and Promoters

Joshua R. Werkman, Sitakanta Pattanaik, and Ling Yuan

Abstract

Unlike rational protein engineering, directed evolution provides an a priori approach toward the engineering of improved proteins and novel promoters. This minimally recursive technique builds upon small improvements by selecting and combining the best changes. Protein–protein/DNA interactions, catalytic efficiency, or resilience to inhibitors can be improved by thousands of times. By working within a subspace of homologous sequences, DNA shuffling recombines that subspace. Individuals are screened for a particular trait or two and selected for when they meet a set threshold. Here we explain basic principles to follow and provide procedures for the preparation, fragmentation, efficient size fractionation, and purification of parental material, as well as for the reassembly and rescue polymerase chain reactions (PCRs).

Key words: Directed evolution, DNA shuffling, mutagenesis, protein engineering.

1. Introduction

Directed evolution is a concept taken from nature. Stemmer (1, 2) was the first to establish an in vitro method for directed evolution that capitalized upon the recombination of linear sequences found in DNA, RNA, and proteins and to demonstrate its power to make greater advances than cassette mutagenesis or error-prone PCR. Typically, it involves DNA shuffling as the means to generate a library of DNA molecules with a great amount of variation that are then screened under extreme conditions. DNA molecules that do not present an advantage to the chosen system will perform poorly, while individuals that do present an advantage survive to be selected for subsequent rounds of DNA shuffling and

L. Yuan, S.E. Perry (eds.), *Plant Transcription Factors*, Methods in Molecular Biology 754,
DOI 10.1007/978-1-61779-154-3_19, © Springer Science+Business Media, LLC 2011

selection. DNA shuffling performs the function of sexual recombination through the fragmentation of DNA molecules from one or more parents. These fragments are then reassembled through the annealing step in the polymerase chain reaction and because they are fragmented, fragments from different parents can anneal at points of sequence homology. Repeated annealing and extension cycles reassemble DNA molecules similar to the parents used, which are then amplified further to build up enough material for a library that can be transferred into expression or reporter vectors. An overview of the major steps involved is shown in **Fig. 19.1**.

Two key aspects drive success in a directed evolution endeavor. The first is determining an efficient scheme to provide selection pressure for the identification of improved offspring. It is all about numbers – the greater the number of individuals that can be screened, the more likely it is to find improved individuals. Bacterial and yeast hosts offer the ability for one to screen tens of thousands of clones on simple Petri dishes quickly and efficiently.

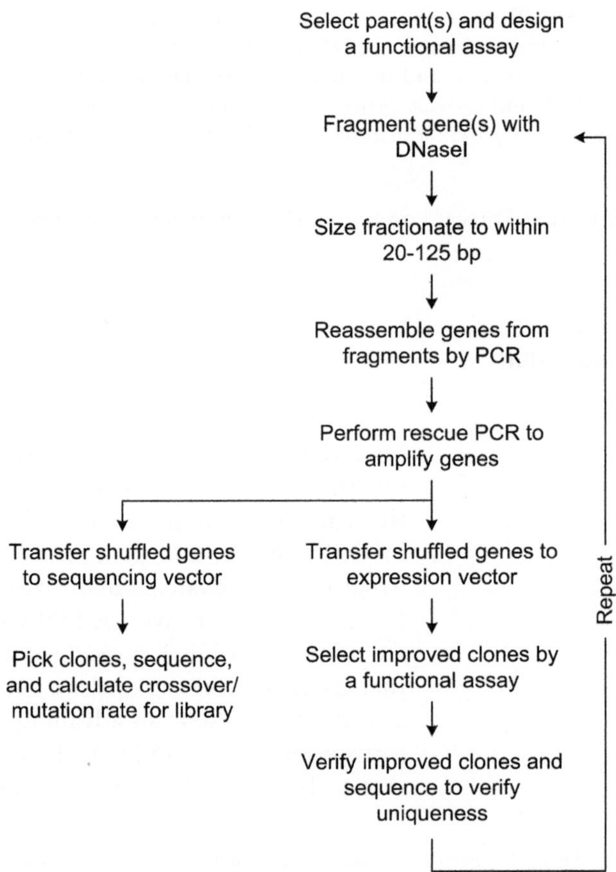

Fig. 19.1. Flowchart summarizing the main steps in a single round of directed evolution via DNA shuffling.

Expression or reporter systems that permit in vivo assays with *yes* or *no* results, directly from a plated transformation, are ideal, but not a necessity.

The second involves parental material selection and the degree to which recombination or mutation is desired. Parents can be anything from promoters to genes or domains of genes. Plasmids and whole bacterial genomes can even be used as parents (3). Single parents are not excluded either. The process of DNA shuffling not only performs recombination but also introduces random mutations, making it suitable for use with a single parent. Once a few improved clones are identified from DNA shuffling of a single parent, they can be used as multiple parents in the next round of DNA shuffling and selection.

With directed evolution, improvements have been made to improve the catalytic efficiency of an enzyme, change the substrate specificity of an enzyme, give an enzyme greater resilience to inhibitors (4), improve a protein's solubility (5), alter color (6), or change a promoter's activity (7). In the Yuan Lab, we have been able to use directed evolution to increase the transactivation of a basic helix–loop–helix (bHLH) transcription factor, Myc-RP from *Perilla frutescens*, toward the *Perilla* dihydroflavonol reductase (*DFR*) promoter by 15-fold in the first round of selection and 70-fold in the second round compared to wild-type Myc-RP (8). Provided here is a method to allow one to pursue an endeavor in directed evolution for their gene or promoter of choice.

2. Materials

1. Gene(s) or promoter(s) of interest: Parental material. Preferably these will already be in the expression or reporter vector.
2. Expression or reporter vector: Must support directional ligations with the insert.
3. High-copy sequencing vector: Preferably with the same restriction sites as used for the expression vector. Directional ligation is not critical.
4. Template-specific primers: Forward and reverse with appropriate restriction sites for directional ligations to the expression or reporter vector (*see* **Note 1**).
5. Sterile Milli-Q-grade water or equivalent.
6. 10× PCR buffer: 200 mM Tris–HCl, pH 8.8, 100 mM $(NH_4)_2SO_4$, 100 mM KCl, 20 mM $MgSO_4$, 1% Triton X-100. Store at −20°C.
7. 50 mM $MgCl_2$ for PCR. Store at −20°C.

8. 100 mM dNTPs (25 mM of each dATP, dCTP, dGTP, and dTTP): Perform a 1:10 dilution for 10 mM dNTPs into 1× TE buffer (item 15 provides the recipe for a 10× stock). Aliquot both and store at −20°C to prevent degradation of the dNTPs.
9. Bovine serum albumin (BSA) (10 mg/mL): Restriction digest grade.
10. *Taq* DNA polymerase (5 U/μL): Use of a low-fidelity DNA polymerase is acceptable as it can add to the variation desired for directed evolution. The use of error-prone PCR can also aid in adding variation with or without DNA shuffling.
11. Thermocycler with a heated lid.
12. Agarose, low-melting, electrophoresis grade.
13. 50× Tris–Acetate–EDTA (TAE) buffer: 242 g/L Tris base, 100 mL/L of 0.5 M EDTA, pH 8.0, and 57.1 mL/L glacial acetic acid (*see* **Note 2**). Prepare with Milli-Q-grade water or equivalent. Store at room temperature. Dilute also with Milli-Q-grade water or equivalent.
14. 1% (w/v) Ethidium bromide (EtBr) in water. Wear gloves when handling. Store at 4°C.
15. 10× Tris–EDTA (TE) buffer: 100 mM Tris–HCl, pH 7.5, 10 mM EDTA, pH 8.0. Prepare with Milli-Q-grade water or equivalent. Autoclave and store at room temperature. Dilute also with Milli-Q-grade water or equivalent.
16. DNA ladders (100 bp and 1 kb) and loading dye: 30% Glycerol in 10× TE buffer will work if the dyes interfere with the observation of DNA fragments. Use 1–2 μL per 10 μL of sample. Mix well.
17. Agarose gel purification kit.
18. DNase I (RNase-free): Bovine pancreatic DNase I (31–39 kDa) possesses random, nonspecific endonuclease activity toward both double- and single-stranded DNA in the presence of magnesium and calcium cations (required).
19. 10× DNase I reaction buffer: 400 mM Tris–HCl, pH 8.0, 100 mM $MgSO_4$, 10 mM $CaCl_2$. Salt concentrations greater than 100 mM reduce DNase I activity, as well as magnesium concentrations of less than 5 mM.
20. DNase I stop solution: 200 mM EDTA, pH 8.0.
21. Centrifugal filter devices: Amicon Ultra-0.5, Ultracel-100 membrane, 100KDa (Amicon 100K) and the Amicon Ultra-0.5, Ultracel-10 membrane, 10 kDa (Amicon 10K) (Millipore, USA) (*see* **Note 3**).

22. Restriction endonucleases with buffer and compatible to restriction sites chosen in expression or reporter vector and insert.
23. T4 DNA ligase (400 U/μL) with 10× T4 DNA ligase reaction buffer: 500 mM Tris–HCl, pH 7.5, 100 mM $MgCl_2$, 10 mM ATP, 100 mM dithiothreitol (DTT). Make aliquots of the buffer and store at −20°C to prevent degradation of the ATP.
24. Luria–Bertani (LB) medium: 10 g/L NaCl, 10 g/L Bacto tryptone, 5 g/L Bacto yeast extract. For a solid medium, add 15 g/L Bacto agar. Mix into Milli-Q-grade water or equivalent. Autoclave for 15 min at 121°C and 15–20 psi for volumes less than 300 mL and 20 min for volumes up to 500 mL. Allow medium to cool to between 45 and 65°C before adding any heat-labile, filter-sterilized components (*see* **Note 4**).
25. High-efficiency, competent *Escherichia coli*: Chemically competent, high-efficiency cells are desired in order to get the most out of the library created. Alternatively, electrocompetent cells can be used in conjunction with an electroporator. These two preparations of cells, however, cannot be interchanged. If possible, select *endA1* mutant strains to aid in the recovery of improved clones.
26. Plasmid isolation kit.

3. Methods

3.1. Selection Pressure

Selection is one of the most critical aspects of directed evolution. Directed evolution can be performed by creating mutations, sequencing clones, selecting clones that have mutations, and testing those clones in assays. However, this would be a brute force approach. Combining an in vivo assay provides a means to rapidly remove deleterious mutations, while selecting for mutations that outperform in the assay. This means that more often than not, an expression or reporter vector will be required. Choice of the vector and host will be dictated by the protein or promoter of interest and the screening scheme. These principles will apply regardless:

1. *Determine the more appropriate host*: As in natural evolution, not all individuals carry mutations and not all mutations carry a benefit. Therefore, in order to increase the chances of finding beneficial changes, nature uses two options: small populations selected over many multiple generations or large populations over a couple generations. In the laboratory, using larger numbers of smaller organisms is often better.

Also, consider using host strains with unique mutations or co-expressing heterologous factors. Do not limit oneself to solely *E. coli* or *Saccharomyces cerevisiae*. If there are readily available expression or reporter vectors and a reliable means to transform a particular host, use it if it means the difference between assaying a population instead of individuals.

2. *Determine a strategy for selection*: Is one trying to improve the solubility of a protein, create resistance to a particular inhibitor (chemical or protein), change an enzyme's substrate specificity, or alter a promoter's activity? Altering the temperature could also be used as a means of selection, depending upon one's goals. Finding a way to prevent the growth of the host with an unimproved or non-functional gene will generally be the most efficient means of selection. Thus, one can screen many more individuals over a shorter amount of time (*see* **Note 5**).
3. *Consider how one will recover the gene from potentially improved clones*: For example, in an initial selection of a shuffled library in plant protoplasts, it will be difficult to ensure that an improved gene can be recovered.
4. *For chemical inhibitors, determine the threshold for selection by determining the inhibitory concentration (IC50) for the system.*
5. *Determine the mutation rate for each round of selection*: A portion of the prepared inserts are ligated into a high-copy sequencing vector (**Fig. 19.1**). Clones are sequenced in order to verify quality and to calculate the rate of recombination for a specific library. Calculations can be based upon the number of crossovers per molecule (points at which one can determine part of the sequence is one parent and the other is from another parent) and then averaged among other sequenced clones to get an overall rate. Calculations can also be made based upon point mutations (about 0.1–0.5% of bases), insertions, deletions, etc. Typically, the desired changes are one or two per individual molecule of 1 kb. A mutation rate that is too high can create too many non-functional offspring. If two mutations are made to a gene, one that would have improved its function and another that makes it non-functional, the non-functional mutation will almost always win.
6. *Candidate clones must be retested by streaking clones onto plates with the same selection originally used to screen and discover the clone.*
7. *Candidate clones must be characterized as to the degree of improvement made*: This is most important for schemes where multiple rounds of selection are to occur, because this

information will determine the threshold of selection (e.g., IC50) in the following round of selection.

8. *Sequence the best candidate clones to determine what changes have occurred and that the clones are unique to each other.*
9. *For multiple rounds of selection, select only a few of the most improved clones as parents for the next round of DNA shuffling.*

3.2. Preparation of Parental Material

Parental material can be from either a single sequence or multiple unique sequences. Since these methods are based upon the ability of small DNA fragments to anneal to other small DNA fragments and subsequently lengthen into larger DNA fragments, it is important that there be some homology amongst DNA sequences from multiple parents. In cases where the primary structure is quite divergent, one can consider codon optimization of one of the parental sequences in whole or part:

1. Prepare 100 μL of the following in a 0.2-mL PCR tube on ice: 1× PCR buffer, 0.8 mM total dNTPs, 0.2 μM forward primer, 0.2 μM reverse primer, 0.1 μg/μL BSA, 0.025 U/μL *Taq* DNA polymerase, and 30–50 ng plasmid template. Bring to volume with sterile water. Mix gently (*see* **Note 6**).
2. Load the tubes into a thermocycler and run the following program: Pre-heat 94°C for 2 min; 30 cycles of 94°C for 30 s, *primer* T_m −3°C for 30 s, 68°C for 1 min/kb; final extension at 68°C for 5–10 min. Store at −20°C.
3. Wearing gloves, prepare a gel tray with comb. A 1% (w/v) agarose gel solution is prepared by adding agarose to room temperature 1× TAE buffer and bringing to a boil in a microwave oven. Occasionally stop and swirl to aid in dissolving the agarose and to prevent it from boiling over. Allow the liquid to cool slightly. In a fume hood, add ethidium bromide (EtBr) to a final concentration of approximately 0.00005% (v/v). Discard the EtBr-contaminated tip into the appropriate hazardous waste container (*see* **Note 7**). Pour the gel and allow it to cool in a fume hood for 20–30 min or until evenly solidified. Once cool and solidified, transfer the tray to a gel rig, remove the comb, and immerse the gel in 1× TAE buffer. Mix in gel-loading dye to the PCR product at about 1–2 μL per 10 μL of sample, load the wells with the PCR product, load an appropriate DNA ladder (100 bp or 1 kb), and run at 100–120 V (depending upon the distance between the electrodes; use constant *V*) until the bands are well separated.
4. Using protective eye wear, verify that the band is of the correct size by exposing the gel to UV light (254 nm) and using a clean scalpel or razor blade, excise the band of correct size (*see* **Note 7**). Transfer the gel piece to a

1.5-mL microcentrifuge tube and purify using an agarose gel cleanup system (*see* **Note 8**). Elute with 30 μL of 1× TE buffer to increase the DNA concentration (*see* **Note 9**). Measure the absorbance at 260 and 280 nm and calculate the DNA concentration (*see* **Note 10**). Safely dispose the EtBr-contaminated gel and contaminated buffer according to your institution's hazardous waste procedure.

3.3. Fragmentation of Parental Material

The length of time required to complete digestion must initially be determined empirically due to unique differences in the parental sequence, any carry over contaminants, variations in reaction conditions, such as absolute Mg^{2+} and Ca^{2+} concentrations, and individual technique. Subsequent digestions may follow this determined length of time:

1. Dilute the DNase I to 0.003 U/μL in 1× DNase I reaction buffer and hold on ice.
2. For each time point for each parent, add the following to a 0.2-mL PCR tube on ice: 10 μL of 10× DNase I reaction buffer, 1 μg of gel-purified parental DNA, 3 μL of 0.003 U/μL DNase I, and sterile water to a total volume of 88 μL (*see* **Note 11**). Mix gently.
3. Ready a stopwatch and transfer to a 25°C water bath, heat block, or thermocycler. Start the timer. At predefined timepoints, add 9 μL of DNase I stop solution to one tube per timepoint per parent. Mix well and place tube back on ice until all timepoints are completed (*see* **Note 12**).
4. Heat inactivate the DNase I digests at 75°C for 10 min.
5. Remove a small sample of about 5–8 μL (50 to 80 ng) for each tube per timepoint per parent, add loading dye, include a 100-bp DNA ladder, and run on a 1.8% agarose gel as described in **Section 3.2**, step 3. Be careful that the tracking dye does not overlap with the DNA smear that will occur between 50 and 500 bp.
6. Samples that do *not* show smearing but rather a blurry band around 100 bp can be chosen for size fractionation and purification (**Fig. 19.2a**). Timepoints from the same parent can be combined. Store at −20°C.

3.4. Size Fractionation and Purification of Digested Parental DNA

Purification of the fragments is critical to a true shuffling and reassembly in the offspring. Gel purification of the DNase I digest tends to result in abysmal yields, while unpurified digest can contain much larger fragments and thus serve as templates during the reassembly. This, in turn, reduces the degree of shuffling obtained and often results in many unchanged offspring.

To solve these two problems, Amicon Ultra-0.5 concentrators can be employed to size fractionate the digested DNA fragments

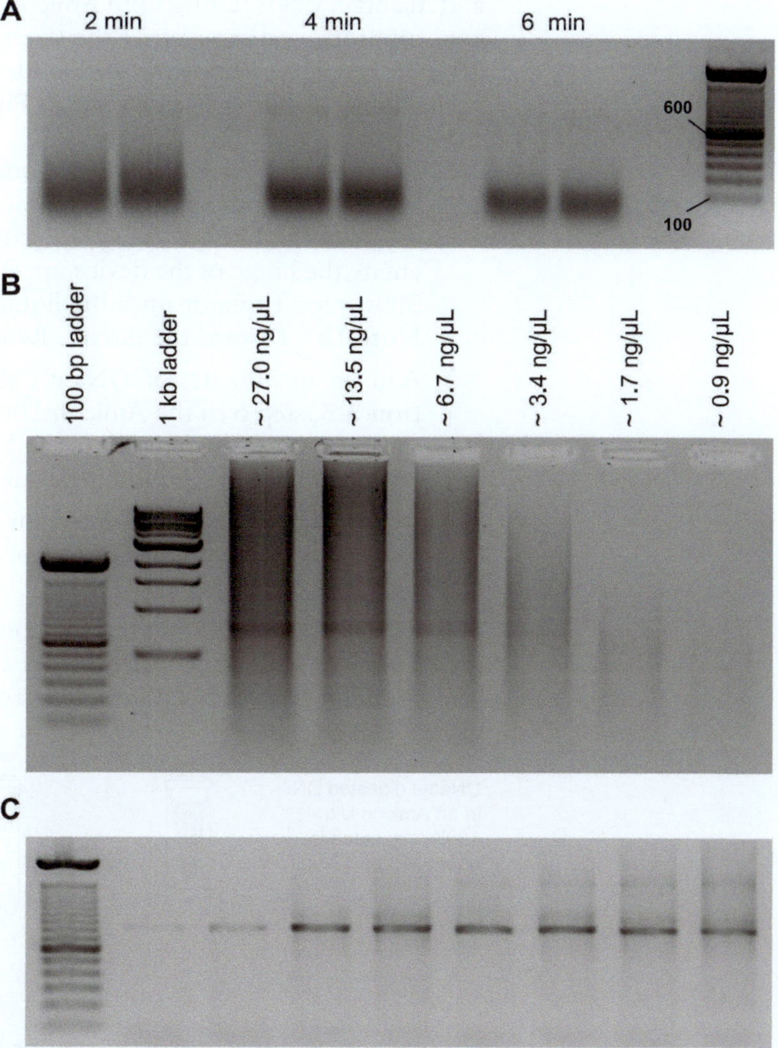

Fig. 19.2. Examples of products derived from different steps of directed evolution. *Arabidopsis* peptide deformylase 1B (*AtPDF1B*; 822 bp) undergoing DNA shuffling. (**a**) Agarose gel of three timepoints for a DNase I digestion of *AtPDF1B*. Note the greater amount of smearing at 2 min versus 6 min. The 6- and 8-min (not shown) timepoints were chosen for size fractionation (*see* **Section 3.3**). (**b**) A gel showing reassembly of the gene over several DNA fragment concentrations (estimated). As reassembly occurs, longer DNA fragments are generated and thus smear on the gel. Here a band also formed near the expected size of the gene (between 700 and 800 bp). Greater amounts of total DNA do not improve the reassembly, while smaller amounts may not yield enough material. Products from 13.5 and 6.7 ng/μL reassembly reactions were chosen for the rescue PCR (*see* **Section 3.5**). (**c**) An agarose gel showing the product of the rescue PCR. Several dilutions of the reassembly PCR reaction were tried. Amplification of the gene was relatively specific and is of expected size (*see* **Section 3.6**). This product was then gel purified, restriction digested, ligated into an expression vector, transformed into *E. coli*, and selected for improved functionality.

and maintain yields. Using both Amicon Ultra-0.5 100K and 10K concentrators, one can restrict the size of the fragments to about 20–125 bp, while removing nucleotides, undigested DNA, and exchanging the buffer. Please refer to **Fig. 19.3** for a summary of these steps:

1. To an assembled Amicon 100K unit, add 200 μL of 1× TE buffer to hydrate and rinse the membrane, close the tube's cap over the filter device, place into a microcentrifuge, orientate the hinge of the device up, and spin at no greater than 500×*g* for 1 min or until the liquid has passed through (*see* **Note 13**). Discard the filtrate. Reuse the collection tube.
2. Add up to 500 μL of DNase I-digested DNA from **Section 3.3**, step 6 to the Amicon 100K filter device. Close the tube's cap over the filter device and spin at 500×*g* for 10 min or until all liquid has passed through the membrane.
3. Remove the filter device from the collection tube and discard. Transfer the collection tube, with the filtrate (flow-through), to ice (*see* **Note 14**).
4. To an assembled Amicon 10K unit, add 200 μL of 1× TE buffer to hydrate the membrane. Close the tube's cap over the filter device, place into a microcentrifuge, orientate the

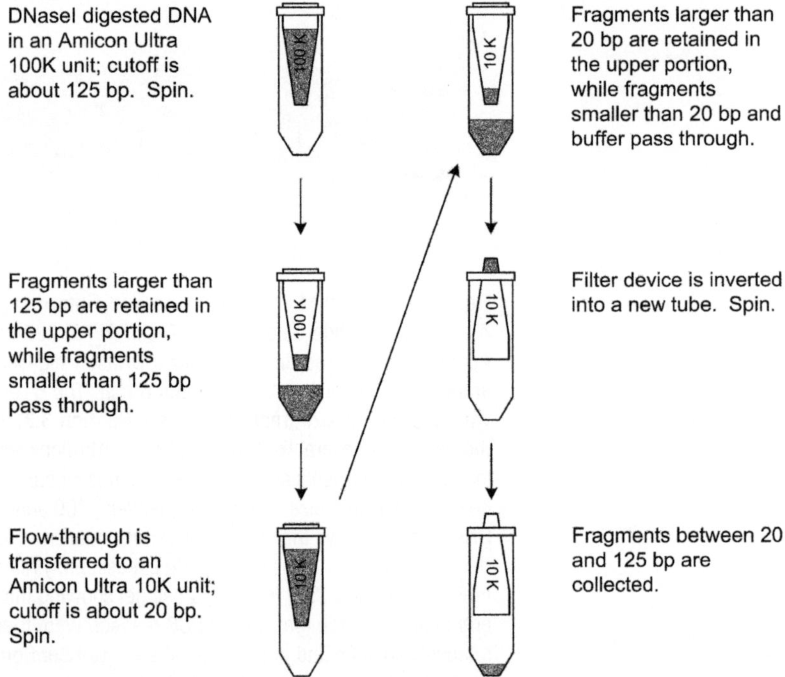

Fig. 19.3. Size fractionation strategy for digested parental DNA. Size fractionation is an effective way to maintain yields of the small DNA fragments created by DNase I, while removing larger DNA fragments and inhibiting salts. It is important to follow the location of the desired fragments throughout this procedure (*see* **Section 3.4**).

hinge of the device up, and spin at no greater than 14,000 × *g* for 1 min or until the liquid has passed through.

5. Transfer the Amicon 100K filtrate (flow-through) to the membrane of the Amicon 10K, close the tube's cap over the filter device, place into a microcentrifuge, orientate the hinge of the device up, and spin at no greater than 14,000 × *g* until all the liquid has passed through, but not to dryness. Discard the 10K filtrate (flow-through) and keep the collection tube for reuse.
6. Perform a buffer exchange by adding 300 μL of 1× TE buffer to the membrane of the Amicon 10K filter device and reduce the volume as in the previous step (*see* **Note 15**).
7. Measure the weight of a new collection tube and record.
8. Add 1× TE buffer to reach the 100 μL mark of the filter device. Gently pipet up and down a small volume to help lift the DNA fragments from the membrane. Remove the Amicon 10K filter device and with it remaining upright, place the new collection tube from the previous step over the filter device, and then invert the entire assembly. The filter device should now be upside down in the collection tube. Carefully place the assembly into the microcentrifuge and pulse briefly to transfer the concentrate to the collection tube.
9. Remove the Amicon 10K filter device, measure the weight of the collection tube plus the concentrate, and place the concentrate tube on ice or store at −20°C.
10. Subtract the weight of the collection tube from the weight of the collection tube plus the concentrate to determine the volume of concentrate recovered. Measure the absorbance at 260 and 280 nm and calculate the DNA concentration (*see* **Note 16**). Multiply the concentration by the volume of concentrate recovered to determine the total amount of DNase I-digested parental fragments available. For the reassembly PCR, a total of 1–2 μg of DNase-I digested fragments, at a concentration between 30 and 50 ng/μL, will be required for reassembly (**Table 19.1**). If multiple parents are to be used, they can be tallied into the total amount of DNA required.

3.5. Reassembly PCR Reaction (Primerless PCR)

For a single parent, the total DNA required will be solely from one parent. For multiple parents, a mixture must be made that represents the parents equally. Subsequent rounds of shuffling from a single parent would be considered a multiple parent case.

When multiple parents are reassembled, use equal amounts of size-fractionated, purified parents to obtain a stock of size-

Table 19.1
Serial dilution strategy for size-fractionated DNA fragments. Recovery of size-fractionated DNA fragments should be at a concentration of between 30 and 50 ng/μL. Find the "initial DNA concentration" that closest matches the concentration of your size-fractionated DNA fragments. Following the serial dilution procedure listed in Section 3.5, one will obtain the estimated final DNA concentrations listed by following right of that number. The total amount of DNA required to perform the serial dilution is also given. An example result can be seen in Fig. 19.2b

Initial DNA concentration (ng/μL)	Final DNA concentrations in PCR Rxn (ng/μL)					Total DNA required (μg)
	Tube 1	Tube 2	Tube 3	Tube 4	Tube 5	
50	40	20	10	5.0	2.5	2.0
45	36	18	9	4.5	2.3	1.8
40	32	16	8	4.0	2.0	1.6
35	28	14	7	3.5	1.8	1.4
30	24	12	6	3.0	1.5	1.2

fractionated DNA fragments near one of the listed initial DNA concentrations (**Table 19.1**; *see* **Note 17**):

1. PCR reaction volumes will be 25 μL and done in a serial dilution fashion with respect to the final DNA concentration in the reassembly reaction. Two master mixes will be made: "No DNA" and "Tube 1." **Table 19.1** provides an estimate of the final DNA concentrations dependent upon the initial DNA concentration (*see* **Note 18**).
2. For the "No DNA," prepare 100 μL of the following in a 0.2-mL PCR tube on ice: 1× PCR buffer, 2 mM $MgCl_2$ (in addition to what is present in the PCR buffer), 0.1 μg/μL BSA, 0.4 mM dNTPs, and 0.05 U/μL *Taq* DNA polymerase. Use 1× TE buffer to bring the mixture to volume. Mix gently.
3. For the "Tube 1," prepare 50 μL of the following in a 0.2-mL PCR tube on ice: 1× PCR buffer, 2 mM $MgCl_2$ (additional), 0.1 μg/μL BSA, 0.4 mM dNTPs, 0.05 U/μL *Taq* DNA polymerase, and 24–40 ng/μL size-fractionated DNA (final concentration; **Table 19.1**, column 2). Use 1× TE buffer to bring the mixture to volume. Mix gently (*see* **Note 19**).
4. Setup four additional 0.2-mL PCR tubes on ice labeled 2 through 5. Transfer the "No DNA" master mix in the following amounts:
   24.5 μL to Tube 2,

23.5 μL to Tube 3,

21.5 μL to Tube 4, and

17.5 μL to Tube 5.

5. Transfer 24.5 μL of "Tube 1" master mix to Tube 2. Mix gently. Discard the pipette tip. Master mix "Tube 1" will now be Tube 1.
6. Transfer 23.5 μL of Tube 2 to Tube 3. Mix gently. Discard the pipette tip.
7. Transfer 21.5 μL of Tube 3 to Tube 4. Mix gently. Discard the pipette tip.
8. Transfer 17.5 μL of Tube 4 to Tube 5. Mix gently. Discard pipette tip (*see* **Note 20**).
9. Load the tubes into a thermocycler and run the following program: pre-heat at 94°C for 2 min; 40 cycles of 94°C for 20 s, 50°C for 30 s (−0.2°C/cycle), 68°C for 30 s (+1 s/cycle for templates under 1 kb, +2 s/cycle for templates between 1 and 1.5 kb, or +3 s/cycle for templates 1.5–2.5 kb); no final extension.
10. Remove a small sample of about 5 μL, add loading dye, include a DNA ladder, and run on a 1% agarose gel as described in **Section 3.2**, step 3. The reassembly PCR product should show a significant amount of smearing with possibly a faint band near the expected size of the full-length, parent DNA fragment (**Fig. 19.2b**). Store reactions at −20°C.

3.6. Rescue PCR Reaction

Using a template-specific primer pair, full-length DNA fragments will be recovered from the reassembly PCR. The primer pairs can be specific to each parent and used as a mixture or primer pairs for a specific parent can be used alone (*see* **Note 1**):

1. Prepare 50 μL of the following in a 0.2-mL PCR tube on ice: 1× PCR buffer, 0.8 mM total dNTPs, 0.2 μM forward primer(s), 0.2 μM reverse primer(s), 0.1 μg/μL BSA, 0.025 U/μL *Taq* DNA polymerase, and 1–5 μL of a 1/100–1/50 dilution of the reassembly PCR (*see* **Note 21**). Use sterile water to bring the mixture to volume. Mix gently.
2. Load the tube(s) into a thermocycler and run the following program: pre-heat at 94°C for 2 min; 30 cycles of 94°C for 30 s, *primer* T_m − 3°C for 30 s, 68°C for 1 min/kb; final extension at 68°C for 5–10 min.
3. Remove a small sample of about 5 μL, add loading dye, include a DNA ladder, and run on a 1% agarose gel as described in **Section 3.2**, step 3. The rescue PCR product should show a strong band near the expected size of the full-length, parental DNA fragment (**Fig. 19.2c**). Store reactions at −20°C.

3.7. Insert Preparation, Ligation, and Transformation

Once the DNA fragment of parental size has been rescued, the rescue PCR reaction is buffer exchanged in an Amicon 100K, the fragment is then digested with the appropriate restriction endonucleases, gel purified, and ligated into the desired expression or reporter vector. The ligation reaction is transformed into an appropriate strain and plated onto plates with the appropriate selection agents for the vector (e.g., ampicillin or kanamycin) and for the selection of improved individuals.

Alternative methods for insertion of the rescued DNA fragment into a plasmid include the use of Gateway recombination technology and TA cloning, in which case both can avoid problems related to finding compatible restriction sites between multiple parents and vector:

1. Buffer exchange the rescue PCR reaction using an Amicon 100K with a pre-rinse of 200 μL of 1× TE, followed by the rescue PCR reaction, and one to two washes with 300 μL of 1× TE buffer all at 500×*g* in a microcentrifuge. Do not spin to dryness. Resuspend the concentrate in 1× TE buffer and transfer the concentrate from the filter device (*see* **Note 22**). The concentrate's total volume should be near 50 μL. Measure the absorbance at 260 and 280 nm and calculate the DNA concentration.
2. Using 0.5–1.0 μg buffer-exchanged, rescue product, perform a restriction digest with the appropriate restriction endonucleases in its respective 1× buffer and at the respective incubation temperature (*see* **Note 23**). Also, set up a separate restriction digest using the same enzymes for the expression or reporter vector using 1.0 μg of plasmid (*see* **Note 24**).
3. Prepare an agarose gel, include a DNA ladder, resolve the digested rescued DNA fragment and the backbone portion of the vector, purify, and quantify as in steps 3 and 4 of **Section 3.2**.
4. Perform a ligation of the prepared insert and vector at a 2:1 or 3:1 molar ratio (insert to vector) with 1 μL of 400 U/μL T4 DNA ligase in 1× T4 DNA ligase buffer. The total amount of DNA should not exceed 100 ng in a 20 μL reaction volume. Incubate the reaction overnight at 16°C (*see* **Note 25**). Store reactions at −20°C.
5. For *E. coli* expression vectors, transform ice-thawed, high-efficiency, chemically competent cells with 5 μL of the ligation reaction per 50 μL of cells by incubating cells plus DNA for 20–30 min on ice, heating the cells plus DNA at 42°C for 30 s, cooling the cells plus DNA for 2 min on ice, and then resuspending in 250 μL of LB broth. Grow the cells at 37°C for 30–60 min and plate out onto the appropriate selection medium decided upon in **Section 3.1** (*see* **Note 26**).

3.8. Recovery of Improved Offspring

Depending upon the selection strategy, candidate clones would be, for example, colonies growing strongly on a chemical selection medium (i.e., in similarity to antibiotic selection) or colonies differing through a color indicator (i.e., in similarity to blue/white selection). In some cases, it may not be possible to directly select from a plated transformation. These cases could involve those that would require a β-galactosidase assay using colony lift and filter paper, such as for yeast, or cases that would require checking the solubility of a protein by determining whether it does not form inclusion bodies or whether it can be eluted from a column. Here one can still process a significant number of individuals and still have successes. If available, the use of robotic rearray machines and liquid handlers can overcome problems where clones need to be assayed individually.

3.9. Preparation for the Next Round

Once candidate clones have been identified and confirmed to perform as when originally discovered, prepare glycerol stocks of the clones for preservation, prepare DNA from the clones for sequencing, sequence the clones, confirm that the clones are indeed unique, determine the new thresholds of selection for the next round of selection, choose one to four of the best performing clones to use as templates in **Section 3.2**, step 1, and repeat **Sections 3.2**, **3.3**, **3.4**, **3.5**, **3.6**, **3.7**, **3.8**, and **3.9** until sufficient progress has been made.

For clones with multiple changes, the impacts of individual changes can be teased out by "backcrossing" to one parent. This simply involves performing a round of DNA shuffling between the clone in question and the parent.

4. Notes

1. If a region of the gene is desired to be mutated, such as in the N-terminus of the protein, and it overlaps where the primers anneal, then the primers used to rescue the gene must be moved outside of that region to prevent reversion back to the parental sequence.
2. Disodium EDTA has a low solubility in water and requires base (NaOH) to bring it into solution. Do not make stock solutions of EDTA above 0.5 M. Wear personal protective equipment (PPE) when using strong acids or bases. Dispense and add glacial acetic acid in a fume hood, adding *acid to water* in order to avoid injury.
3. These are replacements for the discontinued Microcon YM-10 (dsDNA cutoff of 20 bp) and YM-100 (dsDNA cutoff of 125 bp) centrifugal filter devices from Millipore.

4. Some solvents are naturally antimicrobial and thus will not require filter sterilization through a 0.2-μm filter.
5. Adhere to any safety precautions and hazardous waste disposal procedures designated by your institution concerning chemicals used for selection purposes.
6. Depending upon amplification efficiency, more than one 100-μL reaction may be needed in order to obtain the required amounts of DNA for subsequent steps.
7. Both EtBr and UV are mutagenic and hence carcinogenic to living organisms. Do not boil solutions with EtBr.
8. We use the Wizard SV Gel and PCR cleanup system (Promega, USA) for gel purification.
9. Elution in sterile, nuclease-free water is also permissible. The use of a slightly alkaline buffered solution, such as 1× Tris–EDTA (TE), pH 8.0, helps to prevent chemical degradation of the DNA and improves the stability of DNA molecules, which improves spectrophotometric measurements. The presence of EDTA helps prevent errant DNases from destroying one's DNA sample. Eluting with larger volumes will generally recover more DNA; however, it will be of a lower concentration. Elution volumes can be adjusted according to downstream applications. Higher concentrations of macromolecules tend to store better than in dilute form. Plus, it is usually easier to dilute a sample than it is to concentrate it.
10. We use a Nanodrop device to minimize the sample size volume and to avoid having to dilute samples, which increases the quality of the measurements.
11. Using larger amounts of starting material may seem excessive; however, it provides against losses due to absorption by filters, samples taken to monitor product quality, etc. Additionally, subsequent steps also require optimization through empirical testing. Final concentrations of free Mg^{2+} and Ca^{2+} should be 10 and 1 mM, respectively.
12. Typical timepoints include 10, 20, and 30 min for a first run and then 4, 6, and 8 min for a second run. Less DNA can be used for a first run. When the DNase I stop solution is added, the final concentration of EDTA will be 18 mM and thus 7 mM in excess of Mg^{2+} and Ca^{2+} combined. Final DNA concentration of parental DNA will be 10 ng/μL.
13. This step hydrates the membrane and helps to reduce the trace amount of glycerin present in the membrane. Do not touch the pipette tip to the filter membrane at any time. Excessive *g* forces and spin times can generate holes in the

membrane. Do not spin to dryness. Once hydrated, do not allow the membrane to dry. Use a new unit for each parent.

14. A sample of about 5 μL can be taken from the filtrate (flow-through) and saved. A dilution of 1/50 or so can be made and PCR performed using a small amount of this template and the primer pair used for the rescue PCR. If the full-length fragment is found when the PCR product is run on an agarose gel, then undigested material has contaminated the filtrate. If no full-length fragment is found, then it is reasonably safe to assume that no full-length fragment is present in the filtrate, which is the desired outcome. Additionally, if one is curious, take the Amicon 100K filter device, add a small amount (~100 μL) of 1× TE buffer to the membrane, remove to a new tube, perform a dilution, and test for the presence of full-length fragment as for the filtrate. One will most likely find full-length fragment.
15. The buffer exchange step is needed to reduce the amount of EDTA and EGTA to noninhibitory levels for PCR. Additionally, the calcium ions required by DNase I can inhibit *Taq* DNA polymerase.
16. With the presence of some protein from DNase I carryover, the calculated DNA concentrations may be off, but this is only to be an estimation of the DNA concentration.
17. Parents will likely be of similar size and thus each parent should contribute in an equimolar fashion.
18. Do not be overly concerned on the accuracy of the values. By using the serial dilution, this will allow one to choose among reactions that worked and did not work.
19. The reassembly PCR reaction does *not* contain any custom, synthetic primers, but rather the fragmented, parental DNA serves as both primers and template. During the reassembly, single-stranded fragments anneal to other single-stranded fragments at regions of local homology. As this process is repeated, the overall fragment size found in the reaction increases. Inclusion of specific primers at this point would skew and poison this process and likely result in dead genes with concatemers.
20. Tube 5 will have a larger total volume compared to the rest.
21. One may want to try a couple dilutions to find a concentration of reassembled template that yields a clean band when visualized on an agarose gel (**Fig. 19.2c**).
22. Performing a buffer exchange of the rescue PCR reaction serves several purposes. Though some restriction endonucleases have activity in 1× PCR buffer, enzyme can be diverted to primer dimers and thus reduce the overall ability for

your DNA fragment to be cut. Significant amounts of left-over primers can be removed using these filters. From personal experience, the yields are much higher from centrifugal membrane filter devices than from column devices. By placing the gel purification step after the restriction digest, one can obtain cleaner product for ligation and maintain higher yields.

23. To save time, a double digest can be performed. Check with the enzymes' manufacturer for your specific restriction endonucleases.
24. One may want to digest and purify more vector so as not to need to process more in subsequent rounds.
25. Though ligations can be performed in just an hour or two, an overnight incubation allows for the reaction to go to completion. This is essential to capture the diversity of the shuffled library.
26. If the ligation is of sufficient efficiency and the selection strategy appears to be working correctly, repeat the transformation step with the remaining ligation reaction. One can also return to the reassembly PCR product and repeat the subsequent steps to find more clones.

References

1. Stemmer, W. P. C. (1994) Rapid evolution of a protein in vitro by DNA shuffling. *Nature* **370**, 389–391.
2. Stemmer, W. P. C. (1994) DNA shuffling by random fragmentation and reassembly: in vitro recombination for molecular evolution. *Proc. Natl. Acad. Sci.* **91**, 10747–10751.
3. Zahradka, K., Slade, D., Bailone, A., Sommer, S., Averbeck, D., Petranovic, M., Lindner, A. B., and Radman, M. (2006) Reassembly of shattered chromosomes in *Deinococcus radiodurans. Nature* **443**, 569–573.
4. Fan, Z., Yuan, L., Jordan, D. G., Wagschal, K., Heng, C., and Braker, J. D. (2010) Engineering lower inhibitor affinities in β-D-xylosidase. *Appl. Microbiol. Biotechnol.* **86**, 1099–1113.
5. Nasreen, A., Vogt, M., Kim, H. J., Eichinger, A., and Skerra, A. (2006) Solubility engineering and crystallization of human apolipoprotein D. *Protein Sci.* **15**, 190–199.
6. Sawano, A. and Miyawaki, A. (2000) Directed evolution of green fluorescent protein by a new versatile PCR strategy for site-directed and semi-random mutagenesis. *Nucleic Acids Res.* **28**, e78.
7. Lee, S. K., Chou, H. H., Pfleger, B. F., Newman, J. D., Yoshikuni, Y., and Keasling, J. D. (2007) Directed evolution of AraC for improved compatibility of arabinose- and lactose-inducible promoters. *Appl. Environ. Microbiol.* **73**, 5711–5715.
8. Pattanaik, S., Xie, C. H., Kong, Q., Shen, K. A., and Yuan, L. (2006) Directed evolution of plant basic helix–loop–helix transcription factors for the improvement of transactivational properties. *Biochim. Biophys. Acta* **1759**, 308–318.

Index

L. Yuan, S.E. Perry (eds.), *Plant Transcription Factors*, Methods in Molecular Biology 754,
DOI 10.1007/978-1-61779-154-3, © Springer Science+Business Media, LLC 2011

MIX
Papier aus verantwortungsvollen Quellen
Paper from responsible sources
FSC® C105338

If you have any concerns about our products,
you can contact us on
ProductSafety@springernature.com

In case Publisher is established outside the EU,
the EU authorized representative is:
Springer Nature Customer Service Center GmbH
Europaplatz 3, 69115 Heidelberg, Germany

Printed by Libri Plureos GmbH
in Hamburg, Germany